Inhibitors in Patients with Haemophilia

Inhibitors in Patients with Haemophilia

Edited by

E.C. Rodriguez-Merchan
MD, PhD
Consultant Orthopaedic Surgeon
Service of Traumatology and Orthopaedic Surgery
Haemophilia Centre
La Paz University Hospital
Madrid, Spain

C.A. Lee
MA, MD, DSc(Med), FRCP, FRCPath
Professor of Haemophilia
Haemophilia Centre and Haemostasis Unit
Royal Free Hospital NHS Trust
Hampstead
London, UK

Blackwell
Science

© 2002 by Blackwell Science Ltd
a Blackwell Publishing Company
Editorial Offices:
Osney Mead, Oxford OX2 0EL, UK
 Tel: +44 (0)1865 206206
Blackwell Science, Inc., 350 Main Street, Malden, MA 02148-5018, USA
 Tel: +1 781 388 8250
Blackwell Science Asia Pty, 54 University Street, Carlton, Victoria 3053, Australia
 Tel: +61 (0)3 9347 0300
Blackwell Wissenschafts Verlag, Kurfürstendamm 57, 10707 Berlin, Germany
 Tel: +49 (0)30 32 79 060

The right of the Authors to be identified as the Authors of this Work has been asserted in accordance with the Copyright, Designs and Patents Act 1988.

All rights reserved. No part of this publication may be reproduced, stored in a retrieval system, or transmitted, in any form or by any means, electronic, mechanical, photocopying, recording or otherwise, except as permitted by the UK Copyright, Designs and Patents Act 1988, without the prior permission of the publisher.

First published 2002

Library of Congress Cataloging-in-Publication Data

Inhibitors in patients with haemophilia / edited by E.C. Rodriguez-Merchan, C.A. Lee.
    p. ; cm.
  Includes bibliographical references and index.
  ISBN 0-632-06477-3 (hardback)
    1. Hemophilia—Pathophysiology. 2. Hemophilia—Treatment—Complications. 3. Hemophilia—Immunological aspects.
  [DNLM: 1. Hemophilia A—immunology. 2. Factor VIII—antagonists & inhibitors. 3. Factor XI—antagonists & inhibitors. 4. Hemophilia B—immunology. WH 325 155 2002] I. Rodriguez-Merchan, E.C. II. Lee, Christine A.
  RC 642 .I547 2002
  616. 1′572—dc21   2001008760

ISBN 0-632-06477-3

A catalogue record for this title is available from the British Library

Set in 9/12 Sabon by Graphicraft Limited, Hong Kong
Printed in Great Britain by MPG Books Ltd, Bodmin, Cornwall

For further information on Blackwell Science, visit our website:
www.blackwell-science.com

Contents

Contributors, vii
Preface, x

Part 1: Haematology
1 Natural history of inhibitors in severe haemophilia A and B: incidence and prevalence, 3
  J.M. Lusher
2 Characterization of inhibitors in congenital haemophilia, 9
  K. Peerlinck and M. Jacquemin
3 Incidence and prevalence of inhibitors and type of blood product in haemophilia A, 14
  T.T. Yee and C.A. Lee
4 Genetic basis of inhibitor development in severe haemophilia A and B, 21
  J. Oldenburg and E. Tuddenham

Part 2: Management of treatment
5 Methods: plasmapheresis and protein A immunoadsorption, 29
  E. Berntorp
6 Venous access in children with inhibitors, 36
  R.C.R. Ljung

Part 3: Immune tolerance
7 Immune tolerance: high-dose regimen, 45
  H.H. Brackmann and T. Wallny
8 Immune tolerance: low-dose regimen, 49
  E.P. Mauser-Bunschoten
9 Immune tolerance and choice of concentrates, 55
  W. Kreuz
10 Immune tolerance: The North American Immune Tolerance Registry, 59
  D.M. DiMichele and B.L. Kroner

Part 4: Medical management of bleeding episodes
11 The treatment of bleeding episodes in children, 69
  J.M. Tusell
12 General medical management of bleeding episodes: haemarthroses, muscle haematomas, mucocutaneous bleeding and haematuria, 74
  M. Quintana-Molina, V. Jimenez-Yuste, A. Villar-Camacho and F. Hernandez-Navarro
13 Life- and limb-threatening episodes and intracranial bleeds, 78
  S. Schulman, E. Santagostino, R. Grossman, J.M. Lusher and the rFVIIa-CI Group

Part 5: Inhibitors in haemophilia B, inhibitors in mild and moderate haemophilia A and acquired inhibitors to FVIII

14 Factor IX inhibitors and anaphylaxis, 87
   I. Warrier
15 Inhibitors in mild and moderate haemophilia A, 92
   C.R.M. Hay and C.A. Lee
16 Acquired inhibitors to factor VIII, 98
   C.M. Kessler and L. Nemes

Part 6: Musculoskeletal issues

17 Pathogenesis of musculoskeletal complications of haemophilia, 115
   E.C. Rodriguez-Merchan
18 Orthopaedic management of haemarthroses, 120
   E.C. Rodriguez-Merchan
19 Chemical synoviorthesis, 126
   H.A. Caviglia, F. Fernandez-Palazzi, C. Pascual-Garrido, N. Moretti and R. Perez-Bianco
20 Radiosynoviorthesis in patients with haemophilia and inhibitors, 129
   C.J. Petersson and E. Berntorp
21 Musculoskeletal magnetic resonance imaging, 132
   R. Nuss, R.F. Kilcoyne and J.D. Wiedel
22 Treatment of iliopsoas haematomas and compartment syndromes in patients with haemophilia who have a circulating antibody, 139
   M. Heim, M. Warshavski, Y. Amit and U. Martinowitz
23 A rational approach to the treatment of haemophilic blood cyst (pseudotumour) in patients with inhibitors, 142
   M.S. Gilbert and A. Forster
24 Haemophilia patients with inhibitors: a rheumatologist's point of view, 146
   J. York
25 Rehabilitation of patients with haemophilia and inhibitors, 149
   F. Querol, J.A. Aznar, S. Haya and A.R. Cid
26 Physiotherapy in the management of patients with inhibitors, 160
   K. Beeton and B. Buzzard
27 Elective orthopaedic surgery in haemophilia patients with high-responding inhibitors, 169
   I. Hvid, K. Soballe and J. Ingerslev

Part 7: General surgery

28 General and emergency surgery in patients with high-responding inhibitors, 179
   B. White and O.P. Smith
29 Dental extraction in patients with haemophilia and inhibitors, 183
   E.A. Rey, S.A. Puia and W. Castillo

Part 8: Psychosocial issues

30 Psychosocial impact of inhibitors on haemophilia patients' quality of life, 187
   E. Remor, P. Arranz and R. Miller

Part 9: General strategy for management of inhibitor patients

31 General strategy for management of inhibitor patients (summary), 195
   I. Scharrer
32 Immunoadsorption with anti-immunoglobulin antibodies using Ig-TheraSorb columns, 199
   M. von Depka and A. Huth-Kuehne
33 Standard, high and mega bolus doses of rVIIa or recombinant VIIa: comparison of three different treatment regimens, 206
   G. Kenet and U. Martinowitz

Index, 211

Contributors

Y. Amit MD
Israel National Haemophilia Centre; Department of Orthopaedic Surgery, Sheba Medical Centre, Tel Hashomer Hospital, Israel

P. Arranz PhD
Consultant Psychologist, Haemophilia Centre, La Paz University Hospital, Madrid, Spain

J.A. Aznar MD
Director, Unit for Congenital Bleeding Disorders, La Fe University Hospital, Valencia, Spain

K. Beeton
Senior Lecturer, Department of Physiotherapy, University of Hertfordshire, College Lane, Hatfield, Hertfordshire, UK; and Honorary Lecturer, Royal Free and University College Medical School, University College, London, UK

E. Berntorp
Professor, Department for Coagulation Disorders, Malmö University Hospital, SE205 02 Malmö, Sweden

H.H. Brackmann MD
Institute of Experimental Haematology and Transfusion Medicine, University of Bonn, Germany

B. Buzzard
Superintendent Physiotherapist, Newcastle Regional Haemophilia Comprehensive Care Centre, Royal Victoria Infirmary, Newcastle-upon-Tyne, UK

W. Castillo MD
Consultant Odontologist, Department of Oral Surgery, Argentinian Haemophilia Foundation, Soler 3485, CP 1425, Buenos Aires, Argentina

H.A. Caviglia MD
Consultant Orthopaedic Surgeon, Argentinian Haemophilia Foundation, Soler 3485, CP 1425, Buenos Aires, Argentina

A.R. Cid MD
Consultant Haematologist, Unit for Congenital Bleeding Disorders, La Fe University Hospital, Valencia, Spain

D.M. DiMichele MD
Associate Professor of Clinical Paediatrics; Director, Regional Comprehensive Hemophilia Diagnostic and Treatment Center; New York Presbyterian Hospital, Weill Medical College of Cornell, 525 East 68th Street, P-695, New York, NY 10021, USA

F. Fernandez-Palazzi MD
Consultant Orthopaedic Surgeon, Blood Bank, Caracas, Venezuela

A. Forster
International Haemophilia Treatment and Training Center and The Leni and Peter May Department of Orthopaedic Surgery, Mount Sinai Hospital, New York, 1065 Park Avenue, New York, NY 10128, USA

M.S. Gilbert MD
International Haemophilia Treatment and Training Center and The Leni and Peter May Department of Orthopaedic Surgery, Mount Sinai Hospital, New York, 1065 Park Avenue, New York, NY 10128, USA

R. Grossman MD
Central Laboratory of the University Medical Centre, University of Würtzburg, Würtzburg, Germany

C.R.M. Hay MD, FRCP, FRCPath
Department of Haematology, Manchester Royal Infirmary, Manchester, UK

S. Haya MD
Consultant Haematologist, Unit for Congenital Bleeding Disorders, La Fe University Hospital, Valencia, Spain

M. Heim MD
Professor, Department of Orthopaedic Surgery and Orthopaedic Rehabilitation, Israel National Haemophilia Centre, Sheba Medical Centre, Tel Hashomer Hospital, Israel

F. Hernandez-Navarro MD, PhD
Professor and Chairman, Service of Haematology and Haemophilia Centre, La Paz University Hospital, Madrid, Spain

A. Huth-Kuehne MD
Kurpfalz-Hospital, Heidelberg, Germany

CONTRIBUTORS

I. Hvid MD
Professor, Children's Hospital, Department of Orthopaedics E, University of Aarhus, Denmark

J. Ingerslev MD
Director, Haemophilia Centre, University Hospital of Aarhus, Denmark

M. Jacquemin MD
Center for Molecular and Vascular Biology and Haemostasis Treatment Center, UZ Gasthuisberg-University of Leuven, Herestraat 49, 3000 Leuven, Belgium

V. Jimenez-Yuste MD
Consultant Haematologist, Haemophilia Centre, La Paz University Hospital, Madrid, Spain

G. Kenet MD
The Israeli National Haemophilia Centre, Tel Hashomer 52621, Israel

C.M. Kessler MD
Professor of Medicine and Pathology, Lombardi Cancer Center, Georgetown University Medical Center, Division of Hematology-Oncology, 3800 Reservoir Road, NW, Washington, D.C. 2007, USA

R.F. Kilcoyne MD
Professor Emeritus, Department of Radiology and Orthopaedics, University of Colorado Health Sciences Center, Denver, Colorado, USA

W. Kreuz MD
Centre of Paediatrics III, Department of Haematology and Oncology, Comprehensive Care Centre for Thrombosis and Haemostasis of the Johann Wolfgang Goethe University Hospital, Theodor-Stern-Kai 7, D-60596 Frankfurt am Main, Germany

B.L. Kroner PhD
Senior Epidemiologist, Research Triangle Institute International, Rockville, Maryland, USA

C.A. Lee MA, MD, DSc(Med), FRCP, FRCPath
Professor of Haemophilia, Haemophilia Centre and Haemostasis Unit, Royal Free Hospital NHS Trust, Hampstead, London, UK

R.C.R. Ljung MD
Consultant Paediatrician, Departments of Paediatrics and Coagulation Disorders, Lund University, University Hospital, SE-205 02 Malmö, Sweden

J.M. Lusher MD
Division of Haematology/Oncology, Department of Paediatrics, The Children's Hospital of Michigan, 3901 Beaubien Boulevard, Detroit, Michigan 48201, USA

U. Martinowitz MD
Director, Israeli National Haemophilia Centre, Tel Hashomer 52621, Tel Aviv, Israel

E.P. Mauser-Bunschoten MD
Van Creveldkliniek, University Medical Center Utrecht, C01-425, PO 85000, 3508 CX Utrecht, The Netherlands

R. Miller
Social Worker/Family Therapist and Honorary Senior Lecturer, Haemophilia Centre and Hemostasis Unit, Royal Free Hospital and Royal Free and University College Medical School, London, UK

N. Moretti
Physiotherapist, Argentinian Haemophilia Foundation, Soler 3485, CP 1425, Buenos Aires, Argentina

L. Nemes MD
Director, National Haemophilia Center, National Institute of Haematology and Immunology, Daroci u. 24, 1113 Budapest, Hungary

R. Nuss MD
Associate Director, Mountain States Regional Haemophilia and Thrombosis Centre, Associate Professor, Department of Paediatrics, University of Colorado Health Sciences Centre, Denver, Colorado, USA

J. Oldenburg MD
Institute of Transfusion Medicine and Immune Haematology of the DRK Blood Donor Service Hessen, Frankfurt; and Institute of Experimental Haematology and Transfusion Medicine, University of Bonn, Germany

C. Pascual-Garrido MD
Argentinian Haemophilia Foundation, Soler 3485, CP 1425, Buenos Aires, Argentina

K. Peerlinck MD
Center for Molecular and Vascular Biology and Haemophilia Treatment Center, UZ Gasthuisberg-University of Leuven, Herestraat 49, 3000 Leuven, Belgium

R. Perez-Bianco MD
Consultant Haematologist, Argentinian Haemophilia Foundation, Soler 3485, CP 1425, Buenos Aires, Argentina

C.J. Petersson MD, PhD
Department of Orthopaedics, Lund University, Malmö University Hospital, Malmö, Sweden

S.A. Puia MD
Consultant Odontologist, Argentinian Haemophilia Foundation, Soler 3485, CP 1425, Buenos Aires, Argentina

F. Querol MD, PhD
Consultant Rehabilitation Specialist, Unit for Congenital Bleeding Disorders, La Fe University Hospital, Valencia; and Lecturer, Physiotherapy Department, University of Valencia, Avenida de Campaner 21, 46009 Valencia, Spain

M. Quintana-Molina MD
Consultant Haematologist, Haemophilia Centre, La Paz University Hospital, Madrid, Spain

E. Remor PhD
Research Associate Psychologist, Haemophilia Centre, La Paz University Hospital, Madrid, Spain

CONTRIBUTORS

E.A. Rey MD
Consultant Odontologist, Argentinian Haemophilia Foundation, Soler 3485, CP 1425, Buenos Aires, Argentina

E.C. Rodriguez-Merchan MD, PhD
Consultant Orthopaedic Surgeon, Service of Traumatology and Orthopaedic Surgery, Haemophilia Centre, La Paz University Hospital, Madrid, Spain

E. Santagostino MD
A. Bianchi Bonomi Hemophilia and Thrombosis Centre, IRCCS Maggiore Hospital, Milan, Italy

I. Scharrer MD
Medizinische Klinik 1, Haemophilie-Ambulanz, Universitätskliniken Frankfurt, Theodur-Stern-Kai 7, D-60590 Frankfurt am Main, Germany

S. Schulman MD
Coagulation Unit, Department of Haematology, Karolinska Hospital, S-171 76 Stockholm, Sweden

O.P. Smith MA, MB, BA Mod. (Biochem), FRCPI, FRCPEdin, FRCPE Lon, FRCP Glasg, FRCPCH, MRCPath
Consultant Paediatric Haematologist, National Centre for Hereditary Coagulation Disorders, St James's Hospital, Dublin 8, Ireland

K. Soballe MD
Consultant Orthopaedic Surgeon, University Hospital of Aarhus, Denmark

E. Tuddenham MD
MRC Clinical Sciences Centre, Imperial College Faculty of Medicine, Hammersmith Hospital, Du Cane Road, London W12 0NN, UK

J.M. Tusell MD
Consultant Paediatric Haematologist, Vall D'Hebron University Hospital, Barcelona, Spain

A. Villar-Camacho MD
Consultant Haematologist, Haemophilia Centre, La Paz University Hospital, Madrid, Spain

M. von Depka MD
Hannover Medical School, Hannover, Germany

T. Wallny MD
Orthopaedic Hospital, University of Bonn, Germany

I. Warrier MD
Children's Hospital of Michigan, Wayne State University, 3091 Beaubien Avenue, Detroit, Michigan 48501, USA

M. Warshavski MD
Department of Orthopaedic Rehabilitation, Sheba Medical Centre, Tel Hashomer Hospital, Israel

B. White MB, MD, MSc (Mol. Med), MRCPI, MRCPath
Consultant Haematologist, National Centre for Hereditary Coagulation Disorders, St James's Hospital, Dublin 8, Ireland

J.D. Wiedel MD
Professor, Department of Orthopaedics, University of Colorado Health Sciences Centre, Denver, Colorado, USA

T.T. Yee MBBS, MSc (Med), MRCP
Haemophilia Centre and Haemostasis Unit, Royal Free Hospital NHS Trust, Hampstead, London, UK

J. York MD
Clinical Associate Professor, Consultant Emeritus, Department of Rheumatology, The Royal Prince Alfred Hospital, QE II Building, 59 Missenden Road, Camperdown, New South Wales, 2050 Australia

Preface

The development of factor VIII and IX inhibitors creates special problems in haemophilic patients. Alloantibodies (inhibitors) against infused exogenous factor VIII or IX are a significant complication in congenital haemophilia, making it difficult to achieve effective haemostasis. This is particularly true for paediatric patients, for whom repeated venous access is a challenge. Adequate venous access is required for treatment, resulting in an increased need for surgical intervention for the placement or removal of central venous access devices.

This book reviews different haemostatic products and protocols for control of haemorrhage, and for surgery in haemophilic patients with inhibitors. The natural history of inhibitors, their characterization in congenital haemophilia and the genetic basis of inhibitors development are discussed in detail. A number of therapeutic products (human F VIII, porcine F VIII, PCC, APCC, rFVIIa) and methods (plasmapheresis and protein A immunoadsorption) are also reviewed. Immune tolerance, the management of bleeding episodes, inhibitors in haemophilia B and some musculoskeletal issues are also reviewed. Finally, general surgery issues, psychosocial issues and general strategies for the management of haemophilia patients with inhibitors are considered.

The development of factor concentrates in the 1960s ushered in a new era of orthopaedic treatment for the musculoskeletal manifestations of haemophilia. Synovectomy, joint arthroplasty and arthroscopic surgery could be used to correct deformity, relieve pain and increase function. However, these procedures were not available for the patient with an inhibitor. Advances in haemostatic control have now transformed our approach to surgery in haemophilic patients with inhibitors.

We are indebted to our colleagues who have contributed chapters. We have tried to provide an in-depth analysis of the clinical issues involved in the treatment of inhibitors, and we hope that this book will provide information that will be helpful for those treating inhibitors in people with haemophilia.

E.C. Rodriguez-Merchan
C.A. Lee

PART 1

Haematology

CHAPTER 1

Natural history of inhibitors in severe haemophilia A and B: incidence and prevalence

J.M. Lusher

It is difficult to describe the 'natural history' of inhibitor antibodies in severe haemophilia, as prior to the mid-1980s most patients were evaluated for inhibitors quite infrequently. Beginning in the late 1980s, a number of prospective multicentre multinational studies of new products were designed and conducted, with inhibitor assays being carried out at specified intervals after each patient's first dose of the new product. However, therapeutic interventions (such as immune tolerance induction regimens, prophylaxis, or fairly frequent 'on demand' treatment with clotting factor) no doubt interfered with the inhibitor's 'natural history'. None the less, numerous observations and several well-conducted prospective studies [1–9] have demonstrated the following:

1 most inhibitors occur in persons with severe haemophilia A or B (in those with baseline levels of factor VIII (FVIII) or factor IX (FIX) of <0.01 U/mL);
2 most develop after relatively few (median 9–11) exposure days to FVIII or FIX;
3 most develop in early childhood;
4 some patients are 'high responders' with a brisk anamnestic response to FVIII, while others are 'low responders' [10] whose inhibitor concentration never exceeds a few Bethesda units (BU) (Fig. 1.1);
5 some inhibitors (10–30%) are transient and thus can be missed if testing is not carried out at frequent intervals;
6 many (especially FVIII inhibitors) can be completely suppressed or eradicated with an appropriate immune tolerance induction (ITI) regimen.

It has become apparent that certain genetic factors have a role in inhibitor development—certain gene defects cause an individual's haemophilia A or B [11–21] and, at least in haemophilia A, a close family member with an inhibitor [21–23]—and race, with a higher percentage of persons of African descent developing FVIII inhibitors [4–6,8,24,25]. Thus, in reviewing the literature concerning FVIII and FIX inhibitor development and 'natural history', one should critically look at the demographics of the study population, the frequency of inhibitor testing and the frequency of replacement therapy with FVIII or FIX.

As noted by Scandella, for clinical purposes, FVIII inhibitors are often thought of as one entity. However, she has shown that these inhibitors represent numerous antibodies directed against different portions of the FVIII protein. In general, major epitopes which FVIII inhibitors recognize are within the A_2, C_2, and A_3/C_1 domains, with >80% of patients having antibodies to two or more FVIII domains [26].

It is also important to define the terms incidence and prevalence. In the context of inhibitors, incidence refers to the percentage of individuals who develop an inhibitor over a specified period of time. Prevalence refers to the percentage of individuals who have an inhibitor at a certain point in time. For example, over a 6-year period, 30 of 100 persons with severe haemophilia A may have developed inhibitors (for an incidence of 30% over this time period); however, 15 of these persons may have subsequently lost them (as a result of spontaneous disappearance, or ITI) by the end of the observation period. Thus, the prevalence at that time would be 15%.

As most individuals with haemophilia (A or B) who develop an inhibitor to FVIII or FIX do so relatively early in life, the FVIII/FIX Subcommittee of the International Society on Thrombosis and Haemostasis' (ISTH) Scientific and Standardization Committee (SSC) has recommended that studies of antigenicity of new FVIII products be conducted in previously treated patients (PTPs) who have had >150 treatment exposure

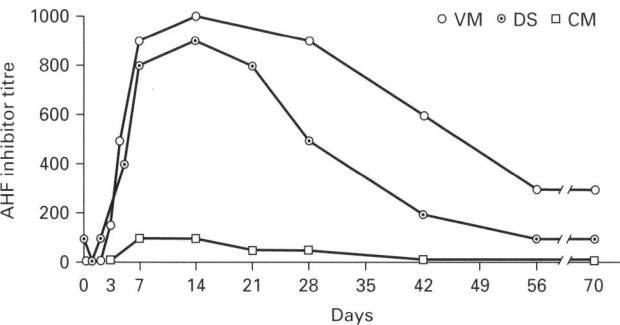

Fig. 1.1 Inhibitor profiles in three haemophiliacs following a single dose of FVIII. Inhibitor titre is expressed as the greatest dilution of patient's plasma in which antibody could be detected. ○ indicates values in patient VM, ⊙ in patient DS, and □ in CM. Note that VM and DS are 'high responders', exhibiting a marked rise in inhibitor titre, whereas CM is a 'low responder', with only a slight increase in inhibitor titre. (Reproduced with permission from [10].)

Table 1.1 FVIII inhibitor incidence, prevalence in representative prospective studies in patients with severe haemophilia A.

Reference	Product (S)	Inhibitor testing interval	Median ED until inhibitor detection	Incidence (%)	High/low responders	Prevalence at end of reporting period (%)	Duration of study reporting period (years)
Ehrenforth et al. [1]	Predominantly intermediate purity plasma-derived	Every 20 ED	11.7	52	12/2	9	15
Lusher et al.	rFVIII (Kogenate™)	3 months	9	29	10/8	17	6
Gruppo et al. [4]	rFVIII (Recombinate™)	3 months	10	30.5	7/15	11.1	6
Rothschild et al. [6]	rFVIII (Kogenate™ and Recombinate™)	3–6 months	18	28.8	5/10	19.0	3
Lusher et al. [28]	rFVIII (BDDrFVIII–ReFacto™)	3 months	9.5	32.0	16/16	8	4
Kreuz et al. [29]	rFVIII-FS (Kogenate™ FS)	Every 3–4 ED for 20 ED; every 10 ED from 21–50 ED (or every 3 months)	3–16	15	5/9	9.7	18 months*

*Early study results (study ongoing).

days (ED) to FVIII, rather than in previously untreated patients (PUPs) [27]. The reasoning being that new inhibitor development would be expected to be relatively uncommon in such PTP groups; thus, a higher rate of new inhibitor formation with a new product would be highly suggestive that the new product was responsible.

Similarly, in an attempt to look at the natural history of FVIII (and FIX) inhibitors, carefully designed and conducted prospective longitudinal studies in PUPs are most useful, despite the previously mentioned limitations related to therapeutic interventions. The first such study, conducted by Ehrenforth et al. [1] in Frankfurt, Germany, provided much interesting and valuable information. Most patients in this study received intermediate purity plasma-derived FVIII concentrates. Patients followed up included not only severely affected individuals, but some with moderate (baseline FVIII 1–5%) haemophilia. The authors obtained FVIII inhibitor assays (Bethesda method) every 20th ED as well as whenever an inhibitor was suspected. They calculated incidence and prevalence of those individuals who were severely affected separately, and reported that 52% of the latter developed inhibitors over the study period of 15 years (individual patients were followed from 0.08 to 15.25 years; median 8 years). The median number of ED until inhibitor development was 11.7 days. Eighty per cent of the patients developing inhibitors were high responders; there were no significant differences between those with high or low titre inhibitors in terms of age or number of ED when the inhibitor was first detected, or in the amount or source of FVIII received before inhibitor detection [1].

While prospective studies of new plasma-derived FVIII (and FIX) concentrates conducted in the last half of the 1980s included inhibitor assays, these PUP studies were predominantly aimed at determining hepatitis (and human immunodeficiency virus) safety. Thus, the studies were short term, and inhibitor assays were generally carried out at 6-month intervals. As a result, transient inhibitors may have been missed, as no doubt were inhibitors which developed slightly later in these young children.

However, when the first recombinant (r) FVIII concentrates entered prelicensure clinical trials, and were shown to produce no more inhibitors in PTP study cohorts than had been seen with plasma-derived FVIII (generally 0–1 new inhibitor per PTP study) prospective clinical trials in PUPs began. These included frequent testing for inhibitors. The multicentre multinational trials in PUPs with the original two rFVIII products (Kogenate™ (Bayer Corp., Berkeley, California) and Recombinate™ (Baxter–Hyland, Glendale, California)) included inhibitor assays (Bethesda method) at three-month intervals, and more often if an inhibitor was suspected. Inhibitor assays were performed in a central laboratory and, if the assay was positive, the patient was recalled for a repeat inhibitor assay and an FVIII recovery and modified half-life were feasible. It is noteworthy that the incidence of inhibitor development with these two rFVIII products was quite similar, with 28–30.5% of PUPs developing inhibitors over the study period of 5 years or 100 ED [2–5]. It is also noteworthy that the median number of ED until inhibitor development was 9–10 days, quite similar to that seen by Ehrenforth et al. in their predominantly plasma-derived FVIII treated patients. Furthermore, in each of these three studies, some patients lost their inhibitors, despite intermittent exposure to FVIII. (Others responded well to ITI regimens, generally employing frequent large doses of the study product.) Thus, the inhibitor prevalence at the end of the study periods was considerably less than the percentage of patients who had developed inhibitors [4,5,11] (Table 1.1).

In the PUP studies with Kogenate™ and Recombinate™, PUPs of African descent had a higher incidence of inhibitor development than did study PUPs of other racial groups [4,5].

This has also been observed in other studies, with low and high purity plasma-derived FVIII products and rFVIII products [6,11,24,25]. It has also been noted that inhibitors may be low titre (generally defined as remaining below 5 or 10 BU) or high titre. It is uncommon to have spontaneous disappearance of a high titre inhibitor (especially those above 10 BU), while approximately 25% of the low titre inhibitors in some studies have disappeared spontaneously (i.e. while the patient was being treated 'on demand' (as necessary)) at intermittent intervals.

An additional 'problem' in looking at the natural history of inhibitors in persons with severe haemophilia who appear to have lost their inhibitor spontaneously, or who appear to have been successfully tolerized with an ITI regimen, is that many then go on prophylactic treatment with FVIII (or FIX). In such individuals, one cannot determine whether or not their inhibitor might have recurred without frequent exposures to FVIII or FIX via prophylaxis.

It should be noted that in each of two other PUP studies conducted with rFVIII, Kogenate™ and Recombinate™—one in France and the other in Germany and Italy—the incidence of inhibitor development was similar to that observed in the multinational trials. Over the first three years of the French study which began in 1993, 15 of 52 (28.8%) severely affected PUPs (baseline FVIII <0.01 U/mL) developed inhibitors (five low titre and 10 high titre). Of the 15 inhibitors five were transient, and two of three black children developed inhibitors [6]. In the German–Italian PUP study, which also began in 1993, six of 29 (20.7%) children with severe haemophilia A developed inhibitors after a median of 9 ED (range 8–35 ED). Four of the inhibitors were low titre (defined here as <5 BU) while two were >5 BU. At the time of reporting, the median number of ED for the cohort was only 15; thus others remained at risk [30].

Until relatively recently, there was also the problem of deciding whether a patient whose Bethesda assay results [31] hovered around or just above the usual cut-off value of 0.6 BU really had a FVIII inhibitor or not. How should we classify the person who reportedly had a single assay value of 0.9 BU, with subsequent assays being <0.6 BU? The advent of the Nijmegen modification [32] of the Bethesda assay, in which the normal substrate plasma is buffered to a pH of 7.4, and FVIII-deficient plasma is substituted for imidazole buffer in the control incubation mix, has helped us to get rid of many 'low titre' false positives, as it prevents a shift in pH and reduction in protein concentration, which can lead to loss of FVIII and a 'positive' result with the original Bethesda assay. An enzyme-linked immunoabsorbent assay (ELISA) performed on the sample in question has also been helpful in determining whether or not an antibody against FVIII is present.

The PUP study with the truncated B-domainless rFVIII [33–35], ReFacto™ (Wyeth-Genetics Institute, Cambridge, MA), began in 1994, and was quite similar in design to the Kogenate™ and Recombinate™ PUP studies. All subjects had severe haemophilia A, and inhibitor assays were performed in one of two central laboratories (one for North American samples and the other for European samples). However, any sample testing positive (≥0.6 BU) by the standard Bethesda assay was tested by the Nijmegen modification; additionally, as in the Kogenate™ and Recombinate™ studies, samples obtained at these same three-month intervals underwent ELISA testing for FVIII antibodies [8].

The 101 PUPs (all severely affected) who were enrolled in this study and were treated with ReFacto™, had a median number of ED of 143 over the 5-year study period (27 September 1994–31 August 1999). Thirty-two of the 101 PUPs (32%) developed inhibitors during the course of the study, after a median of 12 ED (range 3–609). The median age at the time of first infusion for these 32 patients was 8 months (range <1–33 months), and the median age at the time of inhibitor detection was 16 months (range 1–62 months). Of the 32, 16 were high titre (defined here as ≥5 BU) and 16 low titre. Nine of the patients with low titre inhibitors underwent an ITI regimen, with a success rate of 67%. Additionally, six of 13 with high titre inhibitors had a good response to ITI during the study period. Ten inhibitor patients (mostly low titre) who did not undergo ITI had spontaneous disappearance of their inhibitors while being treated 'on demand' (at irregular intervals, as necessary for bleeding episodes) with ReFacto™. As in other study cohorts, there was a higher incidence of inhibitor development in patients of African descent. Three of three black patients and two of two Hispanic patients developed inhibitors (all of which were high titre) [28]. Thus, the types of inhibitors and number of ED until inhibitor development were quite similar to those described in clinical trials with the full-length recombinant [4,5,7,11] and plasma-derived [1,11] FVIII concentrates.

A multicentre multinational prospective study in PUPs (plus a few minimally treated patients (MTPs) with Bayer's Kogenate SF™ (formulated with sucrose rather than albumin)) began more recently (1998); however, results to date appear quite similar to those generated by the previously described trials [29,36].

While analysis of the defects in the FVIII gene responsible for each patient's haemophilia was either not carried out in a particular study, or is incomplete for the cohort, it has become clear from other studies that certain gene defects are associated with a greater likelihood of inhibitor formation [11,13,14].

From the data generated in these prospective trials in severely affected PUPs which incorporated frequent (usually every 3 months) inhibitor assays and observation periods of 5 years or more, or >100 ED, what can we conclude? In terms of the 'natural history' of inhibitors in persons with severe haemophilia A, it is apparent that most, but not all, individuals who develop FVIII inhibitors do so relatively early in life, after a median number of treatment exposures to FVIII of 9–12 (range 2–50 ED in most series, if one excludes the occasional outliers). Roughly 60% of these inhibitors are high titre and 40% low titre, never reaching 10 BU (or 5 BU). Some—particularly low titre inhibitors—appear to be transient [7], being detected on one or more occasions (some over a period of several months), but not thereafter. However, these so-called 'transient' inhibitors are the subject of controversy, in that most patients have

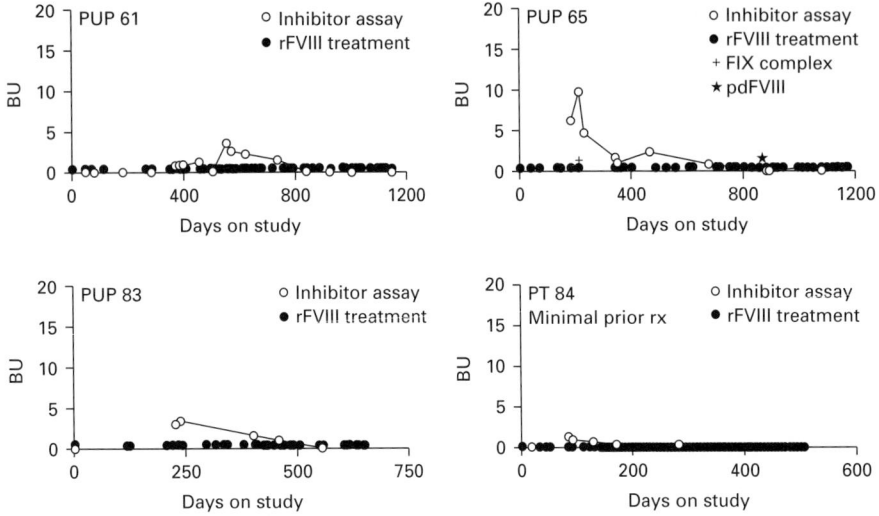

Fig. 1.2 Inhibitor assay patterns in four PUPs who received rFVIII (Kogenate™). Inhibitors were first detected after 4–7 ED. In three of these children inhibitor assays remained low (<5 BU), and then became undetectable despite continued episodic treatment with rFVIII. The fourth PUP (66), had a high titre inhibitor (10 BU) which gradually disappeared despite episodic treatment. However, each of these PUPs may actually have been 'tolerized' with their episodic exposures to rFVIII. (Reproduced with permission from [37].)

received FVIII intermittently for bleeding episodes (Figs 1.2 and 1.3), and some have been on routine prophylactic doses of FVIII. Thus, one could argue that at least some of these 'transient' inhibitors have been suppressed or eradicated by a modified ITI. Additionally, if one starts patients with low titre inhibitors on an ITI regimen too quickly, some inhibitors which may have been transient without ITI may be misclassified. It should also be noted that if a patient with an inhibitor, particularly a low titre inhibitor, does not receive any FVIII-containing product over a period of months, his or her inhibitor level may fall below the level considered to be positive (generally 0.6 BU).

However, most such individuals will have an anamnestic rise in inhibitor titre if they are again exposed to FVIII. These are not transient inhibitors.

The studies described here, as well as others, have documented that patient factors (not only severity of haemophilia, but underlying gene defect, race and family history of inhibitors) play a part in inhibitor development. Thus, depending on the demographics of the particular patient cohort being reported, the incidence and prevalence of inhibitors may vary. Perhaps because these do not differ that much in the multi-centre multinational studies which have been carried out,

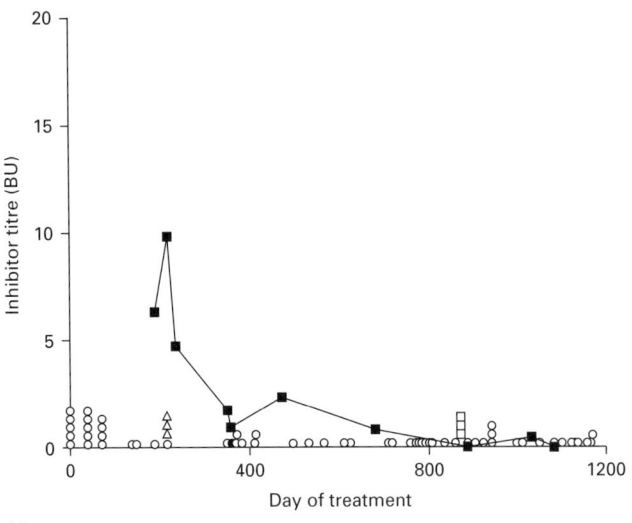
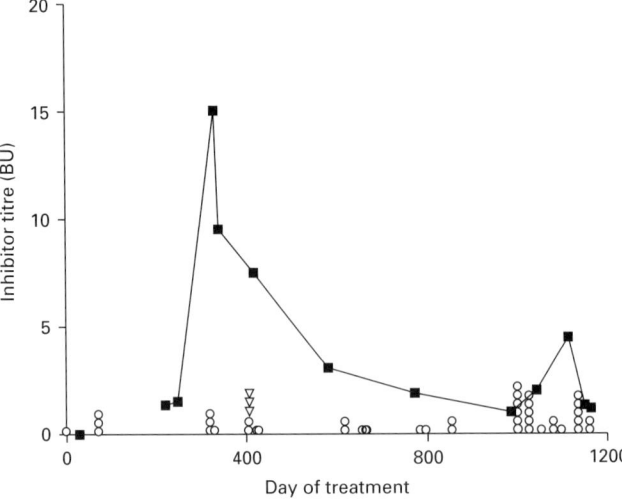

Fig. 1.3 Inhibitor assay patterns in two additional PUPs who received rFVIII (Kogenate™). (a) Shows the course of the inhibitor (first measured at 10 BU) in a child whose inhibitor gradually disappeared with fairly frequent exposures to rFVIII. (b) Shows the course of another child whose inhibitor peaked at 15 BU, but then fell and subsequently remained low despite continued episodic treatment with rFVIII. Day 0 represents the day of initial treatment with rFVIII.

o represents a single day of exposure to rFVIII. △ represents doses of factor IX complex in (a), and porcine FVIII in (b), and □ represents doses of plasma-derived FVIII. (Note the difficulty in deciding whether the child depicted in (a) has a 'transient' inhibitor, or was tolerized by his fairly frequent exposures to FVIII.) (Reproduced with permission from [2].)

incidence and prevalence after 50–100 ED have been fairly consistent in the prospective studies conducted over the past 13 years or so. The percentage of severely affected PUPs developing inhibitors over a period of 5 years or >100 ED has been 28–30% in most series, with the prevalence at the end of these studies varying between 11.1 and 19%. The latter wider range may well reflect the number of inhibitor patients tolerized in some fashion during the course of the various studies. In addition to patient genetic factors, immunological and environmental factors are thought to have a role, although this has not been substantiated.

While most FVIII inhibitors develop early in life, a few develop later [24]. Thus, there is probably no age or number of ED which make an individual completely safe from developing an inhibitor.

The situation in haemophilia B

The natural history of inhibitors in haemophilia B is in some ways easier to define in that they are relatively uncommon [20,38–40]. High [16] has speculated on the reasons for the lower percentage of persons with haemophilia B who develop inhibitors. In various cohorts, 0–3.8% of persons with severe haemophilia B have been reported to have inhibitors [1,38–40]. Because haemophilia B is much less frequent than haemophilia A, many large haemophilia centres have only one or two (or no) patients with FIX inhibitors. Perhaps because of the small numbers of patients affected, or perhaps reflecting a real difference, there have been no reports indicating that race plays a part here as it does in FVIII inhibitor development. However, there is considerable evidence that the underlying gene defect causing the individual's haemophilia B is an important determinant of inhibitor formation [16–19,41,42]. In the world database, 26 of 244 (11%) of persons with severe haemophilia B who had large deletions, nonsense or frameshift mutations developed inhibitors, whereas none of 265 with missense mutations did so. While deletions account for only 1–3% of mutations seen in haemophilia B, they account for 50% of the mutations seen in haemophilia B patients with inhibitors [19].

A unique feature of the inhibitors seen in severe haemophilia B is that FIX inhibitor development is associated with anaphylaxis or severe allergic manifestations on exposure to any FIX-containing product in approximately 50% of patients [41]. Most of these patients have complete gene deletions [42]. Because inhibitor development and severe allergic manifestations are temporally related, it has been possible to estimate the number of ED prior to inhibitor development. It appears that most FIX inhibitors (as FVIII inhibitors) occur relatively early in life, after a median of 9–11 ED [41]. The natural history of inhibitors associated with severe allergic reactions to any FIX-containing product is rather uncertain, as subsequently most of these patients have been treated with a product which contains no FIX (recombinant FVIIa) [43,44].

References

1 Ehrenforth S, Kreuz W, Scharrer I *et al*. Incidence of development of factor VIII and factor IX inhibitors in haemophilias. *Lancet* 1992; **339**: 594–8.

2 Lusher JM, Arkin S, Abildgaard CF, Schwartz RD and the Kogenate Previously Untreated Patient Study Group. Recombinant factor VIII for the treatment of previously untreated patients with hemophilia A. *N Engl J Med* 1993; **328**: 453–9.

3 Bray GL, Gomperts ED, Courter S *et al*. A multicenter study of recombinant factor VIII (Recombinate): safety, efficacy and inhibitor risk in previously untreated patients with hemophilia A. *Blood* 1994; **83**: 2428–35.

4 Gruppo R, Bray GL, Schroth P, Perry M, Gomperts ED for the Recombinate PUP Study Group. Safety and immunogenicity of recombinant factor VIII (Recombinate) in previously untreated patients (PUPs): a 6.5 year update. *Thromb Haemost* 1997; **162** (June Suppl.): Abstract PD-663.

5 Lusher JM, Arkin S, Abildgaard CF, Hurst D and the Kogenate PUP Study Group. Recombinant FVIII (Kogenate) treatment of previously untreated patients (PUPs) with hemophilia A: update of safety, efficacy and inhibitor development after seven study years. *Thromb Haemost* 1997; **162** (June Suppl.): Abstract PD-664.

6 Rothschild C, Laurian Y, Satre EP *et al*. French previously untreated patients with severe hemophilia A after exposure to recombinant factor VIII: incidence of inhibitor and evaluation of immune tolerance. *Thromb Haemost* 1998; **80**: 779–83.

7 Brown DL, Bray GL, Scharrer I. Transient inhibitors in patients with hemophilia A. *Thromb Haemost* 1999 (August Suppl.); **573**: Abstract 1804.

8 Lusher JM, Gringeri A, Hann I, Rodriguez D. Safety, efficacy and inhibitor development in previously untreated patients (PUPs) treated exclusively with B-domain deleted recombinant FVIII (BDD rFVIII). *Blood* 1999; **94** (Suppl. 1): 23A (Abstract 1037).

9 Lusher JM, Kreuz W, Gazengel C *et al*. and the International Kogenate-FS PUP/MTP Study Group. Inhibitor formation monitoring in pediatric patients with severe hemophilia A receiving a second-generation rFVIII concentrate formulated with sucrose. *Blood* 1999; **94** (Suppl. 1): 237A (Abstract 1050).

10 Lusher JM. Etiology, natural history and management of FVIII inhibitors. *Ann NY Acad Sci* 1987; **509**: 89–102.

11 Scharrer I, Bray GL, Neutzling O. Incidence of inhibitors in haemophilia A patients: a review of recent studies of recombinant and plasma-derived factor VIII concentrates. *Haemophilia* 1999; **5**: 145–54.

12 Hay CRM. Review: why do inhibitors arise in patients with haemophilia A? *Br Haematol* 1999; **105**: 584–90.

13 Kemball-Cook G, Tuddenham EGD. The factor VIII mutation database on the World Wide Web: the haemophilia A mutation, search, test and resource site HAMSTeRS update (version 3.0). *Nucleic Acids Res* 1997; **25**: 128–32.

14 Tuddenham EGD, McVey JH. The genetic basis of inhibitor development in haemophilia A. *Haemophilia* 1998; **4**: 543–5.

15 Schwaab R, Brackmann HH, Meyer C *et al*. Haemophilia A: mutation type determines risk of inhibitor formation. *Thromb Haemost* 1995; **74**: 1402–6.

16 High KA. Factor IX: molecular structure, epitopes and mutations associated with inhibitor formation. In: Aledort LM, Hoyer LW, Lusher JM, Reisner HM, White GC, eds. *Inhibitors to Coagulation Factors*. New York: Plenum Press, 1995: 79–86.

17 Ljung RCR. Gene mutations and inhibitor formation in patients with hemophilia B. *Acta Haematol* 1995; **94** (Suppl. 1): 49–52.

18 Giannelli F, Green PM. The molecular basis of haemophilia A and B. *Baillière's Best Prac Res Clin Haematol* 1996; **9**: 211–28.

19 Giannelli F, Green PM, Sommer SS *et al*. Haemophilia B (6th edn): a data base of point mutations and short additions and deletions. *Nucleic Acids Res* 1996; **24**; 103–18.

20 Shapiro AD. Unique aspects of inhibitors in patients with hemophilia B. In: Shapiro A, ed. *Inhibitors in Hemophilia: Current Perspectives and Future Directions*. National Hemophilia Foundation: New York, 2001: 7–15.

21 White GCII. Predicting inhibitor development. In: Shapiro A, ed. *Inhibitors in Hemophilia: Current Perspectives and Future Directions*. National Hemophilia Foundation: New York, 2001: 43–52.

22 Astermark J, Berntorp E, White GC, Kroner BL and the MIBS Study Group. The Malmö International Brother Study (MIBS): further support for genetic predisposition to inhibitor development. *Haemophilia* 2001; **7**: 267–72.

23 Gill GM. The role of genetics in the development of factor VIII inhibitors. In: Shapiro A, ed. *Inhibitors in Hemophilia: Current Perspectives and Future Directions*. National Hemophilia Foundation: New York, 2001: 31–42.

24 Gill GM. The natural history of factor VIII inhibitors in patients with hemophilia A. *Prog Clin Biol Res* 1984; **150**: 19–29.

25 Addiego JE, Kasper C, Abildgaard C *et al*. Increased frequency of inhibitors in African American hemophilia A patients. *Blood* 1994; **84** (Suppl. 1): 239 [Abstract].

26 Scandella D. Epitope specificity of anti-factor VIII antibodies. In: Shapiro A, ed. *Inhibitors in Hemophilia: Current Perspectives and Future Directions*. National Hemophilia Foundation: New York, 2001: 17–29.

27 White GC, DiMichele D, Mertens K *et al*. Utilization of previously treated patients (PTPs), noninfected patients (NIPs), and previously untreated patients (PUPs) in the evaluation of new factor VIII and factor IX concentrates: recommendation of the Scientific Subcommittee on factor VIII and factor IX of the Scientific and Standardization Committee of the International Society on Thrombosis and Haemostasis. *Thromb Haemost* 1999; **81**: 462.

28 Lusher JM, Shapiro A, Gruppo R, Bedrosian CL, Nguyen K and the ReFacto PUP Study Group. Safety and efficacy in previously untreated patients (PUPs) treated exclusively with B-domain deleted factor VIII (BDD rFVIII). *Thromb Haemost* 2001; Congress Suppl.: 2558 [Abstract].

29 Kreuz W, Manco-Johnson M, Gazengel C *et al*. Recombinant FVIII formulated with sucrose (rFVIII-FS) is safe and efficacious in pediatric patients with hemophilia A. *Blood* 2000; **96**: 226A [Abstract].

30 Gringeri A, Kreuz W, Escuriola-Ettinghausen C *et al*. Anti-FVIII inhibitor incidence in previously untreated patients (PUPs) with hemophilia exposed to Kogenate (German–Italian PUP Study on Inhibitors). *Thromb Haemost* 1997; **648** (Congress Suppl. June): Abstract PS-2642.

31 Kasper CG, Aledort LM, Counts RB *et al*. A more uniform measurement of factor VIII inhibitors. *Thromb Diath Haemorrh* 1975; **34**: 869–72.

32 Verbruggen B, Novakova I, Wessels H *et al*. The Nijmegen modification of the Bethesda assay for factor VIII inhibitors: improved specificity and reliability. *Thromb Haemost* 1995; **73**: 247–51.

33 Sandberg H, Almstedt A, Brandt J *et al*. Structural and functional characteristics of the B-domain deleted recombinant proteins, r-VIII SQ. *Thromb Haemost* 2001; **85**: 93–100.

34 Berntorp E. Second generation, B-domain deleted recombinant factor VIII. *Thromb Haemost* 1996; **78**: 256–60.

35 Lusher JM. Recombinant clotting factor concentrates. *Baillière's Best Prac Res Clin Haematol* 1996; **9**: 291–303.

36 Kreuz W, Gazengel C, Gorina E, Kellermann E and the European PUP/MTP Study Group. Eighteen months experience with a sucrose-formulated full-length rFVIII (rVIII-SF) in previously untreated patients (PUPs) and minimally treated patients (MTPs) with severe hemophilia A. *Ann Hematol* 2001; **80**: A37 [Abstract].

37 Lusher JM. Summary of clinical experience with recombinant factor VIII products: Kogenate. *Ann Hematol* 1994; **68**: 53–6.

38 Sultan Y and the French Hemophilia Study Group. Prevalence of inhibitors in a population of 3435 hemophilia patients in France. *Thromb Haemost* 1992; **67**: 600–2.

39 Katz J. Prevalence of factor IX inhibitors among patients with haemophilia B: results of a large-scale North American survey. *Haemophilia* 1996; **2**: 28–31.

40 Soucie JM, Evatt B, Jackson D. Occurrence of hemophilia in the United States. The Hemophilia Surveillance System Project Investigators. *Am J Hematol* 1998; **59**: 288–94.

41 Warrier I, Ewenstein BM, Koerper MA *et al*. Factor IX inhibitors and anaphylaxis in hemophilia B. *J Pediatr Hematol Oncol* 1997; **19**: 23–7.

42 Thorland EC, Drost JB, Lusher JM *et al*. Anaphylactic response to factor IX replacement therapy in haemophilia B patients: complete gene deletions confer the highest risk. *Haemophilia* 1999; **5**: 101–5.

43 Lusher JM. Inhibitor antibodies to factor VIII and factor IX: management. *Semin Thromb Haemost* 2000; **26**: 179–88.

44 Lusher JM. Inhibitors in young boys with haemophilia. *Baillière's Best Prac Res Clin Haematol* 2000; **13**: 457–68.

CHAPTER 2

Characterization of inhibitors in congenital haemophilia

K. Peerlinck and M. Jacquemin

The first occurrence of inhibitor following transfusion of factor VIII (FVIII) in a haemophilia A patient was reported by Lawrence and Johnson in 1941 [1]. The circulating anticoagulant was detected by its ability to prolong the clotting of normal blood. It was only several years later that the nature of this novel type of inhibitor was established. It was first demonstrated that the inhibitor activity was associated with the gammaglobulin fraction of serum [2]. In 1947, Craddock and Lawrence [3] finally established that the anticoagulant identified was an antibody recognizing the protein called at that time 'antihaemophilic globulin', and later FVIII. Indeed, they demonstrated that gammmaglobulin present in the patient's plasma precipitated a protein in FVIII concentrate. This phenomenon, called a 'precipitin reaction', was then considered as the hallmark of the presence of antibody. Craddock and Lawrence therefore concluded that inhibitor antibodies appeared in response to exposure to the protein lacking in the plasma of haemophilia A patients [3]. Reports by several groups further confirmed the presence of anticoagulants, which gave precipitin reaction with antihaemophilic globulin [4,5].

The isotypic distribution of anti-FVIII antibodies was reported much later, at the end of the 1960s. Among the first patients reported, most had inhibitor antibodies of the IgG isotype only [6,7]. The subclass distribution of such antibodies was studied by immunodepletion experiments. Inhibitor titres were measured before and after removal of each immunoglobulin G (IgG) subclass by immunoprecipitation with specific antiserum. Andersen and Terry [8] reported in *Nature* in 1968 the preponderance of the IgG4 subclass within inhibitor antibodies. This was the first demonstration that an IgG4 antibody could bind to a specified antigen. Indeed, at that time it was well established that IgG1, IgG2 and IgG3 were specific for particular antigens [9], but the low plasma concentration of IgG4 (a few per cent of total IgG) had precluded the study of its characteristics. Several studies using different methods confirmed these early observations on the nature of circulating anticoagulants in haemophilia A patients treated with FVIII concentrates [10,11]. Although the mechanisms regulating the production of IgG4 antibody are still incompletely understood, it is noteworthy that this IgG subclass is elicited by chronic exposure to antigens [12].

Quantification of inhibitor antibody

The Oxford and Bethesda methods

The first quantitative method for evaluating inhibitor potency in a mixing assay, the Oxford method, was developed by Biggs and Bidwell [13] in the late 1960s. The method was based on the evaluation of the rate of FVIII inactivation upon incubation of FVIII for a limited period of time with an excess of inhibitory antibody. Characterization of a series of inhibitor antibodies indicated, however, that the method did not provide quantitative results in all cases. In addition, because the assay was performed in the presence of an excess of inhibitory antibody, the method was not sufficiently sensitive enough to assess clinically relevant low titre inhibitor. The assays to quantify anti-FVIII antibodies were therefore abandoned until the characterization of the mode of action of inhibitor antibodies led to the development of more generally useful quantitative assays [14,15].

In the New Oxford method, described by Rizza and Biggs [16] in 1973, FVIII was added in excess over antibody in the assay and the incubation with inhibitor antibodies was prolonged up to 4 h. This extended incubation period allowed accurate detection of inhibitor antibodies characterized by a slow reaction rate with FVIII. The concept at the basis of this method was that, for many antibodies, there is a linear relationship between the inhibitor concentration and the log of residual FVIII following incubation with the inhibitor for a given period of time.

The principle of the New Oxford method was largely adopted for the measurement of FVIII inhibitor potency. However, differences in methodology, notably with regard to the incubation time and the source and amount of FVIII used for the assay, brought about difficulties in comparing results obtained in different centres. A standardized procedure, called the Bethesda method, was therefore proposed in 1974 by a group of investigators led by K. Kasper [17]. An incubation mixture constituted by one part citrated patient (diluted or undiluted) plasma and by an equal part of pooled normal plasma as a source of FVIII is left at 37°C for 2 h together with a control consisting of one part normal pooled plasma and one part buffer. Residual FVIII activity of both incubation mixtures is assayed; residual

FVIII activity of the control mixture is used as 100% reference in this assay. A dilution of the test mixture containing 50% of FVIII activity relative to this reference after 2 h of incubation is defined to contain one Bethesda unit (BU) of inhibitor per mL. The amount of inhibitor can be read directly from a semilogarithmic graph. This quantification method is best suited to measure inhibitors arising in haemophiliacs; the designation of BUs in an inhibitor plasma does not imply that an amount of FVIII can be calculated from this that would neutralize the patients' circulating inhibitor [17].

Quantification of inhibitors in a standardized way is useful for a meaningful comparison between laboratories: a lower inhibitor titre may give some indication as to which patients can still benefit from treatment with FVIII; a low inhibitor titre at start of immune tolerance therapy is an important predictor for treatment outcome [18,19]; and quantification of inhibitor titre can be used to study the formation and disappearance of inhibitor in the individual patient. The sensitivity and the specificity of the Bethesda assay are strongly dependent on the pH of the incubation mixtures (residual FVIII activity in the mixtures decreases with increasing pH). Therefore it is important to use buffered pooled normal plasma [20] or to add 0.1 M imidazole pH 7.4 to the normal plasma as described by Verbruggen *et al.* [21].

Detection of anti-FVIII antibodies by ELISA

In enzyme-linked immunoabsorbent assay (ELISA)-based systems both inhibitory and non-inhibitory antibodies to FVIII are detected. In a recent evaluation of a solid-phase ELISA, Martin *et al.* [22] found a sensitivity of 97.7% and a specificity of 78.4% for the detection of FVIII inhibitory antibodies. The implications of non-inhibitory antibodies are not well known, although they might have a role in FVIII clearance and half-life [23]. Although for current patient management antibodies that inhibit FVIII function seem most important, ELISA-based techniques could have a role as a method of rapid and less labour-intensive screening assay for the presence of anti-FVIII antibodies, with an impressive negative predictive value.

Epitope mapping (Fig. 2.1)

In early studies on the epitope specificity of FVIII inhibitors initiated by Fulcher *et al.* [24], highly purified plasma-derived FVIII was used. Binding sites for FVIII inhibitors were found on the FVIII light chain and on a fragment corresponding to the A2 domain. In later studies using different techniques [25–27] detailed inhibitor epitopes were specified in the A2 and C2 domains and additionally in the A3-C1 domain [28,29]. In most patients FVIII inhibitors consist of a heterogeneous mixture of antibodies that react with multiple epitopes present on FVIII [30].

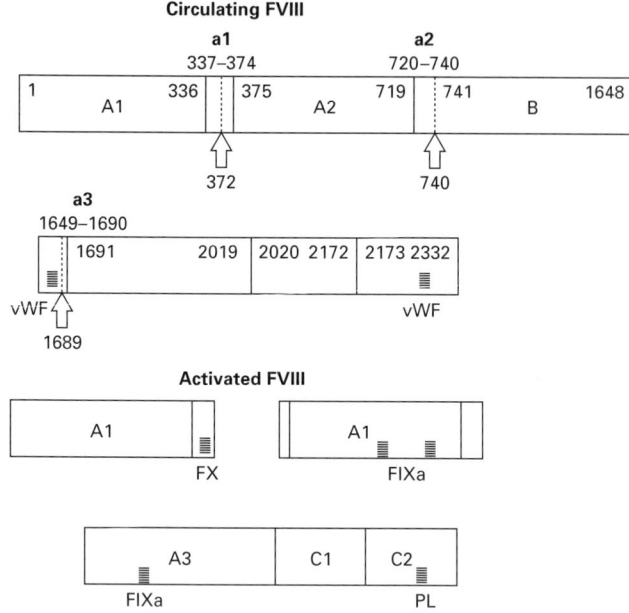

Fig. 2.1 Structure–function relationship of the FVIII molecule. This schematic representation of the FVIII structure is based on sequencing the FVIII gene and biochemical analysis of purified FVIII. Each domain of FVIII is identified by a letter followed by a numeral. The first and last amino acid residues of each domain are indicated. The small acidic regions are indicated by a1, a2 and a3, respectively. The striped boxes correspond to the binding sites for vWF, FIXa, FX and phospholipids as indicated. The thrombin cleavage sites involved in FVIII activation are indicated by arrows.

Mechanisms of FVIII inhibition

The mechanisms of FVIII inhibition were initially studied using inhibitor plasma with restricted epitope specificity. Several mechanisms by which inhibitors interfere with FVIII activity have been identified. Anti-A2 antibodies reduce the catalytic activity of the factor X (FX) activating complex by a mechanism which is not yet fully elucidated [31]. Anti-A3 antibodies prevent factor IXa (FIXa) interaction with FVIIIa [32,33]. The majority of anti-C2 antibodies prevent FVIII binding to phospholipids and to von Willebrand factor (vWF) [34,35]. However, certain rare inhibitors to the C1 or C2 domains stabilize the binding of FVIII to vWF, thereby preventing FVIII binding to phospholipids [36].

Competition with vWF for binding to FVIII provides a likely explanation for the partial FVIII inactivation observed with certain inhibitor antibodies (type II inhibitor) [37]. Indeed, upon purification from plasma and in the absence of vWF, the majority of these inhibitor antibodies completely inactivate FVIII, whereas in the presence of vWF the inactivation remains partial, like in plasma [37].

Interestingly, studies on the mechanisms of action of inhibitor antibodies have provided important insights in the understanding of FVIII structure–function relationship. Using patients' anti-FVIII antibodies, Kemball-Cook *et al.* [38]

obtained data suggesting that the FVIII binding site for phospholipids was located on the light chain. These data confirmed the results of previous experiments using recombinant FVIII fragments [39]. Arai et al. [34] then reported that inhibitor antibodies recognizing the C2 domain prevented FVIII binding to phospholipids. This observation suggested that the FVIII phospholipid-binding site was located within the C2 domain, which was later confirmed by several groups using synthetic peptides and recombinant FVIII fragments [40,41].

Patients' anti-C2 antibodies have also increased our understanding of how FVIII interacts with vWF. The observation that inhibitor antibody to C2 prevented FVIII binding to vWF suggested that the FVIII C2 domain was involved in FVIII binding to vWF [35]. This was later confirmed by demonstrating the binding of a recombinant FVIII C2 domain to vWF [42].

An additional mechanism of FVIII inactivation has recently been described for inhibitor antibodies. Anti-FVIII antibodies, purified by immunoadsorption from plasma of three haemophilia A patients with inhibitor, inactivated FVIII by proteolysis [43]. This discovery may ultimately reconcile opposing hypotheses formulated in the 1960s about the nature of FVIII inhibitor. Most investigators then believed that inhibition of FVIII was stoichiometric [13,44], while some authors suggested that the inactivation was mediated by an enzyme [45]. It would be interesting to determine the incidence of catalytic antibodies in plasma of patients with inhibitor and to compare the rate at which they can hydrolyse FVIII in a purified system [43] and in plasma.

Characterization of human monoclonal antibodies to FVIII

Anti-FVIII inhibitor antibodies proved difficult to characterize with regard to their mechanism of FVIII inhibition because of their polyclonal nature and therefore the multiple epitopes on the FVIII molecule. In addition, certain antibodies do not inhibit FVIII function [11,46]; and anti-idiotypic antibodies neutralizing FVIII inhibitors can be present in patient plasma, notably following successful desensitization by the administration of high doses of FVIII [47]. It is therefore difficult to establish a relationship between epitope specificity and the mechanisms of FVIII inactivation. To circumvent the difficulties inherent in the use of polyclonal antibodies, human monoclonal antibodies directed towards FVIII and representative of patients' pathogenic antibodies were produced by immortalization of the patients' B lymphocytes [48,49] or by phage display [50–52].

Immortalization of B lymphocytes of a severe haemophilia A patient with inhibitor with the Epstein–Barr virus allowed the production of the first human monoclonal antibody to FVIII, BO2C11 [48]. This antibody belongs to the IgG4 subclass and inhibits FVIII activity very efficiently (14 000 BU/mg). This high inhibitor potency explains why patients can have inhibitor titres as high as 3000 BU/mL plasma. For these reasons, BO2C11 is currently being evaluated by the National Institute of Biological Standards and Controls (NIBSC) and the Inhibitor Working Party of the International Society on Thrombosis and Haemostasis Scientific and Standardization Committee (ISTH/SSC) subcommittee on FVIII and FIX, as a potential inhibitor standard for standardization of inhibitor assays such as the Bethesda assay.

BO2C11 recognizes the C2 domain and inhibits FVIII binding to both phospholipids and to vWF. It was therefore of interest to determine the epitope recognized by this antibody. The crystal structure of a Fab fragment of BO2C11 bound to the C2 domain was solved recently [52]. As shown in Plate 2.1 (facing p. 118), BO2C11 makes direct contact with three groups of hydrophobic residues on the C2 domain surface, which were predicted to insert into the phospholipid membrane [53]. Two Asp residues in BO2C11 also interact with two basic residues, which were predicted to interact with charged phospholipid head groups upon binding of the C2 domain to the membrane [53]. The ability of BO2C11 to prevent FVIII binding to phospholipids is therefore easily understandable.

BO2C11 also inhibits FVIII binding to vWF. Residues mediating BO2C11 binding to the C2 domain could also be involved in binding to vWF. However, it cannot be excluded that BO2C11 prevents FVIII binding to vWF by inducing allosteric alteration of the FVIII molecule or by steric hindrance. Nevertheless, the recent observation that mutation of residue 2201 in the C2 domain, as seen in some haemophilia A patients, alters FVIII binding to vWF [54], suggests that BO2C11 binds to residues involved directly in FVIII–vWF interactions.

The recent characterization of another human monoclonal antibody, LE2E9, has resulted in the identification of an unusual mechanism of FVIII inactivation. LE2E9 recognizes the factor VIII C1 domain and inhibits FVIII activity with a high efficacy (7000 BU/mg antibody) [49]. This indicates that the C1 domain is also a possible target for inhibitor antibodies, although the incidence of inhibitors recognizing this domain remains uncertain, because of the difficulty in producing recombinant C1 domain.

LE2E9 is interesting for several additional reasons. It inhibits only partially FVIII activity (type II inhibitor antibody), and interferes with FVIII binding to both vWF and phosphatidylserine. LE2E9 does not completely inactivate FVIII, even when the antibody is present in large excess over FVIII. This partial FVIII inactivation is also observed when LE2E9 is incubated with recombinant FVIII in the absence of vWF [49]. This was unexpected as most type II inhibitor antibodies identified so far inactivate FVIII partially only in the presence of vWF, whereas in the absence of the latter, the inhibition is complete [37].

The observation that LE2E9 inhibits FVIII binding to vWF prompted investigations to determine whether the C1 domain contributes to FVIII binding to vWF. Evaluation of FVIII interaction with vWF indicated that mutations located in the C1 domain and responsible for mild–moderate haemophilia A could indeed reduce FVIII binding to vWF [55–57]. However, it is not yet established whether residues in the C1 domain make direct contacts with vWF or only indirectly contribute to the three-dimensional structure of the light chain, thereby allowing FVIII binding to vWF.

The observation that LE2E9 interfered with FVIII binding to phosphatidylserine was also unexpected because FVIII binding to phospholipids is attributed to the C2 domain [34,40,41]. Recent analysis of the three-dimensional structure of FVIII bound to phospholipids did not indicate a direct role for the C1 domain [58], contrary to the C1 domain of factor V which was recently suggested to contribute to phospholipid binding [59]. It is thus possible that LE2E9 interferes with phospholipid binding by steric hindrance. Further study of LE2E9 interactions with FVIIIa in the tenase complex is expected to provide more information about the mechanism responsible for the partial FVIII inactivation induced by this antibody.

References

1 Lawrence JS, Johnson JB. The presence of a circulating anticoagulant in a male member of a hemophiliac family. *Trans Am Clin Climatol Assoc* 1941; **57**: 223.

2 Munro FL, Munro MP. Electrophoretic isolation of a circulating anticoagulant. *J Clin Invest* 1946; **25**: 814.

3 Craddock CG, Lawrence JS. A report of the mechanism of the development and action of an anticoagulant in two cases. *Blood* 1947; **6**: 505.

4 Frommeyer WB, Epstein RD, Taylor FH. Refractoriness in hemophilia to coagulation-promoting agents: whole blood and plasma derivatives. *Blood* 1950; **5**: 401–20.

5 Conley CL, Rathbun HK, Morse WI, Robinson JE. Circulating anticoagulant agent as a cause of hemorrhagic diathesis in man. *Bull Johns Hopkins Hosp* 1948; **83**: 288–96.

6 Shapiro SS. The immunologic character of acquired inhibitors of antihemophilic globulin (factor 8) and the kinetics of their interaction with factor 8. *J Clin Invest* 1967; **46**: 147–56.

7 Poulik MD, Lusher J. Immunological characterization of an antihemophilic globulin (AHG) antibody. *Fed Proc* 1967; **26**: 312.

8 Andersen BR, Terry WD. Gamma-G4-globulin antibody causing inhibition of clotting factor VIII. *Nature* 1968; **217**: 174–5.

9 Terry WD, Fahey JL. Subclasses of human g2-globulin based on differences in the heavy polypeptide chains. *Science* 1964; **146**: 400–1.

10 Fulcher CA, de Graaf Mahoney S, Zimmerman TS. FVIII inhibitor IgG subclass and FVIII polypeptide specificity determined by immunoblotting. *Blood* 1987; **69**: 1475–80.

11 Gilles JG, Arnout J, Vermylen J, Saint-Remy JM. Anti-factor VIII antibodies of hemophiliac patients are frequently directed towards nonfunctional determinants and do not exhibit isotypic restriction. *Blood* 1993; **82**: 2452–61.

12 Aalberse RC, van der Gaag R, van Leeuwen J. Serologic aspects of IgG4 antibodies. I. Prolonged immunization results in an IgG4-restricted response. *J Immunol* 1983; **130**: 722–6.

13 Biggs R, Bidwell E. A method for the study of antihaemophilic globulin inhibitors with reference to six cases. *Br J Haematol* 1959; **5**: 379.

14 Biggs R, Austen DE, Denson KW, Borrett R, Rizza CR. The mode of action of antibodies which destroy factor VIII. II. Antibodies which give complex concentration graphs. *Br J Haematol* 1972; **23**: 137–55.

15 Biggs R, Austen DE, Denson KW, Rizza CR, Borrett R. The mode of action of antibodies which destroy factor VIII. I. Antibodies which have second-order concentration graphs. *Br J Haematol* 1972; **23**: 125–35.

16 Rizza CR, Biggs R. The treatment of patients who have factor-VIII antibodies. *Br J Haematol* 1973; **24**: 65–82.

17 Kasper CK, Aledort L, Aronson D *et al*. Proceedings: a more uniform measurement of factor VIII inhibitors. *Thromb Diath Haemorrh* 1975; **34**: 612.

18 Mariani G, Ghirardini A, Bellocoo R. Immunetolerance in hemophilia: principal results from the international registry. *Thromb Haemost* 1994; **72**: 155–8.

19 DiMichele DM, Kroner BL and the ISTH Factor VIII/IX Subcommittee. Analysis of the North American Immune Tolerance Registry (1993–97): current practice implications. *Vox Sang* 1999; **77** (Suppl. 1): 31–2.

20 Kitchen S, McCraw A. *Diagnosis of Haemophilia and Other Bleeding Disorders: a Laboratory Manual*. World Federation of Hemophilia, Montreal 2000: 72–6.

21 Verbruggen B, Novakova I, Wessels H *et al*. The Nijmegen modification of the Bethesda assay for Factor VIII:c inhibitors: improved specificity and reliability. *Thromb Haemost* 1995; **73**: 247–51.

22 Martin PG, Sukhu K, Chambers E, Giangrande PLF. Evaluation of a novel ELISA screening test for detection of factor VIII inhibitory antibodies in haemophiliacs. *Clin Lab Haematol* 1999; **21**: 125–8.

23 Dazzi F, Tison T, Vianello F. High incidence of antiFVIII antibodies against non-coagulant epitopes in haemophilia A patients: a possible role for the half-life of transfused FVIII. *Br J Haematol* 1996; **93**: 688–93.

24 Fulcher CA, DeGraaf Mahoney S, Roberts JR. Localization of human factor VIII epitopes to two polypeptide fragments. *Proc Natl Acad Sci USA* 1985; **82**: 7728–32.

25 Scandella D, DeGraaf Mahoney S, Mattingly M. Epitope mapping of human factor VIII inhibitor antibodies by deletion analysis of factor VIII fragments expressed in *Escherichia coli*. *Proc Natl Acad Sci USA* 1988; **85**: 6152–6.

26 Healy JF, Lubin IM, Nakai H. Residues 484–508 contain a major determinant of the inhibitory epitope in the A2 domain of human factor VIII. *J Biol Chem* 1995; **270**: 14505–9.

27 Lubin IM, Healy JF, Barrow RT. Analysis of the human factor VIII, A2 inhibitor epitope by alanine scanning mutagenesis. *J Biol Chem* 1997; **272**: 30191–5.

28 Fijnvandraat K, Celie PHN, Turenhout EAM. A human alloantibody interferes with binding of FIXa to the factor VIII light chain. *Blood* 1998; **91**: 2347–52.

29 Zhong D, Saenko EL, Shima M. Some human inhibitor antibodies interfere with factor VIII binding to factor IX. *Blood* 1998; **92**: 136–42.

30 Prescott R, Nakai H, Saenko EL. The inhibitor antibody response is more complex in hemophilia A patients than in most nonhemophiliacs with factor VIII autoantibodies. *Blood* 1997; **89**: 3663–71.

31 Lollar P, Parker ET, Curtis JE *et al*. Inhibition of human factor VIIIa by anti-A2 subunit antibodies. *J Clin Invest* 1994; **93**: 2497–504.

32 Fijnvandraat K, Celie PH, Turenhout EA *et al*. A human alloantibody interferes with binding of factor IXa to the factor VIII light chain. *Blood* 1998; **91**: 2347–52.

33 Zhong D, Saenko EL, Shima M, Felch M, Scandella D. Some human inhibitor antibodies interfere with factor VIII binding to factor IX. *Blood* 1998; **92**: 136–42.

34 Arai, M, Scandella D, Hoyer LW. Molecular basis of factor VIII inhibition by human antibodies: antibodies that bind to the factor

VIII light chain prevent the interaction of factor VIII with phospholipid. *J Clin Invest* 1989; **83**: 1978–84.
35 Shima M, Scandella D, Yoshioka A *et al*. A factor VIII neutralizing monoclonal antibody and a human inhibitor alloantibody recognizing epitopes in the C2 domain inhibit factor VIII binding to von Willebrand factor and to phosphatidylserine. *Thromb Haemost* 1993; **69**: 240–6.
36 Saenko EL, Shima M, Gilbert GE, Scandella D. Slowed release of thrombin-cleaved factor VIII from von Willebrand factor by a monoclonal and a human antibody is a novel mechanism for factor VIII inhibition. *J Biol Chem* 1996; **271**: 27424–31.
37 Gawryl MS, Hoyer LW. Inactivation of factor VIII coagulant activity by two different types of human antibodies. *Blood* 1982; **60**: 1103–9.
38 Kemball-Cook G, Edwards SJ, Sewerin K, Andersson LO, Barrowcliffe TW. Factor VIII procoagulant protein interacts with phospholipid vesicles via its 80 kDa light chain. *Thromb Haemost* 1988; **60**: 442–6.
39 Bloom JW. The interaction of rDNA factor VIII, factor VIIIdes-797-1562 and factor VIIIdes-797-1562-derived peptides with phospholipid. *Thromb Res* 1987; **48**: 439–48.
40 Foster PA, Fulcher CA, Houghten RA, Zimmerman TS. Synthetic factor VIII peptides with amino acid sequences contained within the C2 domain of factor VIII inhibit factor VIII binding to phosphatidylserine. *Blood* 1990; **75**: 1999–2004.
41 Saenko EL, Scandella D, Yakhyaev AV, Greco NJ. Activation of factor VIII by thrombin increases its affinity for binding to synthetic phospholipid membranes and activated platelets. *J Biol Chem* 1998; **273**: 27918–26.
42 Saenko EL, Shima M, Rajalakshmi KJ, Scandella D. A role for the C2 domain of factor VIII in binding to von Willebrand factor. *J Biol Chem* 1994; **269**: 11601–5.
43 Lacroix-Desmazes S, Moreau A, Sooryanarayana Bonnemain C *et al*. Catalytic activity of antibodies against factor VIII in patients with hemophilia A. *Nat Med* 1999; **5**: 1044–7.
44 Leitner A, Bidwell E, Dike G. An antihaemophilic globulin (factor VIII) inhibitor: purification, characterization, and reaction kinetics. *Br J Haematol* 1963; **9**: 245.
45 Breckenridge R, Ratnoff OD. Studies on the nature of the circulating anticoagulant directed against antihemophilic factor. *Blood* 1962; **20**: 1962.
46 Nilsson IM, Berntorp E, Zettervall O, Dahlback B. Noncoagulation inhibitory factor VIII antibodies after induction of tolerance to factor VIII in hemophilia A patients. *Blood* 1990; **75**: 378–83.
47 Gilles JG, Desqueper B, Lenk H, Vermylen J, Saint-Remy JM. Neutralizing anti-idiotypic antibodies to factor VIII inhibitors after desensitization in patients with hemophilia A. *J Clin Invest* 1996; **97**: 1382–8.
48 Jacquemin MG, Desqueper BG, Benhida A *et al*. Mechanism and kinetics of factor VIII inactivation: study with an IgG4 monoclonal antibody derived from a hemophilia A patient with inhibitor. *Blood* 1998; **92**: 496–506.
49 Jacquemin M, Benhida A, Peerlinck K *et al*. A human antibody directed to the factor VIII, C1 domain inhibits factor VIII cofactor activity and binding to von Willebrand factor. *Blood* 2000; **95**: 156–63.
50 van den Brink EN, Turenhout EA, Bank CM *et al*. Molecular analysis of human anti-factor VIII antibodies by V gene phage display identifies a new epitope in the acidic region following the A2 domain. *Blood* 2000; **96**: 540–5.
51 van den Brink EN, Turenhout EA, Bovenschen N *et al*. Multiple VH genes are used to assemble human antibodies directed toward the A3–C1 domains of factor VIII. *Blood* 2001; **97**: 966–72.
52 van Den Brink EN, Turenhout EA, Davies J *et al*. Human antibodies with specificity for the C2 domain of factor VIII are derived from VH1 germline genes. *Blood* 2000; **95**: 558–63.
53 Spiegel PC Jr, Jacquemin M, Saint-Remy JM, Stoddard BL, Pratt KP. Structure of a factor VIII, C2 domain–immunoglobulin G4kappa Fab complex: identification of an inhibitory antibody epitope on the surface of factor VIII. *Blood* 2001; **98**: 13–9.
54 Pratt KP, Shen BW, Takeshima K *et al*. Structure of the C2 domain of human factor VIII at 1.5 A resolution. *Nature* 1999; **402**: 439–42.
55 Liu ML, Shen BW, Nakaya S *et al*. Hemophilic factor VIII, C1- and C2-domain missense mutations and their modeling to the 1.5-angstrom human C2-domain crystal structure. *Blood* 2000; **96**: 979–87.
56 Gilles JG, Lavend'homme R, Peerlinck K *et al*. Some factor VIII (FVIII) inhibitors recognise a FVIII epitope (s) that is present only on FVIII–vWF complexes. *Thromb Haemost* 1999; **82**: 40–5.
57 Jacquemin M, Lavend'homme R, Benhida A *et al*. A novel cause of mild–moderate hemophilia A: mutations scattered in the factor VIII, C1 domain reduce factor VIII binding to von Willebrand factor. *Blood* 2000; **96**: 958–65.
58 Stoylova S, Villoutreix B, Kemball-Cook G, Mertens K, Holzenburg A. Structure of membrane-bound Factor VIII. *Thromb Haemost* 2001; Suppl. (ISSN 0340-6245)
59 Saleh M, Quinn-Allen MA, Macedo-Ribeiro S *et al*. Identification of a putative phosphatidylserine binding site within the C1 domain of human factor V using alanine scanning mutagenesis. *Thromb Haemost* 2001; Suppl. (ISSN 0340-6245)

CHAPTER 3

Incidence and prevalence of inhibitors and type of blood product in haemophilia A

T.T. Yee and C.A. Lee

Inhibitors and haemophilia treatment

Case reports of patients with haemophilia developing a refractory state to the administration of blood, plasma or its derivatives during the course of their treatment has been described since the early 1940s [1]. The development of a circulating anticoagulant after repeated transfusions was observed in these patients and further transfusion caused neither a reduction in clotting time nor clinical improvement. Detailed studies on the nature of the anticoagulant revealed that the anticoagulant activity was associated with the gammaglobulin fraction of the plasma.

In the 1940s, when the process of Cohn fractionation of plasma was just being developed Davidson *et al.* wrote: 'the refractory state to blood and its derivatives follows the repeated administration of blood, plasma or the antihaemophilic globulin fraction (also known as Cohn Fraction 1) and arises during or promptly after a haemorrhagic episode. The exact nature of this refractory state is still obscure, but recent work has suggested that there may be a production of antibodies to the antihaemophilic substance'. What is more interesting and perhaps prophetic is the footnote that was referenced in this particular passage of the article which states: 'Presently available evidence suggests that this refractory state may occur more frequently following the administration of the antihaemophilic globulin fraction than following administration of blood or plasma. The therapeutic use of the antihaemophilic globulin fraction cannot be advised therefore until further studies have eliminated this hazard' [2].

State of the art replacement products are now made from either plasma-derived or recombinant proteins and the evolution of haemophilia replacement therapy over the years is shown in Table 3.1. There has been a dramatic improvement in the therapeutic products available for patients with haemophilia but the relative risk of inhibitor formation with the different products available today is still an ongoing issue 50 years after the interesting footnote was written.

The spontaneous occurrence of factor VIII (FVIII) inhibitors in patients with haemophilia is very rare and almost always follows replacement therapy either in the form of plasma, cryoprecipitate or specific clotting factor concentrates. Although

Table 3.1 History of haemophilia A treatment.

Decade	Milestone
1840s	First blood transfusion
1940s	Transfusion therapy established
1950s	Fresh frozen plasma; antihaemophilic globulin (Cohn Fraction-1)
1960s	Cyroprecipitate
1970s	Intermediate purity factor concentrates; DDAVP
1980s	Monoclonal antibody-purified and high-purity FVIII concentrates; effective viral inactivation
1990s	Recombinant FVIII
2000s	Second generation (albumin-free) recombinant FVIII products; gene therapy

tremendous achievement in haemophilia management has resulted in the availability of the most technologically advanced and safer clotting factor concentrates, inhibitor development persists as a major complication of treatment for a subset of patients with haemophilia. These inhibitors are highly specific immunoglobulins which rapidly inactivate infused FVIII, making replacement therapy difficult and challenging. The presence of an inhibitor makes treatment of bleeding episodes difficult, often prevents the possibility of starting a child on prophylaxis, often makes surgical procedures high risk, and makes overall management far more time-consuming and expensive.

Current treatment regimens

The mainstay of successful haemophilia therapy for either treatment or prevention of acute haemorrhage is prompt and adequate intravenous replacement of the deficient clotting factor to haemostatic plasma levels. All plasma-derived products have similar haemostatic efficacy and undergo a viral attenuation step during the purification process; however, there is a wide variability among concentrates with respect to final product purity, as defined by units of FVIII-specific activity/mg protein.

Several factors determine the choice of concentrate from the many different products available. Cost of the concentrates is one of the most important issues and has a major impact on haemophilia care. Purity and safety are the other factors influencing the clinician's choice of therapeutic products for haemophilia treatment.

The intermediate purity FVIII concentrates contain FVIII, von Willebrand factor (vWF) and plasma proteins such as immunoglobulins, fibrinogen, immune complexes, cytokines such as transforming growth factor β (TGF-β) and many others. These concentrates are produced predominantly by conventional precipitation techniques and have a FVIII content of 2–5 IU/mg of protein.

High purity FVIII concentrates do not contain other contaminating proteins and are produced either by protein precipitation and chromatographic separation or immuno-affinity chromatography using monoclonal antibodies directed against FVIII or vWF. The former method produces a high purity FVIII product that contains 50–200 IU FVIII/mg protein and a variable amount of vWF. The immunochromatography produces much purer concentrates with FVIII activity of 3000 IU/mg which requires addition of stabilizers such as albumin and sugar to make it suitable for storing and clinical use, resulting in a final FVIII activity of 5–30 IU/mg protein.

The isolation and cloning of the cDNA for human FVIII in 1984 paved the way for the biosynthesis of genetically engineered FVIII in cultured mammalian cells. The currently used recombinant factor VIII (rFVIII) products have been made by the insertion in mammalian cells of the cDNA which encodes the entire FVIII or the B domainless FVIII protein. There are certain differences in the preparation of these recombinant products with respect to the mammalian cell line used to express FVIII. The host cells are Chinese hamster ovary cells for Recombinate™ (Baxter) and rFVIII-SQ (B domainless FVIII product) and baby hamster kidney cells for the Kogenate™ (Bayer) product. FVIII is coexpressed with vWF in Recombinate™ and not with the other two products. The rFVIII which is isolated from the cell line medium and subsequently purified, has a specific activity of 2000–5000 IU/mg and contains traces of hamster proteins (derived from the mamalian cell lines), bovine proteins (from the culture medium) and mouse immunoglobulin G (from the purification step). However, the specific activity of the final recombinant product is much lower (5–20 IU/mg) because of the addition of human albumin to ensure stability. The second generation of rFVIII (rFVIII-SQ) is stable without addition of albumin and the specific activity is extremely high at 15 000 IU/mg compared to the first-generation products. More second-generation rFVIII products without albumin are being manufactured and undergoing clinical safety and pharmacokinetic studies [3].

As the therapeutic options for FVIII has improved substantially over the past few years, one of the most complex and contentious clinical issues currently facing the haemophilia physician is whether the source material and purity of therapeutic concentrates are related to the risk of inhibitor development in recipients.

Table 3.2 Factors influencing the development of FVIII inhibitors.

The patient
Molecular defect
Immunological response characteristics
Other immunological challenges at time of FVIII infusion
(vaccinations, infections, medications such as interferon)
Ethnicity
Family history of inhibitors

Therapy
Type of FVIII product (purity as well as viral inactivation method)
Number of exposures
Pattern of exposures
Cumulative exposure
Effects of exposure to several different products

Pathogenesis of FVIII inhibitors

The pathogenesis of FVIII inhibitors is still not fully understood. The capacity to mount an immune response toward FVIII obviously varies from one individual to another and depends on the interaction of several factors. These can be broadly divided into host-related factors and therapy-related factors. Several patient-related characteristics that render an individual susceptible to develop an anti-FVIII response have been examined to some extent [4,5]. The factor VIII gene mutation is one of the most important risk factors as 90–97% of inhibitors occur in patients with severe haemophilia A. Large databases of FVIII mutations have identified certain genotypes, particularly major gene deletions, intron 22 inversions and nonsense mutations to confer an increased risk of inhibitor formation [6]. Yet the risk of developing an inhibitor cannot be related to the FVIII genotype alone, as patients with haemophilia from the same kindred are often discordant for inhibitor formation, indicating that patients with the same gene defect may respond differently to factor VIII treatment, so the immunological response of the individual characteristics are also important [7,8]. Therapy-related factors (e.g. treatment schedule, FVIII product used, factors that modulate the immune response at the time of FVIII infusions) may also play an important part in the development of inhibitors and evidence is accumulating of the role for the non-genetic factors affecting the immune response [9,10]. Table 3.2 shows the patient- and therapy-related variables that can influence the inhibitor incidence.

Type of FVIII product as a risk factor

There is still a significant amount of disagreement and debate regarding the relative risk of inhibitor formation in conjunction with the use of factor VIII products of different purity. The availability of recombinant concentrates has added an additional dimension to the issue of inhibitor risk of whether the source as well the purity of FVIII product poses an additional

Table 3.3 The differences in product associated inhibitors and the classic inhibitors.

	Product-associated inhibitors	Classic inhibitors
Type of inhibitor	Type 2	Type 1
Detection of inhibitor	>50 exposure days	<50 exposure days
Patients' characteristics	Older age group and PTPs	Younger age and PUPs
Response to treatment	Stop particular product and switch to another concentrate	Immune tolerance
Clinical course	More soft tissue/skin bleeds than joint bleeds	Joint bleeds
Characteristics of the inhibitors	Relatively slow acting inhibitors	Fast acting inhibitors
	IgG accounts for majority of the inhibitory activity	
	Majority of the antibodies are directed against the FVIII light chain	Antibodies can be directed against both the heavy and light chain of FVIII

PTP, previously treated patient; PUP, previously untreated patient.

risk to the development of inhibitors. It is difficult to address the question of whether some FVIII products are safer and less likely to induce inhibitor formation than others as the frequency of inhibitors depends on so many variables.

The first instance of product-related inhibitor development occurred in May 1990, when patients in Belgium and The Netherlands were exposed to a new plasma-derived FVIII product (FVIII:CPS-P in The Netherlands and FVIII:P in Belgium) which was purified by controlled-pore silica adsorption and pasteurized at 60°C for 10 h for viral inactivation. This inhibitor outbreak occurred in 19 patients with haemophilia of whom 15 had ≥200 exposure days to other FVIII products without developing an inhibitor. The majority of the inhibitors arose after a mean of 157 exposure days to the new product FVIII:CPS-P, which is also unusual. These inhibitors disappeared in all but one case after the patients were switched to other FVIII concentrates, demonstrating the direct role of FVIII:CPS-P in the immune response [10,11].

The second outbreak of product-related FVIII inhibitors occurred when a new double virus inactivated FVIII concentrate (Bisanect™ in Belgium and Octavi SDPlus™ elsewhere in Europe) purified by ion exchange chromatography and treated by solvent detergent and pasteurization was introduced in 1995. Similar to FVIII:CPS-P, Bisanect/FVIII:SDPlus induced an inhibitor outbreak in multitransfused haemophilia patients, and the inhibitors disappeared after patients were switched to other FVIII products [12,13]. This category of inhibitors occurring in previously treated patients (PTPs) as a result of exposure to a particular FVIII concentrate have certain characteristics that differ from the FVIII inhibitors that are produced at the beginning of treatment in previously untreated patients (PUPs) (Table 3.3). They developed in an older group of patients who had received >100 exposure days to other FVIII products and had a low risk for inhibitor formation. The majority of these inhibitors also arose after >50 exposure days to that particular pasteurized product, which is also unusual. They had complex inhibition kinetics with an epitope specificity towards the C2 domain of the FVIII light chain [12]. The inhibitors gradually declined in titre when the particular concentrate was stopped, suggesting that successful eradication of the inhibitors was caused by elimination of antigen stimulation rather than induction of immunotelerance. There was also no anamnestic response to FVIII regardless of the type of FVIII product which was subsequently used [10–13].

Factor VIII antigenicity

Product-related outbreaks have attracted more attention toward the evidence that FVIII concentrates could vary in their immunogenicity. Variations in the antigenicity of FVIII can occur between concentrates obtained by different preparation procedures, and also between batches of the same preparation. Indeed, very subtle alterations of FVIII, such as those involving its three-dimensional conformation, can have major conseqences for its antigenicity. As these changes do not necessarily cause significant changes in molecular weight or procoagulant function of the molecule, the current analytical methods commonly used for quality control are not able to detect the batch-to-batch differences. The reasons as to why FVIII can become more immunogenic have been reviewed [14]. FVIII can become more immunogenic by the preparation procedure and/or methods used for viral inactivation as evidenced by the two recent outbreaks of inhibitors. Examples of this are cleavage by proteolytic enzymes and dissociation from vWF, as can result from chromatography over ion exchangers. Novel antigenic determinants can also be created by heating the molecule, which could alter its three-dimensional conformation or add sugar residues by the formation of Schiff's bases. Glycosylation is another parameter which could be important with respect to immunogenicity. The degree of glycosylation in recombinant products, which depends on the expression system used, differ and can have a role in the overall antigenicity of the recombinant molecule.

The influence of factors contaminating FVIII concentrates on the overall antigenicity of the FVIII molecule is little known. A significant amount of biologically active materials such as soluble cytokine receptors, natural antibodies to cytokines, major histocompatibility complex determinants, etc., which are unrelated to FVIII can be found in plasma and can contaminate plasma-derived FVIII concentrates. One of these factors is TGF-β, the level of which in concentrates has been correlated with the so-called 'immunosuppressive' properties of FVIII preparations [15]. A more systemic determination of the presence of these factors in various plasma-derived FVIII concentrates and their role in immunomodulating the FVIII product may clarify some of the issues regarding the risk of inhibitors depending on the purity of the products.

Extensive studies on the FVIII products that gave rise to an outbreak of product-related inhibitors in The Netherlands, Germany and Belgium have been carried out. Biochemical and immunochemical studies of the plasma-derived high purity double virus inactivated FVIII concentrate (marketed as Octavi SDPlus™ in Germany and Bisanect™ in Belgium) revealed the presence of a 40-kDa peptide fragment in some batches. Characterization of this peptide material indicated that it was derived from FVIII, as it reacted in the immunoblot with the monoclonal antibody 530p, which is specific for the FVIII heavy chain (A2 domain) [16]. Depending on the source and the quality of the plasma, the amount of 40 kDa peptide varied. These peptides had resulted from partial autolysis of recovered plasma which occurred while the plasma was awaiting flash-freezing prior to fractionation. It was subsequently shown that there was a strong correlation between inhibitor development and batches containing the 40-kDa marker and elevated markers of coagulation activation (fibrinopeptide A, FPA in particular). The study concluded that the inhibitor potential (neoantigenicity) arose as a result of two effects:
1 degradation of FVIII already present in the source material; and
2 heating of unstable FVIII degradation products [16].

Inhibitor studies

Although there were no formal epidemiological studies of inhibitors in haemophilia in the early days, several investigators have reported data on their prevalence and on some characteristics of the affected patients since the early 1950s [17–19]. In the UK, a database on the complications of haemophilia therapy was started in 1969 by the UK Haemophilia Centre Directors Organization (UKHCDO) [20] and a recent updated report by the UKHCDO on the prevalence of FVIII inhibitors in patients with haemophilia of all severities is approximately 6%, with an annual inhibitor incidence of 3.5 of 1000 severe haemophilia A patients registered with the UKHCDO [21].

Several studies published prior to 1985 indicate that 6–15% of patients with severe haemophilia A develop inhibitors [22–24]. When heat treatment for viral inactivation of FVIII concentrates was introduced in the mid 1980s, there was concern that heat treatment could produce neoantigens on the factor VIII protein that might increase the risk of inhibitor induction. However, no systematic evaluation for possible product-related changes in inhibitor risk was carried out, as concern over HIV and other blood-borne viruses was the most overwhelming issue at that time and heat treatment was essential as a viral inactivation procedure.

When ultrapure FVIII concentrates prepared by monoclonal antibody affinity chromatography became available, similar fears were expressed regarding the increased risk of inhibitor formation with the new highly purified FVIII preparations. Data for FVIII inhibitor incidence in PUPs treated only with high purity FVIII were limited as most studies were small and were designed to evaluate infectious disease test markers rather than the incidence of inhibitors. In the clinical investigation of ultrapure plasma-derived FVIII concentrates Monoclate™ (Armour) and Haemophil M™ (Baxter), cumulative incidences for inhibitor formation were 18% (seven of 38) and 13% (four of 30), respectively [25,26]. There were some suggestions that the incidence rates were high and that the age at onset was shifting towards younger patients compared to historical data on inhibitor risk in patient cohorts treated with intermediate purity products [27].

Since the initiation of rFVIII clinical trials in the mid to late 1980s, much attention has been focused on the potential of these preparations to elicit the formation of inhibitors in haemophilia A recipients. While *in vitro* studies of structure and function suggested that recombinant and plasma-derived FVIII are identical, one cannot ignore the fact that these preparations are not derived from human cells. Thus there is theoretical concern that subtle changes in the post-translational processing and/or tertiary structure of FVIII may result in the formation of neoantigens that might render these preparations more immunogenic than plasma-derived concentrates. Certain features have been incorporated into the design of rFVIII studies of PUPs in an effort to circumvent some of the limitations of prior inhibitor studies, such as regular evaluation for inhibitor development once every three months, the longitudinal evaluation of subjects over at least 100 rFVIII exposure days and assessment of recovery at 30 min postinfusion in study subjects at least once every six months to detect the low level transient inhibitors which might escape detection in a Bethesda assay. Altogether there have been six rFVIII studies in PUPs and a total of 360 severely affected PUPs have been enrolled. Although the trials differed somewhat in design there were important similarities in results to date and these have shown that 22–31% of severely affected PUPs treated with rFVIII develop inhibitors, many of which are low titre and transient [28].

There is a large difference in reported incidence rates of inhibitors in patients treated with virus-inactivated plasma-derived clotting factor concentrates, ranging from 0 to 52%. These conflicting data indicate that the frequency of inhibitor development is influenced by several variables. Most of the studies observing inhibitor development of plasma-derived FVIII concentrates were retrospective studies where all patients with severity of haemophilia, PUPs as well as PTPs, were included and prevalence rather than incidence of inhibitors was reported. Table 3.4 shows 10 predominantly prospective

Table 3.4 Viral inactivated plasma-derived factor VIII products and inhibitor development in severe haemophilia A patients.

Studies	Percentage of inhibitor formation in HA <2 U/dL	Concentrate used
Lusher et al. [25]	18	Monoclate (high purity immunochromatography)
Addiego et al. [26]	13	Haemophil M (high purity immunochromatography)
Guerois et al. [29]	9	High purity FVIII (conventional chromatography)
Ljung et al. [30]	21	Various FVIII products
Ehrenforth et al. [31]	52	Various FVIII products
Addiego et al. [32]	28	Various low purity FVIII
De Biasi et al. [33]	22.9	Various FVIII products
Peerlinck et al. [34]	6	Lyophilized cryoprecipitate
Schimpf et al. [35]	0	Intermediate purity FVIII
Yee et al. [36]	0	BPL8Y (intermediate purity)

studies on inhibitor incidence in severe PUPs treated with viral inactivated plasma-derived FVIII concentrates.

Despite the highly different designs of the 10 studies, they all compare the frequency of inhibitor development in children with severe haemophilia A. Peerlinck et al. [34] reported on a cohort of 67 patients with haemophilia A born between 1971 and 1990, who had been treated exclusively with lyophilized cryoprecipitate. Only four of these PUPs (6%) developed an inhibitor that was transient in three. Guerois et al. [29] studied 56 severe PUPs (FVIIIC <1 U/dL) who received only one brand of plasma-derived highly purified FVIII-vWF concentrate, prepared by conventional chromatography with a solvent detergent method for viral inactivation. Five patients (9%) developed an inhibitor, only one being a high responder. Schimpf et al. [35] described a small cohort of 15 PUPs (FVIIIC <3 U/dL) treated solely with an intermediate purity steam-treated FVIII concentrate (Kryobulin–VH™, Immuno) where there was zero incidence of inhibitors. Still more recently, Yee et al. [36] studied inhibitor development in a cohort of 37 boys with severe haemophilia A PUPs (FVIIIC <2 U/dL) exposed only to a single FVIII concentrate (BPL8Y). Only one boy developed a transient inhibitor over a 10-year period. In contrast to these incidences of inhibitors in PUPs below 10%, other reports arrive at figures ranging from 21 to 52% [30–33] where patients received different FVIII products.

Does repeated switching from one product to another increase the risk of inhibitor production? Was one of the products used in Frankfurt [31], which has the highest inhibitor incidence, more immunogenic, such as the one associated with the epidemic in Belgium and The Netherlands? The two ultrapure FVIII concentrates (FVIII activity of 3000 IU/mg) produced by immunochromatography [25,26] had a higher inhibitor incidence despite being used as a single FVIII concentrate. Studies with the low inhibitor incidence used FVIII products that contained vWF and other contaminating proteins. Residual and contaminating proteins in the intermediate purity products have immunomodulatory effects on the immune system of the recipient, thus reducing the likelihood of inhibitor formation. However, some studies using intermediate purity products [30–33] have shown a higher incidence of inhibitors but the patients in those studies had used more than one product. If neoantigens in the products have a role, every switch from one product to another may be associated with additional risk for inhibitor formation. Regarding the issue of purity of FVIII products as a further risk for inhibitor formation, studies where FVIII concentrates which contained vWF [29,34–36] had a lower rate of inhibitor development when compared to studies of plasma-derived FVIII products without vWF [25,26] or the rFVIII concentrates studies [37,38]. The explanation that has been put forward for this discrepancy is that intermediate purity products contain well-preserved vWF multimers that may block epitopes on the light chain of the FVIII molecule to which most of the alloantibodies to FVIII react [36]. An increasing amount of *in vitro* and *in vivo* evidence suggests that vWF may have a role in the reactivity of FVIII inhibitors with FVIII. Littlewood et al. [37] described variable inhibitor titres using different sources of FVIII. In that study, the intermediate purity concentrates yielded lower inhibitor titres, and a potential role of vWF and/or residual phospholipids in the intermediate purity FVIII concentrates could be postulated. Berntop et al. [38] found that concentrates rich in vWF neutralized the inhibitors tested and yielded higher FVIII:C recovery than highly purified FVIII concentrates containing no vWF or only traces of it. Several recent studies [39–41] have also shown that FVIII bound to vWF is less accesssible to the FVIII light chain-specific antibodies.

Up to the present time, the inhibitor studies that have been carried out have not really addressed the issue of whether the risk of inhibitor development is different for different types of products. The best way to answer these questions is to perform prospective randomized controlled clinical studies in which the inhibitor incidence among recipients of concentrate A vs. concentrate B is compared.

Conclusions

A significant proportion of patients, when provided with FVIII isolated either from natural sources such as human plasma or from recombinant material, respond by producing antibody that inhibits the function of FVIII. The development of inhibitors to FVIII is therefore a universal risk for individuals with haemophilia receiving clotting factor concentrates. The factors that control the response of patients with haemophilia to FVIII are likely to be multiple. Genetic factors both at the locus bearing the haemophilia mutation and the other loci are important factors in the development of inhibitors [6–8]. The above-mentioned studies emphasize, in addition, the potential effect of changes in the biochemical and structural integrity of the FVIII molecule in the induction of inhibitors in patients with haemophilia.

It is difficult to answer the question as to whether some products are safer than others with regard to the reduction of the risk of inhibitor formation, because there are many confounding factors. The issue of purity of products as a risk factor for inhibitor formation can only be addressed by the pursuit of trials where only one therapeutic product with a defined virus inactivation procedure is used, and exposure to any other blood derivative should be excluded. There is the possibility that repeated switching from one FVIII product to another facilitates an immune response. The studies with the highest cumulative inhibitor incidence were those in which the patients regularly switched products. Switching could not only facilitate immune recognition of FVIII, it could make it much more difficult to recognize that specific products are more immunogenic than others.

There are still a number of unresolved issues regarding the type of FVIII concentrate as a potential risk for the development of FVIII antibodies. A number of essential questions still remain to be explored on the immunogenicity of FVIII and the mechanisms that modulate and condition the antigenicity of FVIII products. More studies need to be performed that compare products with respect to the potential for inhibitor induction.

More preclinical assessments on the antigenicity of FVIII concentrates should help anticipate their immunogenicity and provide means to select and design FVIII preparations having the lowest chance of inducing inhibitors. For the present, no *in vitro* or animal model can be used to assure the immunological safety of a new FVIII concentrate. FVIII purification and viral attenuation processing may potentially alter the molecule in such a way that it may provoke inhibitor development as shown by the above two episodes [10,11], illustrating the importance of continued pharmacovigilance with new FVIII concentrates, not only in PUPs but also to conduct clinical trials in PTPs before a product licence is granted.

References

1 Craddock CG, Lawrence JS. Hemophilia: a report of the mechanism of the development and action of an anticoagulant in two cases. *Blood* 1947; **2**: 505.

2 Davidson CS, Epstein RD, Miller GF, Taylor FHL. Hemophilia: a clinical study of forty patients. *Blood* 1949; **2**: 97–119.

3 Abshire TC, Brackmann HH, Scharrer I *et al*. Sucrose formulated recombinant human antihemophilic factor VIII is safe and efficacious for treatment of hemophilia A in home therapy. *Thromb Haemost* 2000; **83**: 811–16.

4 Aledort LM. *A Perspective on Inhibitor Development and Factor VIII Concentrates* (Monograph). Washington DC: American Red Cross, 1992.

5 Lusher JM. *Recent Developments in the Study of Inhibitors to Factor VIII Therapy* (Monograph). Washington DC: American Red Cross, 1993: 9–14.

6 Schwabb R, Brackmann HH, Meyer C *et al*. Haemophilia A: mutation type determines risk of inhibitor formation. *Thromb Haemost* 1995; **74**: 1402–6.

7 Hay CRM, Ollier W, Pepper L *et al*. HLA class II profile: a weak determinant of factor VIII inhibitor development in severe haemophilia A. *Thromb Haemost* 1997; **77**: 234–7.

8 Oldenburg J, Picard J, Schwaab R *et al*. HLA genotype of patients with severe haemophilia A due to intron 22 inversion with and without inhibitors of factor VIII. *Thromb Haemost* 1997; **77**: 238–42.

9 Peerlinck K, Rosendaal FR, Vermylen J. Incidence of inhibitor development in a group of young haemophilia A patients treated exclusively with lyophilized cryoprecipitate. *Blood* 1993; **81**: 3332.

10 Peerlinck K, Arnout J, Gilles JG, Saint-Remy J-M, Vermylen J. A higher than expected incidence of factor VIII inhibitors in multi-transfused haemophilia A patients treated with an intermediate purity pasteurized factor VIII concentrate. *Thromb Haemost* 1993; **69**: 115–18.

11 Mauser-Bunschoten EP, Rosendaal FR, Nieuwenhuis HK *et al*. Clinical course of factor VIII inhibitors developed after exposure to a pasteurised Dutch concentrate compared to classic inhibitors in hemophilia A. *Thromb Haemost* 1994; **71**: 703–6.

12 Peerlinck K, Arnout J, Di Giambattista M *et al*. Factor VIII inhibitors in previously treated haemophilia A patients with a double virus-inactivated plasma derived factor VIII concentrate. *Thromb Haemost* 1997; **77**: 80–6.

13 Laub R, Di Giambattista M, Fondu P *et al*. Inhibitors in German hemophilia A patients treated with a double virus inactivated factor VIII concentrate bind to the C2 domain of the light chain. *Thromb Haemost* 1999; **81**: 39–44.

14 Gilles JG, Saint-Remy JM. Strategy for preclinical evaluation of factor VIII concentrates. *Blood Coag Fibrinolysis* 1995; **6**: 558–61.

15 Wadhwa M, Dilger P, Tubbs J *et al*. Identification of transforming growth factor-beta as a contaminant in factor VIII concentrate: a possible link with immunosuppressive effects in hemophiliacs. *Blood* 1994; **84**: 2021–30.

16 Josic D, Buchacher A, Kannicht C *et al*. Degradation products of FVIII which can lead to increased immunogenicity. *Vox Sang* 1999; **77**: 90–9.

17 Lewis JH, Ferguson JH, Arends R. Hemorrhagic disease with circulating inhibitors of blood clotting: anti-AHF and anti PTC in eight cases. *Blood* 1956; **11**: 846.

18 Margolius A, Jackson DP, Ratnoff OD. Circulating anticoagulants: a study of 40 cases and a review of the literature. *Medicine (Baltimore)* 1961; **40**: 145.

19 Strauss HS. Acquired circulating anticoagulants in hemophilia A. *N Engl J Med* 1969; **281**: 866.
20 Biggs R. Jaundice and antibodies directed against factor VIII and IX in patients treated for haemophilia or Christmas disease in the United Kingdom. *Br J Haematol* 1974; **26**: 313.
21 Rizza CR, Spooner RJD, Giangrande PIF on behalf of the UKHCDO. Treatment of haemophilia in the United Kingdom 1981–96. *Haemophilia* 2001; **7**: 349–59.
22 Brinkhous KM, Roberts HR, Weiss AE. Prevalence of inhibitors in hemophilia A and B. *Thromb Diath Haemorrh* 1974; **51**: 315.
23 Shapiro SS, Hultin M. Acquired inhibitors to the blood coagulation factors. *Semin Thromb Hemost* 1975; **1**: 336.
24 Gill FM. The natural history of factor VIII inhibitors in patients with hemophilia A: progress in clinical and biological research. In: Hoyer LW, ed. *FVIII Inhibitors*. New York: Liss, 1984: 19–29.
25 Lusher JM, Salzman PM and the Monoclate Study Group. Viral safety and inhibitor development associated with factor VIIIC ultra purified from plasma in hemophiliacs previously unexposed to factor VIIIC concentrates. *Semin Hematol* 1990; **27**: 1–7.
26 Addiego JE, Gomperts E, Liu S-L *et al.* Treatment of hemophilia A with a highly purified factor VIII concentrate prepared by anti-FVIIIc immunoaffinity chromatography. *Thromb Haemost* 1992; **67**: 19–27.
27 Bray G. Inhibitor questions: plasma-derived factor VIII and recombinant factor VIII. *Ann Hematol* 1994; **68**: S29–34.
28 Lusher JM. Inhibitor development in prospective clinical trials with recombinant factor VIII preparations in previously untreated patients. *Vox Sang* 1999; **77**: 9.
29 Guerois C, Laurian Y, Rothschild C *et al.* Incidence of factor VIII inhibitor development in severe hemophilia A patients treated only with one brand of highly purified plasma-derived concentrate. *Thromb Haemost* 1995; **73**: 215–18.
30 Ljung R, Petrini P, Lindgren A-C, Tengborn L, Nilsson IM. Factor VIII and IX inhibitors in haemophiliacs [Letter to the editor]. *Lancet* 1992; **339**: 1550.
31 Ehrenforth S, Kreuz W, Scharrer I *et al.* Incidence of development of factor VIII and IX inhibitors in haemophiliacs. *Lancet* 1992; **339**: 594–8.
32 Addiego JE, Kasper CK, Abildgaard C *et al.* Frequency of inhibitor development in haemophiliacs treated with low purity factor VIII. *Lancet* 1993; **342**: 462–4.
33 de Biasi R, Rocino A, Papa ML *et al.* Incidence of factor VIII inhibitor development in hemophilia A patients treated with less pure plasma derived concentrates. *Thromb Haemost* 1994; **71**: 544–7.
34 Peerlinck K, Rosendaal FR, Vermylyn J. Incidence of inhibitor development in a group of young hemophilia A patients treated exclusively with lyophilized cryoprecipitate. *Blood* 1993; **81**: 3332–5.
35 Schimpf K, Schwarz HP, Kunschak M. Zero incidence of inhibitors in previously untreated patients who received intermediate purity factor VIII concentrate or factor IX complex. *Thromb Haemost* 1995; **73**: 553–4.
36 Yee TT, Williams MD, Hill FGH, Lee CA, Pasi JK. Absence of inhibitors in previously untreated patients with severe haemophilia A after exposure to a single intermediated purity factor VIII product. *Thromb Haemost* 1997; **78**: 1027–9.
37 Littlewood JD, Bevan SA, Kemball-Cook G, Barrowcliffe TW. *In vivo* studies of activated porcine factor VIII. *Thromb Haemost* 1996; **76**: 743–8.
38 Berntorp E, Ekman M, Gunnarsson M, Nilsson IM. Variation in factor VIII inhibitor reactivity with different commercial factor VIII preparations. *Haemophilia* 1996; **2**: 95–9.
39 Suzuki T, Arai M, Amano K, Kagawa K, Fukutake K. FVIII inhibitor antibodies with C2 domain specificity are less inhibitory to factor VIII complexed with von Willebrand factor. *Thromb Haemost* 1996; **76** (Suppl.): 749–54.
40 Gensana M, Altisent C, Aznar JA *et al.* Influence of von Willebrand factor on the reactivity of human factor VIII inhibitors with factor VIII. *Haemophilia* 2001; **7**: 369–74.
41 Kallas A, Talpsep T. Von Willebrand factor in factor VIII concentrates protects against neutralization by factor VIII antibodies. *Haemophilia* 2001; **7**: 375–80.

CHAPTER 4

Genetic basis of inhibitor development in severe haemophilia A and B

J. Oldenburg and E. Tuddenham

Alloantibodies (inhibitors) against factor VIII (FVIII) in haemophilia A or against factor IX (FIX) in haemophilia B represent a major complication in patient care because they render classical substitution therapy ineffective. Inhibitors occur at a frequency of 20–30% in severe haemophilia A and 3% in haemophilia B, respectively. Several studies have shown that the type of gene mutation is the most decisive known risk factor for inhibitor formation. Those mutations that prevent endogenous synthesis of FVIII/FIX protein, such as large deletions, nonsense mutations and, in haemophilia A, the prevalent intron 22 inversions, are associated with the highest risk of inhibitor formation. These findings indicate that the presentation of a novel antigen to the patient's immune system is the main driving force of inhibitor formation. Because the mutation type profile in haemophilia A shows a three- to fourfold larger proportion of high-risk mutations compared to haemophilia B this might explain part of the difference in inhibitor frequencies. The homology of FIX to the other vitamin K-dependent proteins may also contribute to the lower prevalence of inhibitors in haemophilia B. Specifically in haemophilia B, a severe allergic reaction to FIX infusion can occur simultaneously with inhibitor onset in some patients, especially those with large deletions. The major histocompatibility complex (MHC) represents a further, but less decisive candidate for the risk of inhibitor formation.

Introduction

Observations in the 1970s and 1980s that the risk of inhibitor formation is influenced by the degree of severity, the history of inhibitors in family relatives, and ethnicity/race pointed to an important role for patient genetics in inhibitor formation (Fig. 4.1). Molecular candidates for a genetic predisposition to inhibitor development are the mutation within the FVIII/FIX gene [1], genes involved in the immune response such as MHC Class I and II loci, and also other proteins participating in the presentation of antigens [2,3]. Most of the data have been obtained in haemophilia A, where inhibitor formation represents a frequent phenomenon affecting about one-third of severe haemophilia A patients [4,5]. Therefore this chapter mainly addresses the inhibitors in haemophilia A and refers to the data on haemophilia B where available.

Although the presentation of a novel or immunologically altered FVIII antigen to the patient's immune system has been identified as the main cause of inhibitor development, its pathogenesis is only partly understood. Why do only some patients with severe molecular defects develop an inhibitor while others do not? What is protecting the latter group against inhibitor formation? Why is inhibitor formation in haemophilia B a rare phenomenon but sometimes associated with a severe allergic/anaphylactic reaction? Which mechanisms determine the antibody epitopes on the surface of the FVIII/FIX protein and what makes an inhibitor become high titre, low titre or transient? This chapter addresses the genetic factors that may contribute to the pathogenesis of inhibitor formation.

FVIII/FIX genotype and risk of inhibitor development

An overview of the risk of inhibitor formation with respect to the underlying FVIII and FIX gene mutation types is shown in Table 4.1. Figure 4.2 shows the data for haemophilia A. In general, two main groups can be formed. One group is composed of severe molecular defects—so called null-mutations because they prevent synthesis of FVIII/FIX protein. These patients with large deletions, nonsense mutations and intron 22 inversions exhibit a high inhibitor prevalence with a range of 21–88% in haemophilia A and 6–60% in haemophilia B. With the exception of the group of large deletions, the prevalence of inhibitors in the high-risk groups is much less for

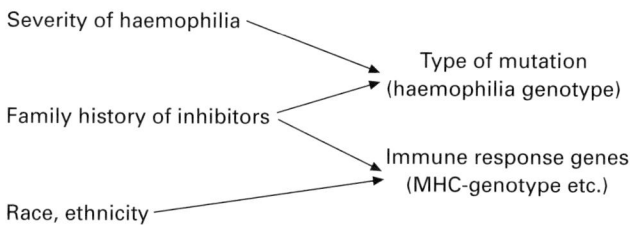

Fig. 4.1 Early indicators of genetic predisposition to inhibitor development in haemophilia.

	Haemophilia A (severe and non-severe)			Haemophilia B (severe)
	Bonn ($n = 533$)	HAMSTeRS ($n = 1022$)	Total (1555)	Haemophilia B database ($n = 663$)
Large deletion	15 (33%)	86 (42%)	101 (41%)	40 (60%)
Multi domain	3 (100%)	23 (87%)	26 (88%)	
Single domain	12 (17%)	63 (25%)	75 (24%)	
Nonsense mutations	45 (42%)	131 (34%)	185 (31%)	184 (6%)
Heavy chain	21 (13%)	61 (18%)	81 (17%)	
Light chain	24 (33%)	70 (42%)	104 (40%)	
Intron 22 inversion	179 (21%)	ND	179 (21%)	
Small deletion/insertion	41 (15%)	115 (16%)	156 (16%)	77 (65%)
Non-A-run	35 (17%)	88 (19%)	123 (19%*)	
A-run	6 (0%)	27 (4%)	33 (3%)	
Missense mutations	243 (1.5%)	669 (6%)	912 (5%)	362 (0.5%)
Non-C1/C2 domain	187 (1%)	431 (4%)	618 (3%)	
C1/C2 domain	56 (4%)	238 (11%)	294 (10%)	
Splice site mutations	10 (0%)	21 (5%)	31 (3%)	ND

Table 4.1 Mutation types and inhibitor prevalence in haemophilia A (Bonn Centre, HAMSTeRS [6]) and B (Haemophilia B Database edition 1998 [7], Ljung [8]). For each mutation type or subtype the number of individuals is given with the relative inhibitor prevalence in brackets.

*Value increases to 21% when in-frame small deletions/insertions are excluded.
ND, not done.

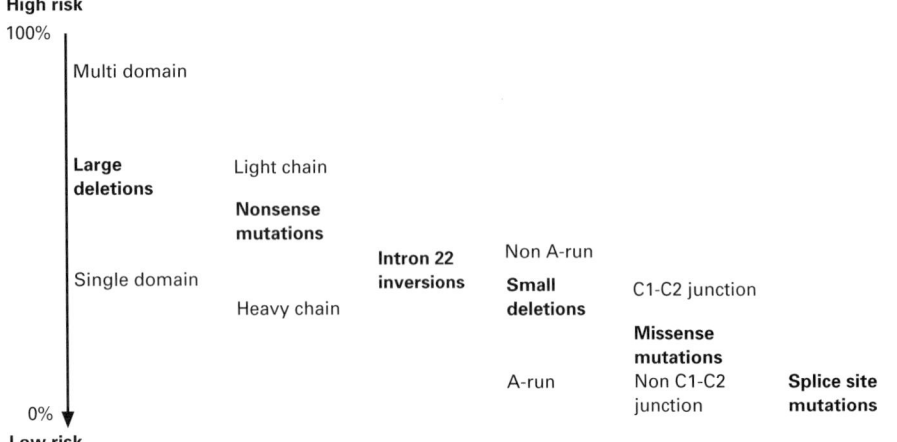

Fig. 4.2 Mutation types/subtypes and inhibitor prevalence in haemophilia A according to the data presented in Table 4.1.

haemophilia B than for haemophilia A. The second group is composed of missense and splice site mutations which result in loss of function, but not complete absence of the FVIII protein, and exhibit an inhibitor prevalence of <10%. The small deletions/insertions belong to the high-risk group in haemophilia B, while in haemophilia A this mutation type exhibits only a moderate risk for inhibitor formation (for detailed discussion see below).

In haemophilia A most of these mutation types can be further subdivided according to their risk of inhibitor formation. Data taken from the HAMSTeRS mutation register (http://europium.csc.mrc.ac.uk) [6] and also data from the Bonn Centre (Table 4.1) revealed that patients with **large deletions**, affecting more than one domain of the FVIII protein, are at the highest risk (about 88%) for developing an inhibitor, which is about three times the risk of single domain deletions. **Nonsense mutations** on the light chain double the risk of inhibitor formation relative to those on the heavy chain, although the predicted truncated proteins would be shorter in patients with stop codons in the heavy chain. At present, no explanation for this observation exists. A further high-risk mutation type in haemophilia A with an inhibitor prevalence of about 21% is the **intron 22 inversion**. This is the most prevalent mutation in severe haemophilia A, especially in the subgroup of patients who develop an inhibitor, where it can be found in up to 60% of patients [9].

Within the group of low-risk mutations, the mutation type of **small deletions/insertions** is of special interest, because it shows an unexpectedly low risk of inhibitor formation in haemophilia A. From the nature of the mutation type—most of the small deletions/insertions lead to a frameshift with a subsequent stop codon—an inhibitor risk similar to the nonsense mutations would be expected. This is exactly the situation observed in haemophilia B (Table 4.1). An explanation for the contrary finding in haemophilia A was reported by Young et al. [10] who discovered endogenous restoration of the reading frame by polymerase errors during DNA replication/RNA transcription in patients with small deletion/insertion mutations that were located at runs of adenines. It is hypothesized that the resulting small amounts of endogenous FVIII protein protect against inhibitor development [11]. Because the two polyadenine runs in the B domain represent mutation hotspots, the high proportion of such small deletions/insertions lowers the overall inhibitor prevalence in this mutation type. In contrast, in haemophilia B no such A runs exist, leading to a similar inhibitor prevalence in small deletion/insertions and nonsense mutations (Table 4.1).

Missense mutations represent the most frequent mutation type overall, accounting for 15% of mutations in severe haemophilia A [12], 80% of the mutations in severe haemophilia B [7] and for almost all mutations in non-severe haemophilia [6]. Patients with missense mutations synthesize some endogenous, although often non-functional, protein that is sufficient to induce immune tolerance in most patients. Consequently, the inhibitor prevalence in this mutation type is as low as 5% for haemophilia A and 0.5% for haemophilia B. In haemophilia A further subgroups of missense mutations can be formed. For patients with missense mutations in the C1 and C2 domains, the risk of inhibitor formation is three times higher (10 vs. 3%) than for missense mutations in other regions, thus suggesting that this part of the FVIII molecule is especially immunogenic. This hypothesis is supported by several studies [13–16]. From these data it can be hypothesized that the C1/C2 domains significantly contribute to the binding of FVIII protein to von Willebrand factor (vWF). During FVIII activation vWF is dissociated and the same C1/C2 region binds to the phospholipid membrane to constitute the 'tenase' complex. Any changes in the three-dimensional structure of this critical part of the FVIII molecule may affects its immunogenicity. The HAMSTeRS data show that the inhibitor prevalence in missense mutations depends more on the location of the mutation than on the degree of severity. There is almost no difference in inhibitor prevalence between severe and non-severe haemophilia A patients with missense mutations, thus indicating that certain regions within the FVIII protein are especially crucial for its antigenic integrity.

Beside the small deletions/insertions at runs of adenine nucleotides, **splice site mutations** represent the second mutation type that are at very low risk for inhibitor formation. The likely explanation for this is that a very small number of FVIII/FIX mRNA molecules are still spliced normally and these few molecules direct synthesis of enough protein to induce immune tolerance.

One of the most important questions for the pathogenesis of inhibitor formation is why do not all patients with a single type of null-mutation develop an inhibitor? The intron 22 inversion in haemophilia A patients provides a good example to examine this issue. It is an almost uniform mutation type (there are proximal and distal crossover points) and, in regard to its pathological mechanism, the intron 22 inversion appears to be a true null-mutation, resulting in no detectable endogenous FVIII protein in the patient's circulation. A speculative reason for non-inhibitor formation is that some maternal FVIII is presented *in utero* to the fetal immune system, thus inducing immune tolerance towards substituted FVIII. The individual characteristics of a patient's polymorphic immune system may also either increase or decrease risk of inhibitor development. Furthermore, it may be speculated that an immune response arises in almost all patients with null-mutations, but is down-regulated in two-thirds of patients by as yet unknown mechanisms. Solving the question of penetrance of inhibitor formation in patients with haemophilia will be the key to understanding the pathological mechanism of inhibitor development.

Why is the inhibitor prevalence in haemophilia B different from haemophilia A?

Although inhibitor formation in haemophilia B is associated with the same mutation types as in haemophilia A, and thus seems to follow similar principles, the overall inhibitor prevalence is much lower (3 vs. 20–30%). A major factor contributing to this difference might be the different proportions of null-mutations; <20% in haemophilia B [7] but >70% in haemophilia A [12]. Also, within the mutation types e.g. nonsense mutations, the inhibitor prevalence is much lower (6 vs. 30%), so beside the mutation profile there must be another reason, which could be the high homology of FIX to the other vitamin K-dependent clotting factors. The finding that there is no major difference in inhibitor prevalence for the mutation type of large deletions, which may be regarded as such severe molecular defects that they cannot be compensated by homologies, argues for this hypothesis. Another remarkable difference between haemophilia B and A is the simultaneous occurrence of severe allergic reactions together with inhibitor onset in haemophilia B patients, especially those with large deletions. Immune tolerance therapy in these patients is complicated by the development of a nephrotic syndrome and a general poor outcome of inhibitor eradication. It is speculated that the much larger amount of circulating protein (5 μg FIX/mL vs. 0.1 μg FVIII/mL) and also the molecular weight difference (FIX 57 kDa vs. FVIII 265 kDa) results in a several hundred-fold larger number of immune complexes in haemophilia B than in haemophilia A inhibitor patients.

Genetics of the immune system and risk of inhibitor development

Information on the role of immune response genes in inhibitor development are at present only available for haemophilia A. Two studies provide indirect evidence that genes participating in the immune response considerably influence the risk of inhibitor development. Scharrer et al. [17] carried out a meta-analysis of three USA studies (Kogenate™ [18], Recombinate™ [19] and a US retrospective study [20]) that clearly demonstrate the influence of race on inhibitor formation. In African Americans the inhibitor incidence in severe haemophilia patients was doubled (51.9%, 14 of 27) when compared to that of white people (25.8%, 51 of 191). Cox Gill [21] compared the incidence of inhibitor formation in haemophilic siblings to more distant haemophilic relatives in the same kindred and found a much higher incidence in siblings (50%) than in more distant haemophilic relatives (9%). Because the genetic defects of the FVIII gene are expected to be similar in both studies (identical within a kindred), the observed difference in inhibitor incidence is most likely caused by genetic variations of the immune system. Candidates for this immunogenetic determination of inhibitor formation are the MHC classes and other polymorphic genes (e.g. cytokines) that participate in the immune response.

The influence of the genes of the MHC classes on inhibitor formation has been addressed in various studies [22–25]. The MHC Class II genes *DQ*, *DR* and *DP* are of particular interest because their function is to present extracellular antigens—such as substituted FVIII—to the patient's immune system. However, with respect to the presentation of endogenously truncated and/or immunologically altered FVIII, the MHC Class I genes that process intracellular antigens may also have an important role in the genesis of inhibitor formation. The results of earlier studies had been inconclusive and sometimes even contradictory. One problem of these studies was that they did not consider the patient's mutation type and, as a consequence, the influence of specific immune response genes on inhibitor formation might have been masked by the strong influence of the FVIII gene defect. In a recent study, the influence of the MHC Class I/II genotype on inhibitor formation was exclusively investigated in patients with the homogenous intron 22 inversion [2]. A condensate of the results is shown in Table 4.2. The MHC Class I/II alleles *A3*, *B7*, *C7*, *DQA0102*, *DQB0602* and *DR15* could be assigned as risk alleles (relative risk 1.9–4.0), because they occurred more often in inhibitor than in non-inhibitor patients. In contrast, the MHC Class I/II alleles *C2*, *DQA0103*, *DQB0603* and *DR13* could be assigned as protective alleles (relative risk 0.1–0.2) because they occurred less often in inhibitor than in non-inhibitor patients. These MHC Class I/II alleles belonged to extended haplotypes (*A3-B7-C7-DQA0102-DQB0602-DR15* and *C2-DQA0103-DQB0603-DR13*) that were also frequent and less frequent, respectively, in the normal population. However, the number of patients was too small to reach a clear statistical significance. Moreover,

Table 4.2 Common and rare MHC Class I and II alleles in severe haemophilia A patients with intron 22 inversion and inhibitor formation according to Oldenburg et al. [2]. For the MHC Class I loci *A*, *B*, *C* the number of individuals and for the MHC Class II loci *DR*, *DQA*, *DQB* the number of chromosomes is given.

Allele	Inhibitors		Non-inhibitors		Relative risk
	No.	%	No.	%	
Common MHC alleles in patients with inhibitor					
A3	11	37.9	9	21.4	2.2
B7	14	48.3	8	19.1	4.0
C7	17	58.6	16	38.1	2.3
DQA0102	20	34.5	16	19.1	2.2
DQB0602	18	31.0	12	14.3	2.7
DR15	19	32.8	17	20.2	1.9
Rare MHC alleles in patients with inhibitor					
C2	1	3.4	6	14.3	0.2
DQA0103	1	1.7	12	14.3	0.1
DQB0603	0	0.0	6	7.1	0.1
DR13	1	1.7	9	10.7	0.1

the inheritance as haplotypes might mask those MHC Class I/II alleles that are decisive for the risk or protection of inhibitor formation. In this context, a finding of Chicz et al. [26] was of interest. They found a 16 amino acid residue peptide from the FVIII light chain (amino acids 1706–1721) that could be eluted from a DR15 cell line. Two functional cleavage sites bounded the peptide, located on the surface of the FVIII molecule. It may be hypothesized that some peptides of the FVIII protein are especially efficient when presented to the patient's immune system, thus leading to the determination of FVIII antibody epitopes. Notably, Hay et al. [3] found the same MHC Class II alleles to be associated at similar frequencies with inhibitor formation in patients with intron 22 inversions (Table 4.3), thus supporting the concept that the MHC is of some significance for the risk of inhibitor formation. However, neither study achieved significant results. Beside the arguments already presented, this may be because additional genes, not only those belonging to the MHC, constitute the immunogenic risk profile for inhibitor formation.

Table 4.3 MHC Class II alleles in severe haemophilia A patients with intron 22 inversion and inhibitor formation according to Hay et al. [3].

Allele	Inhibitors		Non-inhibitors		Odds ratio
	No.	%	No.	%	
Common MHC alleles in patients with inhibitor					
DQA0102	10	43.5	11	20.0	3.1
DQB0602	7	30.4	9	16.4	ND
DR15	8	34.8	10	18.2	2.4

ND, not done.

Patients' genetics and characteristics of FVIII antibody epitopes

Extensive studies on inhibitor epitopes are only available for haemophilia A, where several FVIII antibody epitopes have been characterized. The FVIII A2, A3, C2 domains are the most immunogenic while the A1 and B domains are at best poorly immunogenic. Inhibitory epitopes have been assigned to the ar1 region (aa 351–365) [27], the A2 domain (aa 484–508) [28], the ar3 region (aa 1687–1695) [29,30], the A3 domain (aa 1778–1823) [31] and the C2 domain (aa 2181–2243) [32], (aa 2248–2312) [33]. Inhibitors against the C1 domain have also been reported [34]. Interestingly, the antibody epitopes correlate very well with the functional epitopes of ligands that interact with the FVIII protein (reviewed in [35]).

In severe haemophilia A, little is known about the relationship of the type and site of the mutation and the localization of the FVIII antibody epitopes. Epitopes have been shown to vary among inhibitor patients with respect to their number and specificity [36,37]. Most patients with haemophilia have shown a complex immune response with two or three epitopes contributing significantly to the inhibitor titre [37]. The mechanisms determining the heterogeneous patterns of the various FVIII antibody epitopes are not known. A correlation of the type and site of the mutations to epitope characteristics may help to illuminate and explain this relationship.

In some patients with mild haemophilia A, a direct relationship of the mutation site and the antibody epitope has been clearly established. Fijnvandraat et al. [38] described a patient with mild haemophilia A with an Arg593Cys mutation in whom the FVIII antibodies were directed against substituted FVIII but not against the mutated endogenous FVIII protein. This led to the interesting phenomenon that 1-deamino-8-D-arginine-vasopressin (DDAVP) was able to induce a considerable increase of the FVIII activity while infused replacement FVIII did not. Peerlinck et al. [37] described two patients with mild haemophilia A with a residual Factor VIII:C of 0.09 IU/mL despite inhibitor titres of 300 and 6 Bethesda units, respectively. Both patients had the mutation Arg2150His and produced inhibitory antibodies directed against a FVIII domain encompassing the mutation site. By expression studies and functional tests, the mutation at residue 2150 was shown to lead to reduced FVIII binding to vWF that consequently decreased circulating FVIII:C levels in the patient's plasma [14].

Conclusions

The risk of inhibitor development in haemophilia A and B is determined to a considerable extent by the genetic characteristics of the patients with respect to their underlying mutation and also to a lesser extent by their individual polymorphic immune systems. Some of the differences observed between haemophilia A and B can be explained by different mutation type profiles and protein characteristics, including the molar concentration in plasma and the homologies to other proteins. However, our present knowledge of inhibitor formation is similar to the tip of an iceberg with most of the pathogenesis still undiscovered below the water line. Because the genetics of a patient are—beside some methological challenges—easily assessable and do not change during life, they present an ideal starting point for future studies in this field. Building large and genetically well-characterized patient cohorts will allow the formation of subgroups with similar genetic backgrounds, which may prove necessary for the identification and evaluation of further factors influencing inhibitor development.

References

1 Schwaab R, Brackmann HH, Meyer C et al. Haemophilia A: mutation type determines risk of inhibitor formation. *Thromb Haemost* 1995; **74**: 1402–6.
2 Oldenburg J, Picard J, Schwaab R et al. HLA genotype of patients with severe haemophilia A due to intron 22 inversion with and without inhibitors to factor VIII. *Thromb Haemost* 1997; **77**: 238–42.
3 Hay CR, Ollier W, Pepper L et al. HLA class II profile: a weak determinant of factor VIII inhibitor development in severe haemophilia A: UKHCDO Inhibitor Working Party. *Thromb Haemost* 1997; **77**: 234–7.
4 Kreuz W, Becker S, Lenz E et al. Factor VIII inhibitors in patients with haemophilia A: epidemiology of inhibitor development and induction of immune tolerance for factor VIII. *Semin Thromb Haemost* 1995; **21**: 382–9.
5 Scharrer I, Neutzling O. Incidence of inhibitors in haemophiliacs: a review of the literature. *Blood Coag Fibrinolysis* 1993; **4**: 753–8.
6 Kemball-Cook G, Tuddenham EGD, Wacey AI. The factor VIII structure and mutation resource site: HAMSTeRS version 4. *Nucleic Acids Res* 1998; **26**: 216–19.
7 Gainnelli F, Green PM, Sommer SS et al. Haemophilia B: database of point mutations and short additions and deletions (eighth edition). *Nucleic Acids Res* 1998; **26**: 265–8.
8 Ljung RC. Gene mutations and inhibitor formation in patients with hemophilia B. *Acta Haematol* 1995; **94** (Suppl. 1): 49–52.
9 Oldenburg J, Brackmann HH, Schwaab R. Risk factors for inhibitor development in haemophilia A. *Haematologica* 2000; **85** (Suppl. 10): 7–13.
10 Young M, Inaba H, Hoyer LW et al. Partial correction of a severe molecular defect in haemophilia A, because of errors during expression of the factor VIII gene. *Am J Hum Genet* 1997; **60**: 565–73.
11 Oldenburg J, Schröder J, Schmitt C, Brackmann HH, Schwaab R. Small deletion/insertion mutations within poly-A-runs of the factor VIII gene mitigate the severe haemophilia A phenotype. *Thromb Haemost* 1998; **79**: 452–3.
12 Becker J, Schwaab R, Möller-Taube A et al. Characterization of the factor VIII defect in 147 patients with sporadic hemophilia A: family studies indicate a mutation type-dependent sex ratio of mutation frequencies. *Am J Hum Genet* 1996; **58**: 657–70.
13 Hay CR, Ludlam CA, Colvin BT et al. Factor VIII inhibitors in mild and moderate-severity haemophilia A: UK Haemophilia Centre Directors Organisation. *Thromb Haemost* 1998; **79**: 762–6.
14 Jacquemin M, Lavend'homme R, Benhida A et al. A novel cause of mild/moderate haemophilia A: mutations scattered in the factor

VIII C1 domain reduce factor VIII binding to von Willebrand factor. *Blood* 2000; **96**: 958–65.
15 Liu ML, Shen BW, Nakaya S *et al.* Haemophilic factor VIII C1- and C2-domain missense mutations and their modeling to the 1.5-angstrom human C2-domain crystal structure. *Blood* 2000; **96**: 979–87.
16 Pratt KP, Shen BW, Takeshima K *et al.* Structure of the C2 domain of human factor VIII at 1.5 A resolution. *Nature* 1999; **402**: 439–42.
17 Scharrer I, Bray GL, Neutzling O. Incidence of inhibitors in haemophilia A patients: a review of recent studies of recombinant and plasma-derived factor VIII concentrates. *Haemophilia* 1999; **5**: 145–54.
18 Lusher J, Arkin S, Hurst D. Recombinant FVIII (Kogenate) treatment of previously untreated patients (PUPs) with haemophilia A: update of safety, efficacy and inhibitor development after seven study years. Florence: ISTH. *Thromb Haemost* 1997: Suppl. 162.
19 Gruppo R, Chen H, Schroth P, Bray GL. Safety and immunogenicity of recombinant factor VIII (Recombinate) in previously untreated patients (PUPs): a 7.3 year update. The Hague: WFH. *Haemophilia* 1998; **4**: 228.
20 Addiego J, Kasper C, Abildgaard C. Increased frequency of inhibitors in African American haemophilia A patients. Nashville: ASH. *Blood* 1994: Suppl. 239.
21 Cox Gill J. The role of genetics in inhibitor formation. *Thromb Haemost* 1999; **82**: 500–4.
22 Aly AM, Aledort LM, Lee TD, Hoyer LW. Histocompatibility antigen patterns in haemophilic patients with factor VIII antibodies. *Br J Haematol* 1990; **76**: 238–41.
23 Frommel D, Allain JP, Saint-Paul E *et al.* HLA antigens and factor VIII antibody in classic haemophilia: European study group of factor VIII antibody. *Thromb Haemost* 1981; **46**: 687–9.
24 Lippert LE, Fisher LM, Schook LB. Relationship of major histocompatibility complex class II genes to inhibitor antibody formation in haemophilia A. *Thromb Haemost* 1990; **64**: 564–8.
25 Mayr WR, Lechner K, Niessner H, Pabinger-Fasching I. HLA-DR and Factor VIII antibodies in haemophilia A. *Thromb Haemost* 1984; **51**: 293.
26 Chicz RM, Urban RG, Gorga JC *et al.* Specifity and promiscuity among naturally processed peptides bound to HLA-DR alleles. *J Exp Med* 1993; **178**: 27–47.
27 Foster PA, Fulcher CA, Houghten RA, de Graaf Mahoney S, Zimmerman TS. A murine monoclonal anti-factor VIII inhibitory antibody and two human factor VIII inhibitors bind to different areas within a twenty amino acid segment of the acidic region of factor VIII heavy chain. *Blood Coag Fibrinolysis* 1990; **1**: 9–15.
28 Healey JF, Lubin IM, Nakai H *et al.* Residues 484–508 contain a major determinant of the inhibitory epitope in the A2 domain of human factor VIII. *J Biol Chem* 1995; **270**: 14505–9.
29 Tiarks C, Pechet L, Anderson J, Mole JE, Humphreys RE. Characterization of a factor VIII immunogenic site using factor VIII synthetic peptide 1687–1695 and rabbit anti-peptide antibodies. *Thromb Res* 1992; **65**: 301–10.
30 Barrow RT, Healey JF, Gailani D, Scandella D, Lollar P. Reduction of the antigenicity of factor VIII toward complex inhibitory antibody plasmas using multiply-substituted hybrid human/porcine factor VIII molecules. *Blood* 2000; **95**: 564–8.
31 Fijnvandraat K, Celie PH, Turenhout EA *et al.* A human alloantibody interferes with binding of factor IXa to the factor VIII light chain. *Blood* 1998; **91**: 2347–52.
32 Healey JF, Barrow RT, Tamim HM *et al.* Residues Glu2181–Val2243 contain a major determinant of the inhibitory epitope in the C2 domain of human factor VIII. *Blood* 1998; **92**: 3701–9.
33 Scandella D, Gilbert GE, Shima M *et al.* Some factor VIII inhibitor antibodies recognize a common epitope corresponding to C2 domain amino acids 2248–2312, which overlap a phospholipid-binding site. *Blood* 1995; **86**: 1811–19.
34 Jacquemin M, Benhida A, Peerlinck K *et al.* A human antibody directed to the factor VIII C1 domain inhibits factor VIII cofactor activity and binding to von Willebrand factor. *Blood* 2000; **95**: 156–63.
35 Lenting PJ, van Mourik JA, Mertens K. The life cycle of coagulation factor VIII in view of its structure and function. *Blood* 1998; **92**: 3983–96.
36 Scandella D, Mattingly M, de Graaf S, Fulcher CA. Localization of epitopes for human factor VIII inhibitor antibodies by immunoblotting and antibody neutralization. *Blood* 1989; **74**: 1618–26.
37 Peerlinck K, Jacquemin MG, Arnout J *et al.* Antifactor VIII antibody inhibiting allogeneic but not autologous factor VIII in patients with mild haemophilia A. *Blood* 1999; **93**: 2267–73.
38 Fijnvandraat K, Turenhout EA, van den Brink EN *et al.* The missense mutation Arg593→Cys is related to antibody formation in a patient with mild haemophilia A. *Blood* 1997; **89**: 4371–7.

PART 2
Management of treatment

CHAPTER 5

Methods: plasmapheresis and protein A immunoadsorption

E. Berntorp

The preferred way of repairing the haemostatic defect in haemophilia is to substitute the missing coagulation factors VIII or IX, and such therapy is severely jeopardized if the development of inhibitors occurs. A seemingly attractive therapeutic option is to remove the inhibitors and subsequently administer infusions of factors VIII or IX. However, the anamnestic response to such replacement therapy makes removal of inhibitors feasible only in the short term. The clinical situations in which inhibitor removal, followed by substitution therapy, can be useful are surgery and acute bleeding episodes.

Immune tolerance induction (ITI) is another indication for inhibitor removal when using the Malmö protocol which, in contrast to other ITI treatment regimens, entails not only administration of antigen (factors VIII or IX), but is also supplemented with intravenous immunoglobulin and cyclophosphamide [1]. Antigen excess at the start of treatment is a prerequisite for the Malmö model, and in some patients this can be achieved only if the inhibitor is removed before initiating ITI.

Plasma exchange has been used to remove inhibitors [2], but the results indicate that this method is not very effective. By comparison, extracorporeal adsorption to protein A shows greater promise, although it is not widely applied because of the rarity of the disease being treated and the fact that a team of specially trained caregivers is needed to perform this procedure.

Plasmapheresis

Plasma exchange has been used to treat only a limited number of patients with haemophilia A who also have inhibitors [2–4]. The most extensive work has been carried out by Francesconi *et al.* [2], who carried out 12 plasmaphereses in five patients with factor VIII inhibitors: three with haemophilic antibodies and two with spontaneous antibodies. These investigators observed a marked reduction in levels of the antibodies, and the decrease correlated well with the extent of plasma exchange. About 40 mL plasma per kilogram of body weight (kg bw) had to be removed to reduce the antibody level by half. In the three haemophilia patients, 23–50 mL/kg bw was exchanged and replaced with albumin and/or plasma, and these procedures took about 120–210 min. A rise in factor VIII activity was seen after replacement with factor VIII in two of the three patients with haemophilia as well as in the patients with acquired inhibitors. The antibody titres in the majority of the patients dropped immediately after plasmapheresis but never reached zero. In one of these patients the antibody was quickly eliminated by repeated plasmapheresis and subsequent high-dose treatment with factor VIII. Francesconi *et al.* concluded that plasmapheresis should be considered when patients with antibodies to factor VIII have to be treated for severe bleeding.

Plasmapheresis is a rather inefficient method of lowering inhibitor levels, because it is non-specific, and large volumes of plasma must be removed and replaced to reduce inhibitor concentrations significantly. Furthermore, in high titre inhibitors, it appears to be difficult to lower inhibitor levels close enough to zero to allow successful factor VIII replacement therapy. A better technique for inhibitor removal seems to be extracorporeal adsorption to protein A. If the protein A method is used in parallel with emerging new products that can bypass the coagulation defect during acute bleeding and surgery, such as recombinant VIIa [5] and activated prothrombin complex concentrates [6], plasmapheresis will surely become a low priority therapy in the arsenal of techniques employed to treat patients with inhibitors.

Protein A

Protein A is a cell wall structure that can be isolated from *Staphylococcus aureus* [7,8], and it preferentially reacts with the FC part of human immunoglobulin G (IgG). Protein A has a molecular weight of between 27 and 42 kDa [9] and has five IgG binding epitopes. Protein A can bind all subclasses of human IgG, although interaction with subclass 3 is incomplete. Protein A also reacts to some extent with IgA and IgM. The inhibitors in congenital haemophilia are predominantly IgG antibodies of subclass 4 [10], thus they can be bound to protein A.

Two-column systems for protein A adsorption

The first patient to be treated using extracorporeal adsorption to protein A was a 37-year-old man with haemophilia B and

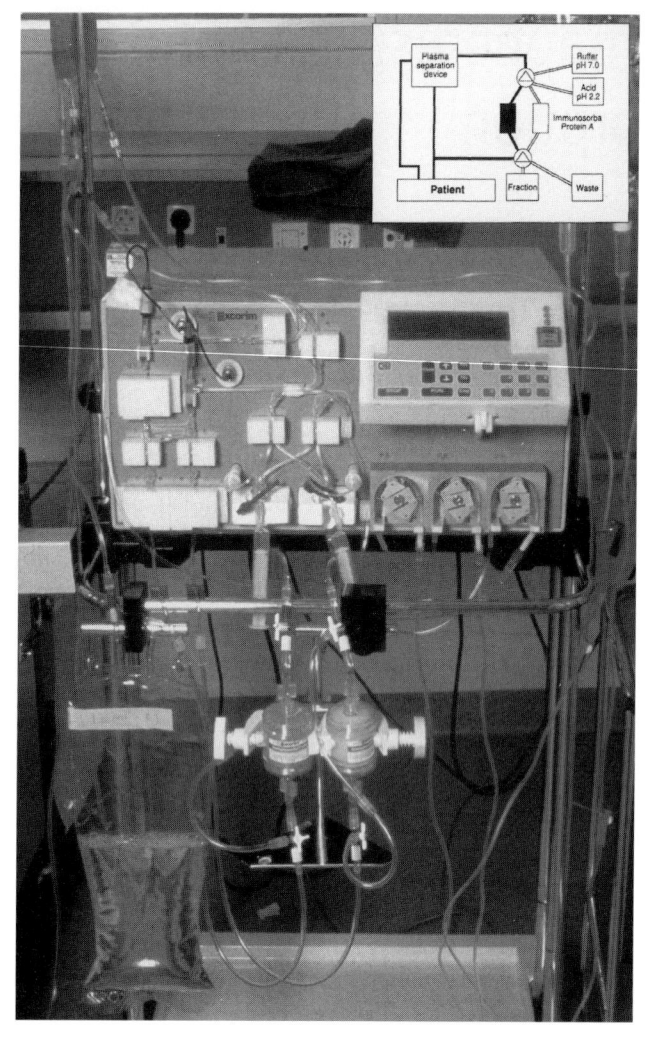

Fig. 5.1 The Citem 10™ system.

sion. Normally 2.5–3 plasma volumes are processed at a flow rate of 50 mL/min daily for two consecutive days, to achieve equilibrium between the inhibitor in the extravascular space and the intravascular compartment. This facilitates elimination of the inhibitor and minimizes the amount of factor concentrates required for inhibitor neutralization [13]. The adsorption procedure requires an experienced team of haematologists, nephrologists and nurses and is therefore not yet widely used to treat haemophilia. However, the method has proven to be very successful for temporary reduction of inhibitors to prepare a patient for subsequent replacement therapy to achieve haemostasis prior to surgery or in case of severe haemorrhage.

Clinical experience of protein A immunoadsorption for removal of inhibitors

Immunoadsorption using the Citem 10/Immunosorba™ system has been carried out at the Malmö Haemophilia Centre since 1980, particularly to treat patients with haemophilia A and B who also have inhibitors. The indication for the adsorption procedure has usually been acute bleeding or surgery, and on some occasions the procedure has been combined with ITI. In a few cases the primary indication for immunoadsorption has been part of ITI treatment administered according to the Malmö protocol (see below). In all, from 1980 to 1995, 10 patients (five with haemophilia A, and five with haemophilia B) were treated with immune adsorption 19 times, and the results have been summarized by Freiburghaus et al. [14]. Data on inhibitor levels, processed plasma volumes and length of treatment in these patients are shown in Table 5.1. Immediately prior to the adsorption therapy, inhibitor levels in the haemophilia A patients ranged from 0.9 to 531 Malmö inhibitor units (MIU), which roughly corresponds to 3–1500 Bethesda units (BU). A total plasma volume of 4.8–52 L was processed during a period of 1–6 days, which is equivalent to 100–183 mL/kg/day. This treatment had a dramatic effect on the inhibitor level in all of the haemophilia A cases. In four of these patients the decline in inhibitors was sufficient to allow neutralization with factor concentrate on all episodes. In one exceptional case, the inhibitor was reduced from 531 to 9.2 MIU (from 1500 to 30 BU) but, despite processing of 42.5 L of plasma, effective substitution with factor VIII could not be accomplished. In most of the patients, haemostasis could be achieved for about 1 week, making surgery possible. In the haemophilia B patients, the inhibitor level before immunoadsorption was 4.5–53 MIU (15–150 BU) and the processed plasma volumes ranged from 2 to 12.6 L, which corresponds to 61–235 mL/kg/day. The treatment was given for 1–2 days, and in all cases the inhibitor level could be effectively reduced and haemostasis achieved for up to about 1 week. The authors concluded that protein A adsorption in a two-column system can rapidly and efficiently remove inhibitors and thereby allow substitution therapy with factor concentrate to control bleeding and permit planned and emergency surgery.

inhibitors. This patient was to undergo surgery for a pseudotumour in his left elbow and, to secure haemostasis during the operation, a two-column device called Citem 5 was developed to increase the adsorption capacity [11,12]. This model was subsequently further developed into the Citem 10/Immunosorba R™ system (Fresenius HemoCare, formerly Excorim), which comprises a plasma separator and a computerized adsorption system made up of two protein A columns and a monitor that oversees the concentration of proteins eluted from the columns (Fig. 5.1). Each Immunosorba column contains 62.5 mL protein A Sepharose™ adsorbent with a total binding capacity of at least 1.2 g IgG. The columns are patient-specific and can be used repeatedly. Blood withdrawn from the patient is immediately citrated (0.113 M sodium citrate, 1 : 10). When the first Immunosorba™ column is saturated with IgG, the plasma flow is switched to the second column, and the first column is regenerated by eluting at low pH (2.2) with a citrate buffer to remove the adsorbed IgG. Blood cells and plasma depleted of IgG are returned to the patient as an infu-

Table 5.1 Immunoadsorption treatment.

Patient	Treatment	Inhibitor start (Miu)	Inhibitor end (Miu)	Processed plasma volume (L)	Length of treatment (days)	Processed vol/kg/day
Haemophilia A						
1A	1	4.7	1.8	5.9	1	102
	2	30	1	6.5	1	112
	3	10	0.7	14.7	2	123
	4	0.9	0.1	10	1	154
2A	1	531	9.2	42.5	5	100
	2	10	0.9	52	6	106
3A	1	11	0	11	2	183
	2	9	0.6	4.8	1	126
4A	1	17	2	19	2	146
5A	1	6	0.1	12	2	133
Haemophilia B						
1B	1	7	1.2	6	2	61
	2	4.5	0.5	4.7	1	94
2B	1	50	2.4		2	
	2	8	0.6	4.8	1	137
3B	1	15	0.1	12.6	2	158
4B	1	10	0	5	2	139
5B	1	30	2	2	1	118
	2	53	0	8	2	235
	3	5	0	5	1	227

Accordingly, the technique can definitely be valuable to patients with initial inhibitor titres that are too high to be neutralized by infusion alone.

Protein A immunoadsorption of inhibitors in haemophilia is clearly effective, but use of this method has been restricted by the fact that only a few centres have access to both the necessary equipment and a team of experienced caregivers to administer the therapy. Moreover, there are some technical and practical aspects that must be taken into consideration when performing immunoadsorption. For instance, efficient blood vessel access is important and can be achieved by withdrawing blood via the radial artery and giving a return infusion via the antecubital vein. In addition, the adsorption is most effective if it is performed on a number of (usually two) consecutive days, which allows overnight equilibrium between intra- and extravascular IgG, and if an average of 3 plasma volumes is adsorbed each day. Adsorbing more than 3 plasma volumes on each treatment occasion does not seem to noticeably increase the efficiency of the IgG removal (Fig. 5.2). Furthermore, Gjörstrup *et al.* [13] have reported data showing that optimal effects were achieved by treatment on two consecutive days. This is also illustrated in Fig. 5.3, which presents results regarding a patient with severe haemophilia and a life-threatening gastrointestinal bleed, who was treated by immunoadsorption and subsequent replacement with factor VIII concentrate at the Malmö centre. In this case, two consecutive adsorption procedures were sufficient to reach haemostatic factor VIII levels in plasma and promptly stop the bleeding.

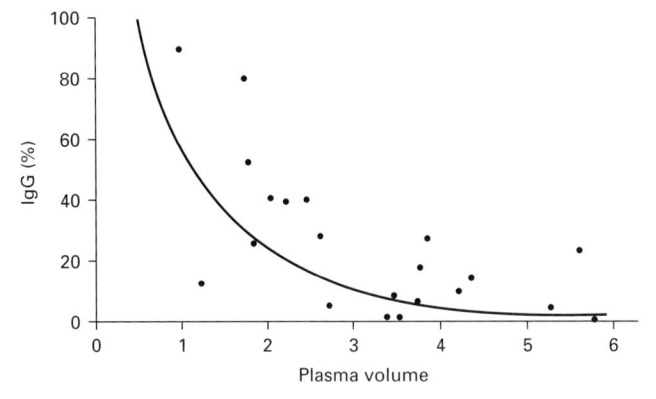

Fig. 5.2 Changes in plasma levels of IgG as an effect of the number of plasma volumes treated on each occasion. Changes are expressed as percent change from initial values and fitted to a logarithmic function (Y, $59.8 - 33.5 \ln X$; r, 0.61; $P < 0.01$; $n = 21$). (From [13] with permission from S. Karger, AG.)

Other methods used for immunoadsorption

A two-column system using antihuman IgG bound to Sepharose™ has also been used quite extensively and has proven to have an even greater adsorption capacity than offered by the Immunosorba™ systems based on protein A. Compared to the Immunosorba™ columns, the Ig-TheraSorb™ columns differ in

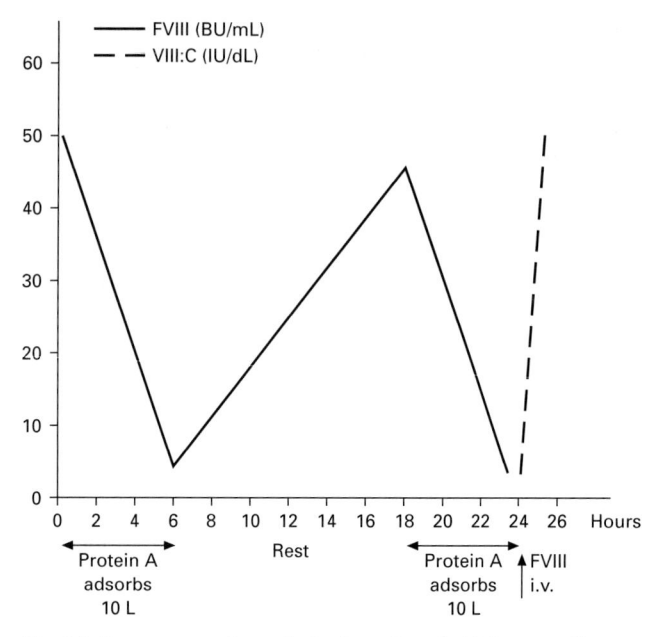

Fig. 5.3 Extracorporeal protein A adsorption of a high-responding inhibitor. After two sessions with a night's rest in between, factor VIII infusion gives a haemostatic VIII:C level.

that they bind all classes and subclasses of immunoglobulins but are essentially equivalent in regard to the plasma volume that can be processed and the possibility to regenerate and reuse the equipment [15,16]. Thus far, clinical experience with Ig-TheraSorb™ columns includes more than 2600 immunoadsorption treatments, but these comprise only five cases of acquired factor VIII inhibitors; two patients with congenital haemophilia A antibodies, and one case of an acquired factor V inhibitor. Considering efficacy in terms of antibody reduction, this system seems to be very similar to the Immunosorba™ protein A columns.

Specific removal of factor IX antibodies in a haemophilia B patient has been reported by Nilsson et al. [17]. In that work, a factor IX preparation with a specific activity of 92 U/mg protein was covalently coupled to Sepharose™ 6 MB and packed in Immunosorba™ columns. The adsorption was performed directly from citrated whole blood, using a minimal extracorporeal volume. The treatment was administered on two consecutive days, after which the anti-factor IX titre decreased from 120 to 7.2 BU/mL of plasma. No signs of activation of the coagulation, fibrinolytic or complement system were detected, nor was there any sign of haemolysis, thus Nilsson et al. concluded that this technique is safe and rapid. The most interesting advantage of this system is that whole blood can be adsorbed on line, using a minimal extracorporeal volume. This specific adsorption procedure has not yet been further developed or applied, but obviously shows great potential for future use.

Role of immunoadsorption in immune tolerance induction

When treating patients who have persistent inhibitors, the ultimate goal should be to try to induce immune tolerance. This is done to eradicate the inhibitors and to make the patient receptive to ordinary replacement therapy, even prophylaxis. In 1977, Brackmann and Gormsen [18] first proposed ITI to be used to treat patients with haemophilia who also had an inhibitor. Since then, a number of protocols have been developed that use a low dose [19,20], an intermediate dose [21,22] or a high dose (often called the Bonn protocol) [23–25] factor VIII regimen, but it has not yet been established which of these is most appropriate.

The procedure referred to as the Malmö protocol also comprises immunomodulation, including the administration of cyclophosphamide and intravenous gammaglobulin [1,26]. To employ this protocol, the concentration of inhibitor at the onset of treatment must be low enough to allow delivery of a saturating dose of factor VIII and it must be possible to achieve free factor VIII levels in the plasma making it possible to adjust the concentration of the factor as needed during the first days of treatment. A similar protocol also exists for induction of immune tolerance in haemophilia B. The desired initial inhibitor level can be reached in two different ways: by introducing a 'waiting period' between the last booster and the start of treatment; or by extracorporeal adsorption of the inhibitor to protein A. The latter is necessary in persons with inhibitors that tend to reach a plateau at a relatively high concentration.

Based on almost 20 years of experience with the Malmö ITI model, the treatment algorithm depicted in Fig. 5.4 has been developed. Briefly, if the inhibitor titre is below 10 BU, tolerance induction is started without protein A adsorption and, if the treatment is successful, the patient can switch to ordinary prophylaxis. On the other hand, if the treatment is not successful, it can be repeated after 6–12 months. At this stage, continuation to a Bonn-like (high dose) regimen is sometimes carried out, although not in the original Malmö model. If the inhibitor

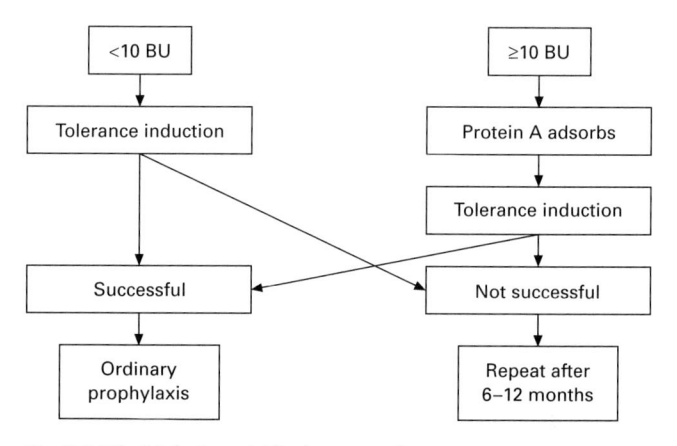

Fig. 5.4 The Malmö model for immune tolerance.

titre is above 10 BU, the patient is subjected to extracorporeal protein A adsorption followed by tolerance induction, according to the following protocol. At the start of treatment, a single dose of corticosteroids 50–150 mg is given to prevent allergic reactions. Thereafter, cyclophosphamide is given intravenously 12–15 mg/kg bw for 2 days, and orally 2–3 mg/kg bw for an additional 8–10 days. Factor VIII (or factor IX for a factor IX inhibitor) is given daily with the intention to maintain a plasma factor concentration of 40–100 U/dL for 2–3 weeks. From the fourth day after the onset of treatment, IgG is given intravenously at daily doses of 0.4 g/kg bw for 5 days. The inhibitor usually reappears after 5–6 days, and the administration of factor concentrates must be intensified. For persistent inhibitors, recombinant factor VIII (NovoSeven™) is preferred for treatment of acute bleeds.

In Malmö, 16 different people with haemophilia A (13 high responders >300 BU) have been treated with this protocol a total of 21 times, and 10 (62.5%) of the patients have become tolerant [27]. On average, the 16 patients received 162 000 U of factor VIII (range 20 000–862 000 U) for a mean of 20 days (range 9–37), and the mean daily doses given per kilogram of bodyweight was 207 U (range 83–441 U). Freiburghaus et al. [27] also reported that six of seven (86%) haemophilia B patients who were treated with the indicated protocol became tolerant. These patients received an average of 219 000 U of factor IX (range 14 000–616 000 U) for a mean of 23 days (range 8–53 days). The mean daily dosage given per kilogram bodyweight was 304 U (range 36–953 U). A modified version of the Malmö protocol with continuous infusion of factor IX at a daily dose of about 300 U/kg bw for 3 weeks has been used to treat two high responders, but this was not successful, although the inhibitor titre declined substantially in one of the patients [28].

Altogether, the Malmö protocol has been used to treat 23 patients on 38 occasions. Because of the inhibitor level, treatment was begun with immune adsorption on eight of the 21 treatment occasions in the patients with haemophilia A and on nine of the 16 treatment occasions in the patients with haemophilia B.

A modified Malmö protocol using plasmapheresis instead of protein A adsorption, and factor VIII given as a continuous infusion instead of intermittent injections, has been employed by Kucharski et al. [29] to treat 15 high responders. A successful outcome was reported for 10 of these patients, and the authors considered the protocol to be superior to their low-dose tolerance induction programme.

People with severe haemophilia usually develop inhibitors soon after initiating treatment. In as much as many of these patients are young children, the immunoadsorption procedure can be technically hazardous, and some physicians are reluctant to use cyclophosphamide. Consequently, today, use of the Malmö protocol is indicated in haemophilia A mainly for patients with long-standing inhibitors. In haemophilia B, the short time required to achieve success with the Malmö model is probably an advantage, because long-term high-dose procedures have been reported to induce nephrotic syndrome in a number of patients with this type of haemorrhagic disorder [30]. To conclude, the quick response to the Malmö protocol should in some cases be more cost-effective than other ITI protocols and might even lead to fewer side-effects in haemophilia B patients as a result of shorter exposure to FIX. However, further studies are needed to confirm the latter suggestion.

Side-effects brought about by protein A immunoadsorption

Extracorporeal protein A immunoadsorption causes few side-effects in haemophilic individuals. The procedure does not seem to induce haemolysis, complement activation or activation of the coagulation system [12]. Theoretically, it is possible that if the volume of blood in the body is small compared to the volume contained in the device (including the centrifuge and tubing), as seen especially in younger children, hypovolaemia and hypotension will occur. However, this can be overcome by priming the adsorption apparatus with, for example, blood from the patient being treated. Furthermore, none of the patients studied in Malmö have actually incurred any volaemic problems. A common adverse effect that can be detected is mild hypocalcaemic symptoms caused by the citrate buffer system used and calcium injections may occasionally be needed. To ensure a sufficiently high rate of blood flow, it is usually necessary to draw blood from an artery, and puncturing arteries can be hazardous in haemophilic individuals, especially those who are very young. This can lead to bleeding complications; indeed, a compartment syndrome caused by bleeding after puncture of the radial arteries in both arms has been reported in a 5-year-old boy [31]. A case of arterial thrombosis also occurred in the patients studied in Malmö: a 14-year-old boy with haemophilia B, who underwent the full Malmö ITI treatment, including protein A adsorption. Two weeks after starting the therapy, this patient experienced a myocardial infarction, and he subsequently developed cardiac insufficiency which caused his death 2 years later. At the time the infarction occurred, the boy had been given a prothrombin complex concentrate and had had measurable factor IX levels. The infarction may have been caused by activation of the coagulation system induced by the concentrate and not by the adsorption procedure *per se*.

Conclusions

Plasmapheresis and extracorporeal immunoadsorption with protein A are two feasible methods for removing inhibitors from the blood of patients with haemophilia. However, the latter procedure is more specific and more efficacious and should therefore be recommended over plasma exchange. Nowadays, antibody adsorption is rarely the method of choice for inhibitor removal carried out before initiating replacement of a missing factor to achieve haemostasis to stop an acute bleed or to cover a surgical procedure. The reason for this is the advent of products which have been shown to secure haemostasis effectively in most patients with inhibitors. Bypassing agents, such as recombinant activated factor VII or activated prothrombin complex concentrate, represent a convenient means of stopping acute bleeds and, more importantly, they can be used for treatment in the home setting [5,6] and are also effective in patients undergoing surgery [32].

Inhibitor removal by plasmapheresis or immunoadsorption to protein A has also been included in ITI protocols, the most well known of which is the Malmö protocol. In that model, if the inhibitor titre is above 10 BU, immunoadsorption is performed at the onset of ITI.

Most of the patients studied in Malmö have had long-standing inhibitors, and today the protocol is seldom used when ITI is to be initiated in very young patients shortly after the detection of an inhibitor. However, in haemophilia B, the Malmö protocol may be an appropriate first-line treatment, because long-term ITI in such patients can lead to nephrotic syndrome [30].

Inhibitor removal has also been employed to treat acquired haemophilia [11,33,34], and in some cases this has been attempted with a modified Malmö protocol that includes immunoadsorption to protein A. The results of ITI seem promising (H.H. Brackmann, personal communication), but the exact role of the adsorption procedure is difficult to elucidate. It appears that more extensive adsorption is necessary in acquired haemophilia than in congenital haemophilia with an inhibitor, because production of factor VIII is intact in the former patients, leading to extensive formation of immune complexes. The extravascularly located complexes are very slowly redistributed to the intravascular space and are therefore not readily available for the adsorption procedure. In aquired haemophilia, there is often substantial production of autoantibodies, jeopardizing effective adsorption, wheras in congenital haemophilia the production can be limited, provided that a waiting period precedes the adsorption procedure, which is recommended in the Malmö protocol.

Immunoadsorption to protein A is an efficient but complicated method requiring a team of experienced specialists to manage both the treatment device and the patient being treated. Therefore, if this procedure, or other adsorption techniques (e.g. use of the Ig-TheraSorb™ columns) are to become feasible and accessible for patients with haemophilia who also have inhibitors, they must be made available at haemophilia centres that are especially focused on the inhibitor problem.

References

1. Nilsson IM, Berntorp E, Zettervall O. Induction of immune tolerance in patients with hemophilia and antibodies to factor VIII by combined treatment with intravenous IgG, cyclophosphamide, and factor VIII. *N Engl J Med* 1988; **318**: 947–50.
2. Francesconi M, Korninger C, Thaler E *et al.* Plasmapheresis: its value in the management of patients with antibodies to factor VIII. *Haemostasis* 1982; **11**: 79–86.
3. Edson JR, McArthur JR, Branda RF, McCullough JJ, Chou SN. Successful management of a subdural hematoma in a hemophiliac with an anti-factor VIII antibody. *Blood* 1973; **41**: 113–22.
4. Cobcroft R, Tamagnini G, Dormandy KM. Serial plasmapheresis in a haemophiliac with antibodies to FVIII. *J Clin Pathol* 1977; **30**: 763–5.
5. Key NS, Aledort LM, Beardsley D *et al.* Home treatment of mild and moderate episodes using recombinant factor VIIa (Novoseven) in haemophiliacs with inhibitors. *Thromb Haemost* 1998; **80**: 912–8.
6. Négrier C, Goudemand J, Sultan Y *et al.* Multicenter retrospective study on the utilization of FEIBA in France in patients with factor VIII and factor IX inhibitors. *Thromb Haemost* 1997; **77**: 1113–19.
7. Jensen K. *Acta Allerg* 1959; **13**: 89–100.
8. Löfkvist T, Sjöquist J. *Int Arch Allergy Appl Immunol* 1963; **23**: 289–305.
9. Moks T, Abrahamsén L, Nilsson B *et al.* Staphylococcal protein A consists of five IgG-binding domains. *Eur J Biochem* 1986; **156**: 637–43.
10. Hoyer LW, Gawryl MS, de la Fuente B. Immunological characterization of factor VIII inhibitors. In: Hoyer, LW, ed. *Factor VIII Inhibitors*. New York: Alan R. Liss, 1984: 73–85.
11. Freiburghaus C. *Extracorporeal immunoadsorption*. Thesis, Lund, Sweden, 1998.
12. Nilsson IM, Jonsson S, Sundqvist SB, Ahlberg Å, Bergentz SE. A procedure for removing high titer antibodies by extracorporeal protein A–Sepharose adsorption in hemophilia: substitution therapy and surgery in a patient with hemophilia B and antibodies. *Blood* 1981; **58**: 38–44.
13. Gjörstrup P, Berntorp E, Larsson L, Nilsson IM. Kinetic aspects of the removal of IgG and inhibitors in hemophiliacs using protein A immunoadsorption. *Vox Sang* 1991; **61**: 244–50.
14. Freiburghaus C, Berntorp E, Ekman M *et al.* Immunoadsorption for removal of inhibitors: update on treatments in Malmö-Lund between 1980 and 1995. *Haemophilia* 1998; **4**: 16–20.
15. Knobl P, Derfler K. Extracorporeal immunoadsorption for the treatment of haemophilic patients with inhibitors to factor VIII or IX. *Vox Sang* 1999; **77** (Suppl. 1): 57–64.
16. Knobl P, Derfler K, Korninger L *et al.* Elimination of acquired factor VIII antibodies by extracorporeal antibody-based immunoadsorption (Ig-Therasorb). *Thromb Haemost* 1995; **74**: 1035–8.
17. Nilsson IM, Freiburghaus C, Sundqvist S-B, Sandberg H. Removal of specific antibodies from whole blood in a continuous extracorporeal system: plasma. *Ther Transfus Technol* 1984; **5**: 127–34.
18. Brackmann JH, Gormsen J. Massive factor VIII infusion in hemophiliac with factor VIII inhibitor, high responder. *Lancet* 1977; **2**: 933.
19. Van Leeuwen EF, Mauser-Bunschoten EP, van Dijken PJ *et al.* Disappearance of factor VIII:C antibodies in patients with haemophilia A upon frequent administration of factor VIII in intermediate or low dose. *Br J Haematol* 1986; **64**: 291–7.
20. Mauser-Bunschoten EP, Nieuwenhuis K, Roosendaal G, van den Berg M. Low-dose immune tolerance induction in hemophilia A patients with inhibitors. *Blood* 1995; **86**: 983–8.
21. Aznar JA, Jorquera JL, Peiro A. Suppression of inhibitors in haemophilia with corticosteroids and factor VIII. *Thromb Haemost* 1983; **49**: 241–3.
22. Ewing NP, Sander NL, Deitrich SL, Kasper CK. Induction of immune tolerance to factor VIII in hemophiliacs with inhibitors. *J Am Med Assoc* 1988; **259**: 658.

23 Brackmann HH. Induced immune tolerance in factor VIII inhibitor patients. *Prog Clin Biol Res* 1984; **150**: 181–95.

24 Brackmann HH, Oldenburg J, Schwaab R. Immune tolerance for the treatment of factor VIII inhibitors: twenty years 'Bonn Protocol'. *Vox Sang* 1996; **70** (Suppl. 1): 30–5.

25 Scheibel E, Ingerslev J, Dalsgaard-Nielsen J, Stenbjerg S, Knudsen JB and the Danish Study Group. Continuous high-dose factor VIII for the induction of immune tolerance in haemophilia A patients with high responder state: a description of eleven patients treated. *Thromb Haemost* 1987; **58**: 1049–52.

26 Berntorp E, Nilsson IM. Immune tolerance and the immune modulation protocol. *Vox Sang* 1996; **70** (Suppl. 1): 36–41.

27 Freiburghaus C, Berntorp E, Ekman M *et al*. Tolerance induction using the Malmö treatment model 1982–95. *Haemophilia* 1998; **5**: 32–9.

28 Tengborn L, Berntorp E. Continuous infusion of factor IX concentrate to induce immune tolerance in two patients with haemophilia B. *Haemophilia* 1998; **4**: 56–9.

29 Kucharski W, Scharf R, Nowak T. Immune tolerance induction in haemophiliacs with inhibitor to FVIII: high- or low-dose programme. *Haemophilia* 1996; **2**: 224–8.

30 Warrier I. Factor IX antibody and immune tolerance. *Vox Sang* 1999; **77** (Suppl. 1): 70–1.

31 Nilsson IM, Berntorp E, Rickard KA. Results in three Australian haemophilia B patients with high-responding inhibitors treated with the Malmö model. *Haemophilia* 1995; **1**: 59–66.

32 Shapiro AD, Gilchrist GS, Hoots WK, Cooper HA, Gastineau DA. Prospective, randomised trial of two doses of rFVIIa (NovoSeven) in patients with inhibitors undergoing surgery. *Thromb Haemost* 1998; **80**: 773–8.

33 Négrier C, Dechavanne M, Alfonsi F, Tremisi PJ. Successful treatment of acquired factor VIII antibody by extracorporeal immunoadsorption. *Acta Haematol* 1991; **85**: 107–10.

34 Taleghani BM, Grossmann R, Keller F, Wiebecke D. Therapy of coagulation factor VIII autoantibodies with long-term extracorporeal protein A adsorption and immunosuppression. *Transfus Sci* 1998; **19**: 39–42.

CHAPTER 6

Venous access in children with inhibitors

R.C.R. Ljung

Treatment of haemophilia A or B with factor VIII (FVIII) and factor IX (FIX) concentrates, respectively, irrespective of whether this is given for a bleed or as a prophylactic infusion, requires uncomplicated venous access. Patients who have developed an inhibitor (antibody) against FVIII or FIX have frequent and life-threatening bleeding which may have to be treated urgently. Ideally, such treatment should be administered immediately at home by the patient or the parents, a situation in which safe and easy access to a vein is mandatory. Immune tolerance programmes for the treatment of patients with inhibitors usually require infusions once or twice daily which may be technically very difficult using a peripheral vein in a small boy and sometimes this is also complicated in adults. Most inhibitors occur in young children after a limited number of infusions of concentrate. The use of a peripheral vein is always the first choice but venous access in very young children may be difficult and require several venepunctures with anxiety and psychological stress for both the child and parent or caregiver.

The good results obtained with various central venous catheters in the treatment of oncological disorders in children [1] prompted the attempt to use central lines in the treatment of haemophilia in the late 1980s and early 1990s [2,3]. Since then several reports have been published with varying experience concerning complications, especially infections [4–11]. For several years there has been an on-going debate among those responsible for the treatment of children with haemophilia concerning the benefits and side-effects of central venous catheters. Some clinicians favour them while others are more critical because of the frequency of complications. This chapter gives an overview of the experience so far with the use of central venous lines in patients with haemophilia, with the focus being on patients who also have inhibitors.

Evaluating the studies on record

There are several reports in the literature on the use of central venous lines for haemophilia treatment [2–6,8–10,12–19]. When evaluating studies on central venous lines for treatment in haemophilia, several aspects need to be focused on.

First, some series include both patients with external catheters and implanted subcutaneous ports [6]. There are various forms of external catheters, such as Broviac and Hickman catheters. Most of the reports on implantable devices have been using the Port-A-Cath™. It has been clearly shown in a large series of non-haemophilic children that implantable systems have a much lower risk of infections compared to external catheters; in one series 0.7 infections/1000 patient days for subcutanous ports vs. 4.3 for external catheters ($P < 0.0001$) [20]. Thus, the first recommendation when considering a central venous line is to use an implantable system.

Secondly, many series are heterogeneous and include patients with haemophilia both with and without inhibitors. It has been shown in many series that the frequency of infection is higher in patients with inhibitors [3,6,10]. In one series with a median follow-up of 30 months, 23% (9 of 39) of patients without inhibitors had complications corresponding to 0.23 complications/1000 patient days. In comparison, 64% (7 of 11) of patients with inhibitors manifested complications, corresponding to 1.9 infections/1000 days [8].

Thirdly, some series include patients with HIV infection who may be immunocompromised and thus naturally have a higher risk of infection than patients who do not have HIV [6]. Series including many HIV-infected persons do not give an accurate risk figure for the non-HIV-infected haemophilic healthy patient considered for a central venous device.

Many series, although usually with the majority of patients affected with haemophilia, also include other coagulation disorders [4,10,15,16]. It is not known if experiences from other coagulation disorders also apply to haemophilia.

Most series using implantable catheters use the Port-A-Cath™ system and almost all experience from larger series in patients with haemophilia use this system. However, some series report also peripheral intravenous access devices almost exclusively used in children (P.A.S. Ports™; SIMS, Deltec Inc.; Slim-Ports™, Bard) [16]. These seem to be well accepted by the children and their parents. In young children it is less threatening to insert a needle into the periphery of the body and thus avoid the visible profile of the port on the chest. However, peripheral ports have been associated with a higher frequency of thrombophlebitis and thrombosis and the average time the patient may benefit from the device is probably shorter [16]. In one series with eight peripheral ports and 32 central ports, the frequency of infection was lower in the peripheral group (0.8 vs. 1.2 episodes/1000 days) but the series was too small to form any firm conclusions [16].

The Percuseal™ device (Percuseal Medical, Huskvarna, Sweden) is a rather novel percutaneous inlet that enables administration without skin puncture in an attempt to combine the benefits from both an external device, involving no skin puncture, and an implanted device, with less impact on daily life and fewer infections. It is made of medical-grade titanium and is implanted into the subcutaneous tissue, with the top portion protruding from the surface of the skin. The implant wall facing the subcutaneous tissue is designed to promote connective tissue ingrowth and prevent epidermal downgrowth, which would lead to rejection of the inlet. A catheter is led into a central vein from the inner end of the inlet. An advantage, apart from the painless clotting factor application without skin puncture, is the immediate recognition of superficial infections, because they grow towards the skin. A pilot study has shown that it is feasible to use the Percuseal™ inlet for frequent FVIII injections, at least on a short-term basis [21]. Although interesting because based on a novel approach, the usefulness of the modified version being introduced needs to be demonstrated in larger series.

This chapter focuses mainly on the Port-A-Cath™ system because almost all data from larger series are from this device.

Infection

Infection is the most frequent complication when using a central venous line. Unfortunately, it is not usually possible to study exclusively a cohort of children with haemophilia who are HIV-negative, or to separate patients with inhibitors from those who do not have inhibitors. Some series separate local infections around the port and proved bacteriaemia or septicaemia. When discussing infections or other complications one should also emphasize that many systems had been used for long periods before the complication occurred.

Table 6.1 shows some of the larger recent studies where it has been possible to study patients with inhibitors separately. It is obvious that the frequency of infection is higher in patients with inhibitors. In the series by Ljung et al. [8] the experience of the Port-A-Cath™ system in 53 children with severe and moderate haemophilia A or B from seven centres was reviewed. Eleven of the 53 children had inhibitors at the time of implantation and three developed during the study period. Table 6.2 shows the complications in the 11 patients who had an inhibitor at the time of implantation (the three who developed inhibitors during the study did not manifest any complications). As can be seen, 45% of the inhibitor patients (5 of 11) manifested infections after 1–47 months of use (mean 15 months, median 15 months). Infections were defined as bacteraemia or septicaemia, and excluding minor skin infections successfully treated at the time of implantation. The rate of infection

Table 6.1 The rate of infection in recent series of patients with haemophilia and inhibitors using implantable (ports) central venous lines (the subgroup in the paper fulfilling these criteria has been extracted from the references).

Study	Total number of patients (*n*)	Number of inhibitor patients (*n*)	Rate of infection/ 1000 patient days	Comments
[6]	19	2	*	*No infection reported in these 2 boys
[16]	30	7	1.2 (central port) 0.7 (peripheral port)	Equal rate inhib/ non-inhibitors 2/7 vs. 7/23
[15]	23	5	4.0	
[8]	53	11	1.9	
[10]	37	14	4.3	
[22]	38	28	0.76 17 of 28 (60%) [6/10 (60%) non-inhibitors]	Equal between inhibitors/ non-inhibitors Median time before infection 211 (range 15–1370) days

Table 6.2 Complications in patients with inhibitors at the time of implantation of the Port-A-Cath™ (From Ljung et al. [8]).

Type of complication	Number of children	Time before complication (months)
Infection	5	1, 4, 15, 18, 47
Leakage	1	5
Fibrin clot in catheter tip	1	6
Total	7	(Mean 15)

corresponds approximately to one infection/17 months or 1.9 infections/1000 patient days for the HIV-negative child with haemophilia and inhibitors.

McMahon et al. [10] studied 86 central venous access devices in 58 children with coagulopathies. Of the 37 children with haemophilia A or B, 14 had inhibitors and 18 Port-A-Caths™ were implanted in this subgroup. Twenty-three infections occurred in these 18 central venous lines, which means one infection/8.5 patient months or 4.3 infections/1000 patient days, which is a considerably higher frequency than in the study by Ljung et al. [8].

In a review, van den Berg et al. [22] found that in various studies 50–83% of patients with inhibitors can be expected to acquire an infection. Collins et al. [15] reviewed 23 patients with bleeding disorders who had 32 central venous lines inserted. There were 25 documented episodes of infection during a median follow-up time of 27 months (range 1–92) which corresponds to one infection/26 patient months at risk. However, the subgroup of five patients with haemophilia who also had inhibitors had a total of 12 infections corresponding to one infection/8.3 months (approx. 4/1000 days) compared to 1/50 months for patients without inhibitors [15].

Perkins et al. [16] presented a series of 30 children with haemophilia of whom seven also had an inhibitor. Two of the seven cases with an inhibitor had infection (positive blood culture) compared to seven of 23 patients who did not have an inhibitor; giving an equal risk of infection between the two subgroups.

An interim report from the Spanish registry of children with haemorrhagic disorders and the use of Port-A-Caths™ ($n = 38$) showed an equal frequency of 60% infection, defined as clinical manifestations and positive blood culture, in the inhibitor group ($n = 28$) and the non-inhibitor group ($n = 10$) [22].

It is obvious that most series show a higher frequency of infection in patients with inhibitors than in patients without inhibitors. One reason for the higher frequency of infections may be that these patients have small bleeds around the port after an injection that may stimulate bacterial growth in the subcutaneous tissue, and subsequent punctures in the area will transfer the bacteria into the port. Another explanation could be that the patient with an inhibitor is often on an immune tolerance induction programme which includes one or two injections/day. The non-inhibitor patient on prophylaxis accesses the port at most every second day and the patient on treatment on-demand even more infrequently. However, in one study no relation could be found between the number of punctures of the Port-A-Cath™ and the frequency of infection [8].

A reasonable conclusion of the different experiences with Port-A-Cath™ in patients with inhibitors is that one may expect one infection/8–17 months of use: in practice, approximately one infection/year/patient. However, these patients need uncomplicated easy venous access both for the treatment of acute bleeds and for immune tolerance induction programmes and the risk of infection has to be judged in that context.

For non-inhibitor patients the need for a port has to be considered together with the risk of complications, which varies between different centres. There seem to be two major experiences concerning infections in non-inhibitor patients: one is approximately 0.2 infections/1000 days [8,9] and the other approximately 1.0 (range 0.7–1.6)/1000 days [6,10,16]. Whether this is an acceptable frequency of infection for this group of patients depends on the situation of the individual patient and the treatment regimen. The child prone to spontaneous bleeds who should start primary prophylactic treatment from the age of 1 year is a greater challenge for venous access than the child with infrequent bleeds on treatment on-demand. Central catheters are only an option for those patients in frequent need of an uncomplicated venous access where the benefits must be calculated to be greater than the risks. The indication for a central line has to be discussed with the parents, and the social situation and the need for home treatment needs to be taken into account. It is always advisable to start treatment, whether prophylaxis or on-demand, by using a peripheral vein in the hope that it will be sufficient.

Pathogens and treatment of infection

In all series, the most frequent pathogens are *Staphylococcus epidermidis* and *S. aureus* [10,16]. However, Gram-negative organisms are also common and in some series they are more likely to have been the cause of removal of the device. In the series by McMahon et al. [10], 68% of the ports infected with Gram-negative organisms had to be removed, compared to 40% of those infected by Gram-positive bacteria. In the immunocompetent patient with haemophilia, *Candida albicans* or other fungal species are rarely found in the catheter.

The use of antibiotics in the peroperative period varies between centres and remains controversial. Most of the experience is from cancer patients and in some studies antibiotics have been of value [24], while in others no such correlation was found [25]. The frequency of infections in the immediate postoperative period varies between series, reflecting that the surgical technique and general aseptic conditions are important. However, these infections are usually treated successfully without the device being removed and seem to have little or no importance for bacteraemia later in the course. More important is probably the overall use of antibiotics in the society and thus the frequency of resistance against antibiotics. Vigorous education in aseptic techniques and follow-up and reactivation of the education are keystones in reducing the risk of infection. It has also been speculated that the common use of EMLA™-anaesthetic cream may have a role [16]. A reduction in infections was found after parents were instructed to scrub the cream off by soap and water and not only by an ordinary medical swab in order to remove the residual lipid from the skin. This is probably an experience to include in the education of aseptic techniques.

It has also been speculated that younger children, below the of age 2–4 years, may be more susceptible to infections because of their immature immune systems. For example, in one series, the children who manifested infections were 3.2 years vs. 6.6 years for those who did not become infected [16]; however, in another study no such age effect was seen [8]. Inhibitors usually develop in small children after less than 20 exposures of concentrate, and immune tolerance induction is thus often performed in small children whose immune system is still developing.

An infection in the central line does not necessarily mean that the catheter has to be removed. There are several reports in the literature that catheters may be used after the treatment of an infection. This is usually the case when the patient has an infection around the port in the postoperative period. It may be advisable not to use the port by repeated punctures in the immediate postoperative period to ensure that no infection develops in the area around the port that may be transferred to the inside of the system. However, many authors seem to have positive experience of accessing the device intraoperatively and use the needle for postoperative infusions of concentrate. There are anecdotal reports of a beneficial effect of the installation of antibiotics in the catheter between injections during treatment of a bacteraemia. After treatment of bacteraemia or septicaemia the system has to be checked repeatedly before it is considered free from bacterial contamination. In many cases it has to be removed and it may be advisable not to insert a new system in the same operative procedure. However, the experiences of the use of a second device if the first fails, encourage the recommendation to implant a second device [8].

Other complications

Infections are the main side-effects when using central venous catheters. However, several other complications have to be considered, as shown in Table 6.3. Thrombosis has been a major concern regarding indwelling catheters. The figures are very low for clinically apparent thrombosis in the larger series on record. No thromboses were seen in several of the larger studies including both inhibitor and non-inhibitor patients [8,10]. However, one should be aware that routine venograms were not performed in these series. In cases where this has been carried out for various reasons, a fibrin clot at the tip of the catheter has sometimes been seen, the clinical significance of which is not clear. There are reports of subclavian thrombosis: 1 in 25 cases of central catheters in the series of Blanchette et al. [6] and in two boys in another report [26]. A recent report presented a child who after an infection with *S. aureus* in the postoperative period was later found to have bacterial endocarditis further complicated by pulmonary embolism [27]. There are reports in the literature even of thrombosis in patients with inhibitors [26] but, despite the cases on record, the risk seems to justify only routine clinical surveillance for this potential complication. It is important that the position of the tip of the catheter is in the right atrium or in the superior vena cava where the flow of blood diminishes the risk of thrombosis compared to when

Table 6.3 Complications other than infections in various studies with an estimate of risk (refers not only to inhibitor patients).

Complication	Frequency	Study
Technical problems (blockade or buckling of catheter, damage from trauma)	3/16 complications. Blockade after a mean of 13 months.	[8]
	Catheter splitting/disconnection after a mean of 2 years	[10]
Short catheter	1/16*	[8]
	1/23*	[6,7]
Erosion of skin over port	0/53*	[8]
	3/86†	[10]
	3%	[16]
Thrombosis	0/53*	[8]
	1/23†	[6,7]
	0/86†	[10]
	0/40†	[16]
	2 case reports	[26]
	1 case report	[27]
Worn membrane	Rare	
Bleed around the port	Not unusual in inhibitor patients	

*Number of patients.
†Number of devices.

placed in a smaller vessel. Routine annual radiography of the position of the catheter tip seems warranted in this context.

The risk of catheter blockage may be diminished by proper education in the use of saline flush and a 'heparin lock' after use (a few mL heparin 100 U/mL, depending on the system used). If not used, the catheter should be flushed and a new heparin lock should be administered every second month at least. Another prophylactic measure is to remove protein film on the inside of the catheter by the routine installation of urokinase or t-PA in the catheter. For example one may use 2–3 mL urokinase 5000 U/mL, or 1.5 mL tissue plasminogen activator (t-PA), Actilyse™, 1 mg/mL, for children >5 years and 0.3 mL/kg for children <5 years for 1–2 h. A small volume syringe (2–5 mL) should be used to avoid high pressure in the catheter during the procedure if there is already a clot in the catheter. How often this has to be performed depends on the use of the catheter, but at least when the injections in the catheter become less smooth. The nurse or doctor should be familiar with the procedure. One should also take into account that the intravascular part of the catheter may be coated on the exterior by plasma proteins which can act as an adhesive for certain bacteria if these are present in the blood.

Erosions of the skin over the port have occurred in a few reported cases and have to be considered as a serious complication. Inhibitor patients with bleeds and subsequent infection around the port seem to be at particular risk. Other complications may be leakage, disconnection between catheter and port or a worn membrane (although these are made to last for several thousand penetrations).

Haemostasis during central line insertion

A central venous access device should be surgically placed by an experienced surgeon during general aneasthesia. Haemostasis in the patient with haemophilia and an inhibitor has to be obtained by the regimen used at the centre, recombinant FVIIa (rVIIa, NovoSeven™), porcine FVIII (Hyate C™) or FVIII bypassing agents (Feiba™, Prothromplex™). Smith et al. [28] reported successful use of rVIIa in four line insertions in patients with inhibitors and an update of the same material had 16 port insertions on record under the cover of rVIIa without bleeds [10]. Another case was reported by Bell et al. [29] using rVIIa. Haemorrhagic complications at the insertion of a central venous device have been reported in several studies, especially in patients with inhibitors. Warrier et al. [5] reported bleeds in 20% of patients and Ljung et al. [3] in 15%. The reader is referred to the current literature for suggestions for dose and dosage interval when using rVIIa. As a general guideline, bolus doses should be 100–200 µg/kg repeated every 2–3 h. However, continuous infusion of rVIIa with the aim of keeping VIIa concentration at a level of 10 U/mL may considerably reduce the factor needed [30]. The duration of treatment is 2–4 days in patients with inhibitors. For doses and dosage interval on FVIII bypassing agents and porcine factor VIII the reader is referred to the manufacturers' recommendations.

Conclusions

In summary, a peripheral vein should always be the first choice for the patient with an inhibitor. In many children a central venous access device is necessary when access to a peripheral vein is difficult or not possible for the modern frequent treatment for induction of immune tolerance, or for the immediate treatment of a serious bleed. Some of the complications on record so far may be reduced by adequate aseptic measures, both during implantation and in the subsequent use, and systematic basic routines for regular surveillance. The final decision to use a central line in a patient has to be a compromise between the medical goal, the medical risk, and the social situation and the risk of complications.

Acknowledgement

This work was supported in part by grants from the Swedish Medical Research Council (no. K20000-71X-13493-01 A and no. 10409), research funds from the University of Lund (ALF), and regional funds from the county of Skåne and Malmö University Hospital, Sweden.

References

1 Wesenberg F, Flaaten H, Janssen CJ. Central venous catheter with subcutaneous port (Port-A-Cath): 8 years clinical follow-up with children. *Paediatr Haematol Oncol* 1993; **10**: 233–9.
2 McWhirter WR, Gray L. Indwelling intravenous catheter in a young haemophiliac [Letter]. *Lancet* 1988; ii (8602): 99–100.
3 Ljung R, Petrini P, Lindgren AK, Berntorp E. Implantable central venous catheter facilitates prophylactic treatment in children with haemophilia. *Acta Paediatr* 1992; **81**: 918–20.
4 Liesner RJ, Vora AJ, Hann IM, Lilleymann JS. Use of central venous catheters in children with severe congenital coagulopathy. *Br J Haematol* 1995; **91**: 203–7.
5 Warrier I, Baird-Cox K, Lusher J. Use of central venous catheters in children with haemophilia: one haemophilia treatment centre experience. *Haemophilia* 1997; **3**: 194–8.
6 Blanchette VS, al Musa A, Stain AM, Filler RM, Ingram J. Central venous access catheters in children with haemophilia. *Blood Coag Fibrinolysis* 1996; **7** (Suppl. 1): S39–44.
7 Blanchette VS, Al Musa A, Stain AM, Ingram J, Fille RM. Central venous access devices in children with hemophilia: an update. *Blood Coag Fibrinolysis* 1997; **8** (Suppl. 1): S11–14.
8 Ljung R, van den Berg M, Petrini P et al. Port-A-Cath usage in children with haemophilia: experience of 53 cases. *Acta Paediatr* 1998; **87**: 1051–4.
9 Miller K, Buchanan GR, Zappa S et al. Implantable venous access devices in children with hemophilia: a report of low infection rates [see comments]. *J Pediatr* 1998; **132**: 934–8.
10 McMahon C, Smith J, Khair K et al. Central venous access devices in children with congenital coagulation disorders: complications and long-term outcome. *Br J Haematol* 2000; **110**: 461–8.
11 Mumtaz H, Williams V, Hauer-Jensen M et al. Central venous catheter placement in patients with disorders of haemostasis. *Am J Surg* 2001; **180**: 503–6.

12 Vidler V. Use of Port-a-Caths in the management of paediatric haemophilia. *Prof Nurse* 1994; **10**: 48–50.
13 Bollard C, Teague L, Berry E, Ockelford P. The use of central venous catheters (Port-A-Caths) in children with haemophilia. *Haemophilia* 2000; **6**: 66–70.
14 Damiano M, Hutter J. Immune tolerance for haemophilia patients with inhibitors: the western United States experience. *Haemophilia* 2000; **6**: 526–32.
15 Collins PW, Khair KS, Liesner R, Hann IM. Complications experienced with central venous catheters in children with congenital bleeding disorders. *Br J Haematol*, 1997; **99**: 206–8.
16 Perkins JL, Johnson VA, Osip JM *et al*. The use of implantable venous access devices (IVADs) in children with hemophilia. *J Pediatr Hematol Oncol* 1997; **19**: 339–44.
17 Schultz WH, Ware R, Filston HC, Kinney TR. Prolonged use of an implantable central venous access system in a child with severe hemophilia. *J Pediatr* 1989; **114**: 100–1.
18 Miser AW, Roach JE, Harmel RP Jr, Sayers MP, Miser JS. Insertion of a central venous catheter for long-term venous access in a child with severe hemophilia and recurrent intracranial hemorrhage. *Clin Pediatr (Phila)* 1984; **23**: 589.
19 Lilleyman JS. Domiciliary desensitization therapy for young boys with haemophilia and factor VIII inhibitors. *Br J Haematol* 1994; **86**: 433–5.
20 Ingram J, Wietzman S, Greenberg ML. Complications of indwelling venous access lines in the pediatric hematology patient: a prospective comparison of external venous catheters and subcutaneous ports. *J Pediatr Hematol Oncol* 1991; **13**: 130–6.
21 Berntorp E, Frick K, Mätzsch T, Carlsson M, Lethagen S. The Percuseal device: a new option for effective continuous prophylaxis? *Haemophilia* 1998; **4**: 184.
22 Tusell J, Villar A, Lopez M *et al*. High incidence of Port-A-Cath (PAC) infections in children with congenital coagulopathies: Spanish registry. *European Society of Paediatric Haematology and Immunology* 2001: Abstract 88.
23 Van Den Berg HM, Fischer K, Roosendaal G, Mauser Bunschoten EP. The use of the Port-A-Cath in children with haemophilia: a review. *Haemophilia* 1998; **4**: 418–20.
24 Lim SH, Smith MP, Machin SJ, Goldstone AH. A prospective randomised study of prophylactic teicoplanin to prevent early Hickman catheter-related sepsis in patients receiving intensive chemotherapy for hematological malignancies. *Eur J Haematol Suppl* 1993; **54**: 10–13.
25 Ranson MR, Oppenheim BA, Jackson A, Kamthan AG, Scarffe JH. Double-blind placebo controlled study of vancomycin prophylaxis for central venous catheter insertion in cancer patients. *J Hosp Infect* 1990; **15**: 95–102.
26 Vidler V, Richards M, Vora A. Central venous catheter-associated thrombosis in severe haemophilia. *Br J Haematol* 1999; **104**: 461–4.
27 Hothi D, Kelsall W, Baglin T, Williams D. Bacterial endocarditis in a child with haemophilia B: risks of central venous catheters. *Haemophilia* 2001; **7**: 507–10.
28 Smith OP, Hann IM. rVIIa therapy to secure haemostasis during central line insertion in children with high-responding FVIII inhibitors. *Br J Haematol* 1996; **92**: 1002–4.
29 Bell BA, Birch K, Glazer S. Experience with recombinant factor VIIA in an infant hemophiliac with inhibitors to FVIII:C undergoing emergency central line placement: a case report. *Am J Pediatr Hematol Oncol* 1993; **15**: 77–9.
30 Schulman S, Bech Jensen M, Varon D *et al*. Feasibility of using recombinant factor VIIa in continuous infusion. *Thromb Haemost* 1996; **75**: 432–6.

PART 3

Immune tolerance

CHAPTER 7

Immune tolerance: high-dose regimen

H.H. Brackmann and T. Wallny

The development of inhibitor antibodies to factor VIII (FVIII) still represents the most serious complication in haemophilia A treatment. Up to 30% of severely affected patients develop antibodies against FVIII, which render FVIII therapy ineffective. The only curative treatment is induction of permanent immune tolerance by down-regulating or eradicating FVIII antibodies. In 1977 Brackmann and Gormsen reported the first curative treatment protocol for inhibitor patients—known as the 'Bonn protocol'—which represented a high-dose regimen designed to induce lifelong immune tolerance towards substituted factor VIII [1,2]. Based on these findings several modifications of the Bonn protocol were reported in the following years [3–8].

Birth of the Bonn protocol

More than 25 years ago, a 1.5-year-old-inhibitor patient came to the Bonn centre with severe bleeding episodes in his right shoulder, right upper arm and right chest. The inhibitor titre at that time was >500 Bethesda units (BU). In 1974 Kurczinsky and Penner [9] reported successful treatment of bleeding episodes in patients with an inhibitor by activated prothrombin complex concentrates; however, this product was not available for the German market at the time. Thus it was decided, therefore, to combine high dosages of factor VIII with a regular prothrombin concentrate. The regimen was given twice daily to the patient, and the bleeding was controlled. After 3 weeks the patient had recovered completely. The inhibitor titre decreased to about 40 BU during this treatment.

As a consequence of this experience the procedure was evaluated in several more inhibitor patients suffering from acute bleeding episodes. An initial booster effect of the inhibitor was observed in some. However in all cases the inhibitor titre decreased after some weeks; it was therefore decided not to stop but to continue the treatment. Following this regimen the inhibitor finally disappeared and the half-life of factor VIII normalized. From this time, the dosage schedule of the Bonn protocol, still used today, was developed [10].

The Bonn protocol

The Bonn protocol initially comprised two treatment phases (Table 7.1). During the first phase, 100 IU FVIII/kg and 50 IU/kg activated prothrombin complex concentrate (aPCC; Feiba™) respectively were given twice daily until the inhibitor decreased to less than 1 BU and the 30 min FVIII recovery became measurable. The rationale of using aPCC application was the prevention of intercurrent haemorrhages in patients of high bleeding risk. In the second phase (inhibitor titre <1 BU) therapy was continued only with factor VIII (150 IU FVIII/kg

Table 7.1 Original Bonn method for introduction of immune tolerance in inhibitor patients with haemophilia A and a tendency to bleed or orthopaedic problems (Schedule I), or no such problems (Schedule II).

Treatment schedule	Agent	Dosage	Duration
Ia	FVIII aPCC (Feiba™)	100 U/kg body weight every 12 h 50 U/kg every 12 h (to prevent intercurrent haemorrhages)	Until inhibitors = <1 BU/mL
Ib	FVIII	150 U/kg body weight every 12 h	Until inhibitors disappear and half-life of factor VIII is normal
II	FVIII	150 U/kg body weight every 12 h	Until inhibitors disappear and half-life of factor VIII is normal

aPCC, activated prothrombin complex concentrate; BU, Bethesda unit; FVIII, factor VIII.

twice daily) until the inhibitor disappeared completely and the FVIII half-life had normalized. Today, complete immune tolerance therapy (ITT) is performed using 150 IU FVIII/kg twice daily. Feiba™ therapy is only used in those patients with high bleeding frequency [2,10].

The Bonn protocol has proven to be highly effective in eliminating inhibitors in the great majority of patients. Patients are more likely to respond if the inhibitor titre is low when ITT is started and when the maximum inhibitor titre has not exceeded 100 BU [11]. In some cases, however, several years of therapy were required to attain immune tolerance [3,5,12,13].

Experiences with the Bonn protocol

The following data describe the results of ITTs performed according to the Bonn protocol in 60 haemophilia A patients with inhibitors (36 high responders, 24 low responders). The overall success rate (by August 1997) was 86.7% (52 patients). ITT was unsuccessful in 13.3% (eight patients), and was only stopped in one patient due to failure. In the remaining seven patients the therapy failed because of death during treatment (four patients), lack of compliance (one patient), movement to another centre (one patient) or serious complications of therapy (endocarditis, one patient) [2].

The success rate observed at the Bonn Centre corresponds to that reported by Mariani *et al.* [11] for high-dosage protocols (>200 IU FVIII/kg/day). Evaluating data from the international registry, Mariani *et al.* found a success rate of 93.3% for low-responder and 84% for high-responder inhibitor patients. In contrast, application of low-dosage protocols to high-responder patients yielded much lower success rates, ranging from 31.2% (<50 IU FVIII/kg/day) to 50% (50–200 IU FVIII/kg/day) in high-responder patients. Overall, the outcome of ITT positively correlated to the factor VIII dosage used with the best success rate found in patients with a dosage >200 IU/kg/day [2,11]. These data suggested application of high-dosage protocols especially in the high-responder patients. Because of the high costs of the Bonn protocol, several modifications were reported in more recent years [3,4,6,8,14].

In order to show the long-term effect of ITT on the clinical outcome, 12 haemophilic patients from the Bonn Centre were followed up regarding their arthropathies in 2001, a median 12.75 years (range 8–21) after onset of the inhibitor. The average age at time of detection of the inhibitor was 2.4 years (9 months–9 years). The inhibitor was treated by ITT for an average 16.2 months (5–34). Two of 12 patients were affected by slight arthropathies in one joint (elbow and ankle) before developing an inhibitor. The elbow showed a flexion contracture of eight degrees with a thickened and scarred capsule; the ankle exhibited a slight synovitis with swelling and heat.

Clinical assessment of these patients in 2001, a median 12.75 years after onset of the inhibitor, revealed that the majority of the patients had maintained their clinical status. Clinical or radiological arthropathic changes could no longer be detected in the formerly affected ankle of the patient mentioned above.

The patient who had slight changes in his elbow at the beginning of inhibitor therapy showed clinically, but not radiologically, a slight increase of arthropathy (flexion contracture of 10°, supination 70°); in daily life, he had no functional restrictions. One further patient, without any arthropathies at the beginning of inhibitor therapy, showed a thickened capsule in the ankle and a slight contracture of the dorsal flexion with pain on loading. The radiological assessment revealed an advanced arthropathy in both ankles with a Petterson score of 3 and 7 points respectively. The elbow of the same patient had a flexion contracture of 10° with no further clinical or radiological references to an arthropathy. Subjectively, he occasionally complained of pain in his ankle but had no functional restriction in daily life.

In conclusion, the results impressively demonstrate that in spite of developing an inhibitor these patients experienced no significant progress in arthropathy over many years, thus showing ITT as a highly effective means of restoring quality of life and life expectancy for haemophilic patients who experience an inhibitor.

Side-effects during immune tolerance induction

Severe side-effects occurred in the past with the use of non-virus-inactivated FVIII concentrates. Until the 1980s these infections were mainly hepatitis B and hepatitis C, whereas HIV infection has dominated the last two decades. Today, due to higher purification and efficient virus inactivation procedures of plasma-derived FVIII and because of the introduction of recombinant FVIII concentrates, factor VIII substitution therapy has become safe with respect to viral infections.

The early concentrates that were of low purity and not blood-group specific, occasionally led to haemolysis when using high doses. Haemolysis can still be observed today when intermediate purified FVIII concentrates are used in ITT, however, it has become a rare phenomenon. Independent of plasma or recombinant origin, allergic reactions to FVIII preparations may occur infrequently.

Another side-effect associated with the use of PCCs and aPCCs, especially at high doses, are symptoms of disseminated intravascular coagulation and thromboembolic events which can complicate the ITT.

Poor venous access especially in very young ITT patients could make the implantation of port or catheter systems necessary. This gives an added risk of infection of the device which may influence duration and outcome of the inhibitor eradication therapy [15].

German recommendations for the immune tolerance induction

Data guidelines of the Bundesärztekammer (German Medical Association) for therapy with blood components and plasma

Table 7.2 Indications and dosage recommendations for treatment of inhibitor patients with haemophilia A according to the German recommendations [1,7].

	Dosage
Treatment of acute bleeding (children and adults)	
Low responder (<5 BE)	⇒ FVIII concentrate until haemostatically active FVIII level is achieved.
	⇒ Activated prothrombin complex concentrate: initial dose, until 100 IU/kg twice daily; or recombinant FVIIa, mean initial dose 90 U/kg
High responder (>5 BE)	a) Activated prothrombin complex concentrate: initial dose, until 100 IU/kg, maintenance dose, until 100 IU/kg twice daily; or recombinant FVIIa, mean initial dose 90 U/kg
	b) Without effectiveness and checking the cross-reactivity porcine factor VIII concentrate (Hyate C) 50–100 IU/kg twice daily
	c) In an emergency and failure of (a) and (b) immune adsorption apheresis
Elimination of inhibitors by induction of immune tolerance	
Children	
Low responder (<5 BE)	⇒ 50–100 IU/kg FVIII concentrate three times weekly until normal recovery and half-time. Inhibitor monitoring required twice weekly, afterwards regular treatment schedule
High responder (>5 BE)	⇒ 100–200 IU/kg FVIII concentrate twice daily until recovery and half-time return to normal for several months, afterwards regular treatment schedule
Reduction of bleeding tendency during the inhibitor elimination	Combination with activated prothrombin complex concentrate ⇒ 50 IU/kg twice daily possible
Adults	
Low responder (<5 BE)	No elimination therapy at regular treatment schedule
	⇒ FVIII concentrate 50 IU/kg three times weekly
High responder (>5 BE)	⇒ 100–150 IU/kg FVIII concentrate twice daily for several months until recovery and half-time return to normal
Reduction of bleeding tendency	Combination with activated prothrombin complex concentrate ⇒ 50 IU/kg twice daily possible

If eradication therapy failed, ITT was usually discontinued after 1 year.
ITT, immune tolerance therapy.

derivatives, recommendations for ITT have been established [1,16]. Table 7.2 shows the main statements of these guidelines.

Conclusion

Nowadays, more than two decades after the first presentation of a treatment regimen by Brackmann and Gormsen, elimination of an inhibitor by immune tolerance induction has become the preferred treatment method. During this time data in treating patients with inhibitors have been gathered, and successes as well as complications have been documented by haemophilia doctors all over the world. At present, little is known about how immune tolerance is achieved by ITT and future immunologically based studies are needed to clarify the pathomechanism of ITT and to define parameters that allow the success of ITT to be predicted [2].

Acknowledgements

The authors thank Dr Anke Tripp from Wyeth Pharma (Münster, Germany) for her technical support.

References

1 Brackmann HH, Gormsen J. Massive factor VIII infusion in a haemophiliac with factor VIII inhibitor, high response. *Lancet* 1977; 2: 933.
2 Oldenburg J, Schwaab R, Brackmann H-H. Induction of immune tolerance in haemophilia A, inhibitor patients by the 'Bonn Protocol': predictive parameter for therapy duration and outcome. *Vox Sang* 1999; 77 (Suppl. 1): 49–54.
3 Ewing N, Sanders NL, Dietrich SL, Kasper CK. Induction of immune tolerance to factor VII in hemophiliacs. *JAMA* 1988; 259: 65–8.
4 van Leewen EF, Mauser-Bunschoten EP, van Dijken PJ, Kok AJ, Sjamsedin-Visset EJM, Sixma JJ. Disappearance factor VIII. C antibodies in patients with haemophilia A upon frequent administration of factor VIII in intermediate or low dose. *Br J Haematol* 1986; 64: 291–7.
5 Manno CS. Treatment options for bleeding episodes in patients undergoing immune tolerance therapy. *Haemophilia* 1999; 5 (Suppl. 3): 33–41.
6 Brackmann HH, Lenk H, Scharrer I, Auerswald G, Kreuz W. German recommendations for immune tolerance therapy in type A haemophiliacs with antibodies. *Haemophilia* 1999; 5: 203–6.
7 Scheibel E, Ingerslev J, Dalsgaard-Nielsen J, Stenberg S, Knudson JB, and the Danish study group. Continuous high-dose factor VIII for the induction of immune tolerance in haemophilia A patients

with high responder state: a description of eleven patients treated. *Thromb Haemost* 1987; **58**: 1049–52.
8 White GC, Roberts HR. The treatment of factor VIII inhibitors—a general overview. *Vox Sang* 1996; **70** (Suppl. 1): 19–23.
9 Kurczinsky EM, Penner JA. Activated prothrombin concentrate for patients with factor VIII inhibitors. *N Engl J Med* 1974; **291**: 164–7.
10 Brackmann HH, Oldenburg J, Schwaab R. Immune tolerance for the treatment of factor VIII inhibitors—twenty years 'Bonn Protocol'. *Vox Sang* 1996; **70** (Suppl. 1): 30–5.
11 Mariani G, Ghirardini A, Bellocco R. Immune tolerance in hemophilia—principal results from the international registry. Report of the factor VIII and IX subcommittee. *Thromb Haemost* 1994; **72**: 155–8.
12 Brackmann HH. Successful treatment of hemophilia A inhibitor patients with induced immunotolerance. In: *Proceedings of the 4th International Symposium on Hemophilia Treatment*. Tokyo 1984: 187–96.
13 Brackmann HH. Induction of Immune tolerance in hemophiliacs with inhibitors to factor VIII and IX. XIIth Congress of the International Society on Thrombosis and Haemostasis, Kyoto 1989.
14 Scharrer I, Neutzling O. Incidence of inhibitors in haemophiliacs. A review of the literature. *Blood Coag Fibrinol* 1993; **4**: 753–8.
15 Brackmann H-H, Schwaab R, Effenberger W, Hess L, Hanfland P, Oldenburg J. Side effects during immune tolerance induction. *Haematologica* 2000; **85** (Suppl. 10): 75–7.
16 Leitlinien Bundesärztekammer (Guidelines of the German Medical Association). www.bundesaerztekammer.de

CHAPTER 8

Immune tolerance: low-dose regimen

E.P. Mauser-Bunschoten

Some patients with inhibitors develop a high titre antibody, having a brisk anamnestic response following infusion with any factor VIII material. In some of these patients the inhibitor disappears spontaneously after stopping factor VIII, but will relapse after the next factor VIII infusion [1,2]. For this reason, until 1980 it was the practice to stop factor VIII therapy as soon as a patient with haemophilia developed an inhibitor.

In 1974 Brackmann *et al.* [3] made the first serious protocol for the eradication of inhibitors in haemophilia A by designing the Bonn protocol for induction of immune tolerance in these patients. The protocol originally consisted of two phases. In the first phase, 100 units/kilogram bodyweight (U/kg bw) factor VIII and 40–60 U/kg bw activated prothrombin complex concentrate (APCC, Feiba™) were infused every 12 h until the inhibitor titre fell under 2 Bethesda units (BU)/mL. In the second phase, 150 U/kg factor VIII was administered every 12 h. The dosage was then slowly reduced to 100 U/kg bw daily until factor VIII half-life became normal. This regimen was successful in patients with low and extremely high titre inhibitors. Since the beginning of the 1980s various regimens for the introduction of immune tolerance have been introduced [4–8]. Sometimes plasmapheresis, cyclophosphamide, gammaglobulin or corticosteroids were added to these regimens [9,10]. However, none of these strategies was based on any quantitative research and the mechanism of successful treatment remains ill understood.

Low-dose immune tolerance therapy

Low-dose immune tolerance therapy (ITT) was developed because this was thought to be less demanding for patients and staff because patients have to be infused only two or three times a week. Also the amount of factor VIII infused is low, making it more attractive for economical reasons.

In The Netherlands this low-dose immune tolerance regimen was first introduced in 1981 in three young patients with life-threatening bleeds.

Dosage regimen

In patients in whom factor VIII was started with the sole aim of obtaining immune tolerance, the factor VIII (FVIII) dosage was 25–50 U FVIII/kg bw every other day or three times a week, independent of the inhibitor titre. Initially, in very young children in whom venous access was difficult, factor VIII was injected twice weekly. Since 1990 in these patients an intravenous catheter (Port-A-Cath™ system, PAC) is implanted in order to obtain adequate venous access.

When factor VIII treatment was started because of surgery or life-threatening bleeding and the inhibitor was less than 10 BU/mL, an initial high dose of factor VIII was given to neutralize the antibodies. The neutralizing dosage was calculated as follows [11]:

$$2 \times BW \frac{80 \times (100 - Ht)}{100} \times I$$

where BW is bodyweight in kilograms, Ht is haematocrit, and I is inhibitor in BU/mL.

The initial high dose was followed by infusion of 25 U FVIII/kg bw twice daily for 1–2 weeks, depending on the clinical status of the patient and the anamnestic response to factor VIII. After this period, factor VIII was continued three times a week or every other day in a dosage of 25–50 U FVIII/kg bw.

Dose adjustment

Once factor VIII antibodies decreased and factor VIII recovery was restored, or when there was no anamnestic response, factor VIII was reduced if the absolute factor VIII recovery was higher than 30% until a standard prophylactic dosage of 10–15 U FVIII/kg bw was reached. Since 1997 the standard prophylactic dose in children could be, depending on its clinical effect, a dose of 25 U FVIII/kg bw three times a week [12].

In patients in whom the inhibitor titre showed no tendency to decrease over a period of 6 months or longer, the factor VIII dosage was increased to 50–100 U FVIII three times a week or every other day. In those patients in whom the inhibitor remained high despite 2 years of low-dose ITT, high-dose ITT was started according to the Malmö protocol [10].

Choice of product

The choice of factor VIII product varied with time. Different factor VIII products were used: cryoprecipitate and intermediate purified product, with and without von Willebrand factor, as well as monoclonal purified and recombinant factor VIII.

Patients

Since 1981 all haemophilia patients visiting the Van Creveldkliniek are tested for antibodies at least once a year and also in cases of clinical suspicion of an inhibitor. When a positive antibody test is obtained, blood samples for repeated antibody testing are taken and a factor VIII recovery study is performed. Patients are considered to have a type A inhibitor when recovery of 50% or less is measured, with or without the clinical evidence of an inhibitor. Patients are considered to have transient (type B) inhibitors when the second sample tests negative for antibodies and a normal recovery takes place [13]. Type B inhibitors were excluded from the study. Twenty-seven patients with persistent antibodies were included in the study.

Informed consent

Informed consent was obtained from all patients.

Laboratory assays

Plasma sampling

Plasma samples for factor VIII and inhibitor assays were collected according to standard techniques; 4.5 mL of venous blood was drawn with a disposable needle into a silicone-treated Vacutainer in which 0.5 mL of 3.8% (0.13 M) sodium citrate was added. Immediately after collection, samples were carefully mixed and centrifuged at $3000 \times g$ for 15 min at 4°C. The platelet-poor plasma was carefully pipetted off and stored in a plastic tube at −20°C.

All samples were analysed at the coagulation laboratory of the University Medical Center, Utrecht (head Professor Dr J.W.N. Akkerman).

Inhibitor assay

Inhibitor measurements were taken using the Bethesda method as described by Kasper et al. [14]. Inhibitor titres of 1 BU/mL or more were considered positive [5]. Since 1997, inhibitor measurements have been performed using the modified Nijmegen assay [15]. Using this method, inhibitors >0.3 BU/mL are considered positive.

Stored blood samples from all patients tolerized before 1997 were tested for inhibitors using the Nijmegen method. The results were comparable with those using the original Bethesda inhibitor assay.

In patients with positive inhibitor tests, blood samples for inhibitor measurement were subsequently taken every 4–8 weeks.

Factor VIII assay

Factor VIII assays were performed using the one-stage method based on the kaolin-activated partial prothrombin time and expressed as a percentage of factor VIII present in pooled human plasma [16].

In vivo recovery

Blood samples for factor VIII assays were taken before and 15 min after infusion with factor VIII. Recovery was defined as the percentage of factor VIII measured and the expected level calculated by the method according to Lee et al. [17].

Definition of success

Originally, immune tolerance was considered to be clinically successful when the inhibitor decreased to <2 BU/mL, with a factor VIII recovery of at least 50% of normal and a half-life of 6 h or more and the absence of an anamnestic response after infusion with factor VIII. Since 1997, after the introduction of the inhibitor measurement according to the modified Nijmegen method, an inhibitor of <0.4 BU/mL was taken as the cut-off point. Clinical success was chosen as an endpoint because these patients can be treated with prophylaxis to prevent bleeds, and bleeds can be treated adequately with factor VIII.

Complete success was defined as absence of inhibitor, normal recovery and half-life time.

Follow-up

After initiation of ITT, patients were seen at least every month. During follow-up visits samples for antibody test were taken. When the inhibitor was <2 BU/mL, recoveries were also performed after infusion with the same dosage of factor VIII the patient was currently using for ITT.

When the inhibitor was <1 BU/mL (Kasper) or <0.4 BU/mL in the Nijmegen assay, and the recovery was >50%, the half-life was performed after infusion with 50 U/kg bw.

Once a patient was tolerized, blood samples for inhibitor measurement and recovery were taken at least twice a year. The date of the last inhibitor assay was taken as the endpoint for evaluation.

Treatment of bleeds during immune tolerance therapy

Bleeds in patients with active inhibitors were treated with 50 U/kg bw APCC (Feiba™) or 90 µg FVIIa (NovoSeven™). When a patient had a factor VIII recovery, bleeds were treated with (increased) doses of factor VIII. Infusion with clotting factor was repeated depending on the clinical situation of the patient.

Statistical analysis

Probabilities of disappearance of the inhibitor over time were estimated using Kaplan and Meier analysis and were compared using the log–rank statistic. The time lapse until disappearance of the inhibitor was also examined by univariate step-wise Cox regression analyses. All variables found to have P-values of less than 0.10 in univariate were considered candidate variables for multivariate analyses.

Results

The group consisted of 27 patients with severe haemophilia A. Patient data are summarized in Tables 8.1 and 8.2. The median age at inhibitor development was 3 years. Inhibitors developed after a median of 34 exposures. The median age at the start of ITT was 13 years. The total period of follow-up since start of ITT was 336 years with a median of 13 years/patient. Two patients were lost for follow-up, and three patients died from AIDS.

In patients 1–11 factor VIII was continued after an inhibitor developed, whereas in patients 12–27 factor VIII was discontinued for at least 1 year.

During immune tolerance induction patients were seen at least every month. We checked the diaries kept by the patients against the amounts of factor VIII supplied to them. Based on these data compliance was almost 100%.

In 23 patients (85%) success was obtained with low-dose therapy, in two patients (patients 9 and 10) dose had to be adjusted to 75–100 U FVIII/kg bw three times a week before success was seen and they were considered to be failures. In two other patients therapy failed completely.

Success was obtained after 2–28 months. The Kaplan–Meier plot of the presence of inhibitor is almost linear in the first 2 years of ITT, indicating a constant chance of disappearance of inhibitor. Even after 3 years of therapy there is a chance the inhibitor will disappear, as shown in patient 9. So far ITT totally failed in two patients after 48 and 50 months, respectively (patients 26 and 27). In these two patients therapy failed even when therapy with high-dose factor VIII, intravenous gammaglobulin and cyclophosphamide as described by Nilsson *et al.* [10] was given. Patient 26 received three courses of this regimen without success and factor VIII therapy was discontinued after 3 years; patient 27 received one course without success.

In patients 9 and 10, in whom factor VIII dosage was adjusted, success was seen after 18 and 36 months, respectively. In patient 9 ITT was started at 50 U FVIII/kg bw twice a week and after an anamnestic response to 220 BU/mL the inhibitor decreased in 24 months to 5 BU/mL when the factor VIII dosage was increased to 50 U/kg bw three times a week. This resulted in clinical tolerance after another 12 months. In patient 10 the inhibitor developed at the age of 6 months. Six months later ITT was started with 50 U FVIII/kg bw, resulting in an anamnestic response with a maximum inhibitor titre of 753 BU/mL. In this patient a Port-A-Cath™ was implanted using continuous infusion with porcine factor VIII and bolus injection with Feiba™. Four months after the start of ITT the dose was increased to 75 U FVIII/kg bw and later 100 U FVIII/kg bw three times a week to try to obtain immune tolerance more quickly. During ITT, four periods with continuous infusion were given because of (re)placement of the Port-A-Cath™. Twenty-four months after starting treatment, ITT was successful in this patient.

Logistic regression analyses revealed (Table 8.3) that there was a relationship between the highest inhibitor level and successful ITT and the time needed before success was obtained. Patients with low inhibitor titres did better. All patients with an inhibitor level of <40 BU/mL were treated successfully. Patients in whom therapy completely failed or in whom the dose was adjusted had inhibitors between 44 and 753 BU/mL. In patients with maximum inhibitor titre of <40 BU/mL, success was obtained within 10 months (median 6 months). In patients with maximum titres >40 BU/mL, the median time before success was obtained was 18 months, but even after 36 months ITT was successful.

Whether therapy was started directly or many years after the inhibitor development did not seem to make much difference in this group of patients. Treatment with a neutralizing dose at the start of immune tolerance induction did not improve the results. Furthermore, no difference was seen in HIV-positive ($n = 6$) and HIV-negative ($n = 21$) patients and no influence of hepatitis B or C infection was observed. Moreover, the type of product used to obtain immune tolerance did not affect the results. Tolerance was also obtained using monoclonal purified and recombinant product (Table 8.2).

Complications

The most frequent complication during ITT was the occurrence of bleeds. In the beginning bleeds were treated with APCC

Table 8.1 Demographic data of patients on immune tolerance therapy (ITT).

	Median	Range
Age at inhibitor development (years)	3	0.5–23
Number of exposures before inhibitor development	34	8–53
Age at start ITT (years)	13	1–43
Period of follow-up (years)	13	4–20

Table 8.2 Course of inhibitor before, during and after ITT.

Pat	First inhibitor in BU/mL	Highest inhibitor level before ITT in BU/mL	Inhibitor level at onset ITT in BU/mL	Highest anamnestic response at ITT in BU/mL	Time needed for success (months)	Product used for ITT	Follow-up since start ITT (years)
1*	1.3	8.7	8.7	None	6	Intermediate purified	15
2*	1.1	1.1	1.1	None	2	Cryoprecipitate	18
3†‡	8.1	8.1	4.0	11	6	Intermediate purified conc.	16
4‡	2.5	2.2	2.2	68	24	Intermediate purified conc.	16
5‡	3.9	3.9	3.9	None	3	Cryoprecipitate	19
6*	4.2	4.5	4.5	4.6	8	Intermediate purified conc.	15
7*	1.1	1.3	1.3	11.0	6	Intermediate purified conc. with vWf	8
8‡	9.7	9.7	9.7	24	6	Intermediate purified conc.	10
9‡	8.0	8.0	8.0	220	36 (dose adjusted)	Intermediate and monoclonal conc.	10
10*	2.0	25	5.0	753	24 (dose adjusted)	Recombinant	5
11*	5.0	5.0	5.0	None	6	Recombinant	4
12*	Pos	23	0.6	2.3	6	Monoclonal	4
13‡‡	Pos	177	0.4	17	12	Intermediate purified conc.	18
14*†	Pos	160	4.6	66	27	Intermediate purified conc.	11
15*†	Pos	3.5	0.7	3.3	10	Intermediate purified conc.	16
16*†	Pos	94	2.8	34	12	Intermediate purified conc.	6
17*†	Pos	15.5	2.5	None	6	Intermediate purified conc.	20
18*	Pos	15.6	1.0	2.7	9	Intermediate purified	20
19*†	Pos	164	1.2	1.6	14	Intermediate purified	16
20*†	52	76	1.7	8.5	18	Intermediate purified	18
21*†	Pos	83	5.7	83	18	Intermediate purified conc.	15
22*	8.0	10.4	0.3	None	3	Cryoprecipitate	19
23*	5.0	64	7.3	31	28	Intermediate purified	13
24*†	3.0	3.0	1.1	None	2	Intermediate purified conc.	4
25*†	4.3	51	1.5	19.3	8	Monoclonal purified	8
26*†	Pos	33	2.8	450	Malmö protocol failure	Intermediate purified and monoclonal	5
27*	Pos	25	0.6	44	Malmö protocol failure	Intermediate purified and monoclonal	7

*Factor VIII 25–50 U/kg bw three times a week or every other day.
†Initial high doses of factor VIII to neutralize the antibodies, followed by 25 U/kg bw factor VIII twice daily for 1–2 weeks.
‡Factor VIII 25–50 U/kg bw twice a week.
BU, Bethesda units; ITT, immune tolerance therapy.

Table 8.3 Success of ITT in relation to inhibitor titre.

	Max titre <40 BU/mL	Max titre >40 BU/mL
Number of patients	15	12
Success	15	8
Success after dose adjustment		2
Failure	0	2
Time needed before success (months)		
median	6	18
range	2–10	12–28

BU, Bethesda units; ITT, immune tolerance therapy.

Table 8.4 Present prophylactic regimen, bleeding frequency, last recovery and half-life time in immune tolerized patients.

Pat	Year of birth	Prophylactic dose (U FVIII/kg bw)	Number of bleeds/year	Number of surgical interventions	Last inhibitor titre (BU/mL)	Recovery (%)	Half-life time (h)
1	1975	1996*		0	Neg	80	12
2	1982	10 U/kg bw 2× week	0–5	0	Neg (0.0)	100	14
3	1983	15 U/kg bw 3× week	0–5	0	Neg (0.0)	110	15
4	1984	10–15 U/kg bw 3× week	0–5	1	Neg (0.0)	110	18
5	1979	10 U/kg bw 3× week	0–5	0	Neg (0.1)	100	12
6	1984	10–15 U/kg bw 3× week	5–10	0	Neg (0.0)	100	12
7	1993	25 U/kg bw 3× week	5–10	1	Neg (0.1)	100	10
8	1989	20 U/kg bw 3× week	5–10	3	Neg (0.1)	110	6
9	1985	15 U/kg bw 2× week	0–5	0	Neg (0.0)	80	8
10	1995	25 U/kg bw 3× week	0–5	2	Neg (0.0)	100	8
11	1996	25 U/kg bw 3× week	0–5	1	Neg (0.1)	100	6
12	1967	1993*		1	Neg	100	12
13	1955	20 U/kg bw 3× week	10–25	5	Neg (0.0)	120	12
14	1967	1993*		3	Neg	100	12
15	1964	10–15 U/kg bw 2× week	0–5	0	Neg (0.0)	120	12
16	1967	1990 lost for follow-up		0	Neg	60	7
17	1967	15 U/kg bw 3× week	0–5	1	Neg (0.0)	130	14
18	1967	10 U/kg bw every other day	0–5	0	Neg (0.0)	90	12
19	1971	10 U/kg bw 3× week	0–5	0	Neg (0.0)	100	19
20	1971	10 U/kg bw 3× week	0–5	5	Neg (0.0)	80	10
21	1975	10–15 U/kg bw 3× week	0–5	0	Neg (0.0)	110	10
22	1976	10–15 U/kg bw 3× week	0–5	1	Neg (0.0)	100	8
23	1967	15–20 U/kg bw 3× week	0–5	0	Neg (0.2)	140	6
24	1974	1985 lost for follow-up		0	Neg	100	12
25	1987	25 U/kg bw 3× week	0–5	3	Neg (0.0)	100	9

* Year of death.

and, since 1998, with factor VIIa. When the inhibitor was low and a factor VIII recovery was measured, bleeds were treated with factor VIII. Venous access was problematic in some small children so Port-A-Cath™ systems were implanted in five patients at the start of ITT when the inhibitor was low, or during ITT under cover of an increased dose of factor VIII, porcine factor VIII, APCC or continuous infusion with factor VIIa. During ITT there were six Port-A-Cath™-related infections in three patients. Four Port-A-Caths™ were replaced for this reason.

Until 1985 seven of 12 patients treated with ITT were infected with HIV and three of them died from AIDS.

In one patient (patient 4) a relapse of the inhibitor was observed 1 year after he was tolerized. During this period a maximum inhibitor titre of 1 BU/mL was found, with absence of factor VIII recovery. This patient was treated with a second course of low-dose ITT with good result. Ten years later he has had no second relapse, is on normal dose prophylaxis and has no spontaneous bleeds.

Present status

At the most recent visit (Table 8.4) all patients were treated with prophylaxis and in most patients bleeds are adequately prevented. In one patient with severe arthropathy a high bleeding frequency was observed, which probably is caused by the poor physical condition of this patient. In young patients with risk behaviour a higher bleeding frequency is seen. However, these patients do not suffer from spontaneous bleeds.

In one patient prophylaxis has been stopped for 10 years without recurrence of the inhibitor.

After successful ITT 27 surgical interventions were performed in 12 patients under factor VIII cover. Bolus injection as well as continuous infusion were used with good haemostatic effect. There was no postoperative bleeding in any of these patients.

Discussion

Low-dose ITT is successful. The duration of treatment is determined by the maximum inhibitor level. Within 10 months after start of ITT 100% success was seen in patients with a maximum inhibitor titre of <40 BU/mL [5,18]. Success was also achieved in patients with a high titre inhibitor. However, in these patients the time needed for success was longer and even after >24 months of treatment success was achievable.

Low-dose ITT is less demanding for a patient and his or her parents. A disadvantage of low-dose ITT may be the longer period of time needed before tolerance is achieved as compared to the ITT regimens that use daily high-dose FVIII [19]. This may be a problem for patients with a high bleeding tendency or for patients in whom surgery is required.

Since the introduction of ITT at the beginning of the 1980s success has been obtained with different dosage regimens. Usually, 100 U FVIII/day is given, with or without the addition of corticosteroids, gammaglobulin or plasmapheresis [4–10]. Success rates varied between 76 and 89% and success was obtained after 1–18 months. In some studies doses were compared. Haya et al. [20] found that doses <100 U FVIII/kg bw/day did better than higher doses but, generally speaking, success rates in the various studies are similar—only the median time to success differs [19]. However, definitive conclusions about the dosage regimen in relation to the inhibitor titre can only be drawn after prospective randomized controlled studies. It is hoped that the study proposed by Hay et al. will answer this [21].

Conclusions

Low-dose ITT (25–50 U FVIII/kg bw three times a week) is appropriate for patients with low titre inhibitors, for young children in whom an inhibitor has just developed and for patients with high titre inhibitors who have a low bleeding frequency. Low-dose ITT may also be indicated for patients in whom ITT has not been attempted because of the high cost [18].

References

1 Bloom AL. The treatment of factor VIII inhibitors. *Thromb Haemost* 1987; **58**: 447–71.
2 Lusher JM. Management of patients with factor VIII inhibitors. *Transfus Med Rev* 1987; **1**: 123–30.
3 Brackman HH, Gormesen J. Massive factor VIII infusion in a haemophiliac patient with factor VIII inhibitor, high responder [Letter]. *Lancet* 1977; **2**: 933.
4 Wensley RT, Burn AM, Redding OM. Induction of tolerance to factor VIII in haemophilia with inhibitors using low doses of factor VIII. *Thromb Haemost* 1985; **54**: 227–30.
5 Mauser-Bunschoten EP, Nieuwenhuis HK, Roosendaal R, van den Berg HM. Low-dose immune tolerance induction in hemophilia A patients with inhibitors. *Blood* 1995; **86**: 983–8.
6 Ewing NP, Sanders NL, Dietrich SL, Kasper CK. Induction of immune tolerance to factor VIII in hemophiliacs with inhibitors. *J Am Med Assoc* 1988; **259**: 65–8.
7 Oldenburg J, Schwaab R, Brackmann HH. Induction of immune tolerance in haemophilia A inhibitor patients by the 'Bonn protocol': predictive parameter for therapy duration and outcome. *Vox Sang* 1999; **77** (Suppl. 1): 49–54.
8 Kreuz W, Mentzer D, Auerswald G et al. Successful immune tolerance therapy of FVIII-inhibitor in children after changing from high to intermediate purified FVIII concentrate. *Haemophilia* 1996; **2** (Suppl. 1): 19A.
9 Aznar JA, Jorquera JI, Peiro A, Garcia I. The importance of corticoids added to continued treatment with factor VIII concentrates in the suppression of inhibitors in haemophilia A. *Thromb Haemost* 1984; **51**: 217–21.
10 Nilsson IM, Berntorp E, Zetterval O. Induction of immune tolerance to factor VIII in hemophiliacs with inhibitors. *N Engl J Med* 1988; **318**: 947–50.
11 Van Leeuwen EF, Mauser-Bunschoten EP, van Dijken PJ et al. Low dose immune tolerance induction in hemophilia A patients with inhibitors. *Br J Haematol* 1986; **64**: 291–7.
12 Van den Berg HM, Fischer K, Mauser-Bunschoten EP et al. Long-term outcome of individualized prophylactic treatment of children with severe haemophilia. *Br J Haematol* 2001; **112**: 561–5.
13 Rosendaal FR, Nieuwenhuis HK, van den Berg HM et al. A sudden increase in factor VIII development in multitransfused hemophilia A patients in the Netherlands. *Blood* 1993; **81**: 2180–6.
14 Kasper CK, Aledort LM, Counts RB et al. A more uniform measurement of factor VIII inhibitors [Letter]. *Thromb Diath Haemorrh* 1975; **43**: 469.
15 Verbruggen B, Novakova I, Wessels H et al. The Nijmegen modification of the Bethesda assay for factor VIII:C inhibitors: improved specificity and reliability. *Thromb Haemost* 1995; **73**: 247–51.
16 Bouma BN, Starkenborg AE. Dilution of haemophilic plasma used as reagent in the determination of anti-haemophilic factor A (factor VIII). *Haemostasis* 1974; **3**: 94–7.
17 Lee ML, Gomperts ED, Kingdom HS. A note on the calculation of recovery for factor VIII infusions [Letter]. *Thromb Haemost* 1993; **69**: 87.
18 El Alfy MS, Tnatawy AAG, Ahmed MH, Abdin IA. Frequency of inhibitor development in severe haemophilia A children treated with cryoprecipitate and low-dose immune tolerance induction. *Haemophilia* 2000; **6**: 635–8.
19 Dimichele DM. Immune tolerance: a synopsis of international experience. *Haemophilia* 1998; **4**: 568–73.
20 Haya S, Lopez MF, Aznar JA, Battle J and the Spanish Immune Tolerance Group. Immune tolerance treatment in haemophilia patients with inhibitors: the Spanish registry. *Haemophilia* 2001; **7**: 154–9.
21 Hay CRM. Immune tolerance induction: prospective trials. *Haematologica* 2000; **85** (Suppl.): 52–6.

CHAPTER 9

Immune tolerance and choice of concentrates

W. Kreuz

The development of inhibitors is currently the main problem in the treatment of children with haemophilia, occurring in 22–52% of patients with severe haemophilia A [1,2]. Previously published prospective studies of previously untreated patients (PUPs) revealed no significant difference regarding inhibitor incidence between recombinant and plasma-derived products [3–5]. However, preliminary data of a prospectively conducted PUP study comparing different products indicate a slightly higher inhibitor incidence in recombinant-treated patients [6]. However, all studies revealed a high incidence of high titre inhibitors around 50%. These patients present with a high titre inhibitor response after repeated administration of factor VIII (FVIII) concentrate so that FVIII concentrates become ineffective even when given at high dosage. These patients present a severe bleeding tendency, with haemorrhagic episodes being particularly difficult to control even with FVIII-bypassing agents. Thus, the best management of inhibitor patients is with rapid immune tolerance induction (ITI). Various therapeutic regimens, such as the administration of high doses of FVIII twice daily [7–9] or lower doses three times weekly [10], have been attempted. Intermediate dose regimen have also been successfully applied [11]. Elimination of inhibitory antibodies from plasma by immune adsorption followed by immunosuppression and intravenous gammaglobulins combined with a high-dose FVIII regimen (Malmö protocol) [12] has also been used. However, the comparative success of the different regimens is difficult to assess, predominantly because of remarkable discrepancies in the definition of success and, to a lesser extent, in the study cohorts (Table 9.1).

In addition to the therapeutic regimen, variables such as inhibitor peak titre at start of ITI, maximum inhibitor titre during ITI, patient's age, number of exposure days from detection of the inhibitor to start of ITI and inflammatory conditions seem to be significant predictors of successful outcome of ITI [7,9,12,14].

With the introduction of purer FVIII concentrates containing very little or no von Willebrand factor (vWF) in the ITI regimen, the question has been raised as to whether the type of product, and particularly the vWF content, has an influence on success. This hypothesis has been corroborated because significantly decreased success rates have been reported by two haemophilia centres when pure FVIII concentrates were used [15 and H.H. Brackmann, personal communication]. Apart from the type of concentrate, no parameter of the treatment regimen

Table 9.1 Different ITI regimens.

	Bonn protocol [7,13]	Van Creveld protocol [10]	Malmö protocol [12]
Regimen	100–150 IU FVIII/kg twice daily (HR) 50–100 IU FVIII/kg daily or every other day (LR)	25 IU FVIII every other day	Elimination of inhibitors from plasma by immune adsorption Immunosuppression IV immunoglobulins High-dose FVIII
Duration of therapy	HR 4 months (0.5–42) LR 1.5 months (0.5–3)	<40 BU 6 months (1–9) >40 BU 19 months (12–27)	9–37 days
Definition of success	No detectable inhibitor, normal recovery and half-life No reappearance of inhibitor	Inhibitor titre <2 BU/mL Recovery >50% of normal Half-life >6 h No anamnestic response	Normal recovery Normal half-life Tolerance to further replacement therapy
Success rate (%)	91–100	87	62.5

BU, Bethesda units; FVIII, factor VIII; HR, high responder; LR, low responder; ITI, immune tolerance induction.

was changed in either centre (Frankfurt and Bonn, Germany). In addition, we have described the successful induction of immune tolerance after a switch from a pure FVIII concentrate to a concentrate with high vWF content in four patients who showed an unsatisfactory treatment response with pure FVIII concentrates. This also suggests the importance of the presence of vWF during ITI [16].

Immune tolerance therapy with different types of concentrates

In 1977 Brackmann and Gormsen [17] published the first successful inhibitor elimination in a high responder using high doses of FVIII concentrate. Since then the Bonn protocol has been established and successfully used. In Germany, the criteria for successful ITI are: no detectable FVIII inhibitor using the modified Bethesda method on three consecutive occasions; normalization of FVIII recovery in at least two consecutive measurements; and normalization of the half-life of FVIII [18]. Taking into account these hard endpoints, high success rates of around 90–100% have been reported [7–9,13]. However, these data are based on the almost exclusive use of plasma-derived concentrates of intermediate purity with a high content of vWF. In this respect, it must be emphasized that the published results of other immune tolerance therapy regimen, such as the low-dose regimen (van Creveld protocol) [10] and the Malmö protocol [12], also refer predominantly to the use of plasma-derived concentrates. Berntorp [19] evaluated the results of 15 patients with haemophilia who also had inhibitors who were treated according to the Malmö protocol. The overall success rate was 67%, i.e. of 15 patients, 10 inhibitors were successfully eliminated with the exclusive use of intermediate plasma-derived concentrates. Focusing on the five patients who had no successful outcome, they reported that both patients who received a monoclonal purified concentrate failed to achieve immune tolerance.

However, there are also numerous reports about the successful use of pure FVIII concentrates, particularly recombinant products. Since the introduction of these concentrates, which include recombinant as well as plasma-derived products, immune tolerance therapy has been performed using this new generation of concentrates. Induction of immune tolerance was achieved, particularly using the Bonn protocol prescribing about 200 IU FVIII given preferably in two doses or the intermediate FVIII dose regimen (50–100 IU/kg bw/day).

In the Kogenate™ (Bayer, Berkeley, CA, USA) PUP study, an inhibitor elimination rate of 63% (five of eight patients) was achieved in high responders using recombinant FVIII (Kogenate™) [3]. Batlle et al. [20] reported another nine high responders who were treated with Kogenate™, of whom seven (77%) achieved complete immune tolerance. The successful outcome was lower in the Recombinate™ (Baxter Hyland, Glendale, CA, USA) PUP study, where three of six (50%) inhibitor patients achieved immune tolerance [4]. Rothschild reported for the French Recombinate™ PUP study [21] that immune tolerance was difficult to achieve; of eight patients who underwent ITI only two patients (25%) achieved immune tolerance. In contrast, Rocino et al. [22] presented data of ITI at a single institution in 12 high-responder haemophilia A patients who received pure FVIII concentrates (recombinant or monoclonal purified) at doses of around 100 IU/kg/day and 10 patients (83.3%) achieved immune tolerance. Data on ITI with the B-domain deleted recombinant FVIII (ReFacto™, Wyeth, Genetics Institute, Cambridge, MA, USA) are still poor. In the ReFacto™ PUP study, immunotherapy was started in 21 patients (14 high responders and seven low responders). Of 14 high responders, 11 patients showed a decrease in their inhibitor titre and had <0.6 Bethesda units (BU)/mL for their latest available test after ITI. Of the seven low responders, six patients had inhibitor titres <0.6 BU/mL. There are no data reported concerning recovery and half-life [5].

Unuvar et al. [23] reported the results of ITI at the Michigan haemophilia treatment centre. Fourteen patients with haemophilia who also had inhibitors (13 HR, one LR) received recombinant or plasma-derived FVIII doses ranging from 50 to 100 IU/kg/day. Five of 10 (50%) patients with haemophilia who were treated with recombinant FVIII achieved complete immune tolerance. Four patients received plasma-derived FVIII concentrates of low or intermediate purity and three of these (75%) achieved complete immune tolerance.

As a summary of all data, efficacy rates ranged from 25 to 100%. However, a direct comparison of the results is impossible because the study cohorts as well as the immune tolerance protocols differ in many respects. In addition, study endpoints are not equally defined. Until now no prospective randomized trial has been conducted focusing particularly on the highly controversial issues such as FVIII dosage, optimal timing and inhibitor level at start of therapy as well as choice of concentrate.

However, clinical data from the authors' centre (Frankfurt) and the Bonn centre collected over a period of more than 20 years have revealed that FVIII concentrates containing high amounts of vWF are more effective in ITI, at least in patients with inhibitors directed against the light chain of FVIII.

Historical review of immune tolerance therapy at the Frankfurt haemophilia centre

From 1979 to 1993, ITI was performed according to the Bonn protocol at the Frankfurt centre using FVIII concentrates with high vWF content (FVIII/vWF) (Table 9.1). Low responders (LR) received 50–100 IU FVIII/kg bw daily or every second day. High responders (HR) were treated with 100–150 IU FVIII/kg bw/every 12 h. According to patients' bleeding tendency, they received concomitantly activated prothrombin complex (Feiba™, Baxter Hyland, Glendale, CA, USA) 50–100 IU/kg bw twice daily. Twenty-one haemophilia A patients with inhibitors aged 0.4–6 years (16 HR and five LR) were treated according to this protocol using exclusively Humate p™ (Aventis Behring, Marburg, Germany), a plasma-derived FVIII concentrate of intermediate purity with high vWF content. Inhibitor elimination was achieved in 19 of 21 (91%) patients after a median time of 4 months (range 0.5–42 months) in HR and 1.5 months (range 0.5–3 months) in LR. Therefore the success rate was 91% for all patients; 100% for LR and 88% for HR, respectively [7].

Both the number of FVIII exposures from inhibitor detection until the start of ITI and the interruption of ITI therapy showed a significant negative correlation with a successful outcome. Both parameters also correlated with a longer duration of time needed to induce immune tolerance.

With the introduction of purer FVIII concentrates containing very little or no vWF in the ITI regimen, the success rate decreased significantly to 29% when pure FVIII concentrates were exclusively used (Table 9.1). Therefore, patients were again switched to FVIII/vWF concentrates.

From 1993 to 2000 ITI was invariably performed according to the Bonn protocol but using different types of FVIII concentrates. In all cases we used the same concentrate for each patient that resulted in inhibitor development: 16 patients (15 HR, one LR), one adult and 15 children, aged 0.1–50 years (median 1 year) underwent ITI receiving: plasma-derived FVIII/vWF ($n = 2$, Humate p™); plasma-derived FVIII with or without trace amounts of vWF ($n = 8$); and recombinant FVIII ($n = 6$). Both patients who were treated with plasma-derived FVIII/vWF as well as four patients receiving FVIII concentrates without vWF showed a significant decline in inhibitor titre or even achieved immune tolerance. However, in the remaining 10 patients receiving FVIII concentrates without vWF, the course of ITT was unsatisfactory. After a median therapy duration of 3 months (range 1.4–18 months) they showed no decline or even a rise in inhibitor titres. Therefore, these non-responders were switched to intermediate and high purity plasma-derived concentrates with high vWF content (Humate P™; Immunate™, Baxter Hyland, Glendale, CA, USA; Haemoctin SDH™, Biotest, Dreieich, Germany; Profilate™, Fanhdi™, Grifols, Barcelona, Spain).

The course of the inhibitor (BU) showed a rapid decline in inhibitor titre in nine out of 10 patients after changing the concentrate; however, complete immune tolerance was achieved in eight out of 10 HR after a median of 17 months of treatment (range 5–36 months). In comparison to our previous experience with FVIII/vWF concentrates where high titre inhibitors were completely eliminated after a median time of only four months, inhibitor elimination time increased significantly.

The inhibitor reappeared in two patients after switching them to a high purity FVIII concentrate devoid of vWF. Remarkably, the inhibitor disappeared almost immediately after returning them to a regimen with FVIII/vWF concentrate.

The results of our longitudinal evaluation on ITI in patients with inhibitors implies that the type of concentrate used in the induction regimen has an important role. This issue has been corroborated, as our findings have been confirmed by the experience of the Bonn haemophilia centre [14]. Also in this centre, the success rate declined from 90 to 60% after the introduction of pure FVIII concentrates for ITI as the unique parameter changed in the treatment regimen. This clinical observation again indicates the importance of the choice of concentrate in ITI [14].

Impact of the choice of concentrate—hypothetical considerations

Admittedly, our data are based on a clinical observation of a small number of patients. However, there are some theoretical considerations which might substantiate our findings.

The epitope specificity of the inhibitor might play a crucial part, in particular in the choice of concentrate for ITI. The spectrum of epitope specificity includes the light chain and the heavy chain of FVIII. Patients with haemophilia usually have at least two different inhibitor antibodies. However, epitope specificity seems to be dependent on the type of concentrate administered until inhibitor development. Patients with haemophilia who received plasma-derived FVIII (low or intermediate purity) until inhibitor detection develop predominantly inhibitors against the light chain (C2 domain). In contrast, those patients receiving recombinant FVIII showed inhibitor activity against the light chain (C2 domain) and against the heavy chain (A2 domain) [24].

This observation could be important because one may assume that vWF protects FVIII against degradation by proteinases (inhibitors) by steric inhibition, as suggested by two publications by Berntorp *et al.* [25] who reported a higher *in vivo* recovery in HR patients against C2 domain after administration of FVIII/vWF concentrates. In addition, patients' plasma samples were tested against different concentrates showing a lower inhibitory activity against FVIII concentrates with high amounts of vWF. Suzuki *et al.* [26] showed *in vitro* that FVIII inhibitor antibodies with C2-domain specificity were less inhibitory to FVIII complexed with vWF compared to recombinant FVIII products. No difference was seen with both types of concentrates in A2-domain inhibitor plasma samples.

In patients with a predominant C2-specific inhibitory activity the vWF, which is complexed to FVIII in FVIII/vWF concentrates, can mask the epitope (C2 domain) to which the anti-light-chain inhibitor is directed. Therefore degradation of FVIII is temporarily inhibited and the antigen (FVIII) can be longer exposed to the immune system, which might have a positive impact on inducing immune tolerance.

Another factor in the success of ITI is that low purity concentrates might have an immunomodulatory effect [19] and therefore be beneficial in ITI. Until 1993, we exclusively used FVIII concentrates of intermediate purity. Thereafter high purity concentrates with high content of vWF were used. Thus we conclude, supported by both clinical observations and theoretical considerations, that vWF has a central role in the success of ITI. However, the benefit is only likely to be applicable for inhibitors with C2-domain specificity.

Conclusions

With the introduction of pure FVIII concentrates containing very little or no vWF in the ITI regimen, a significantly decreased success rate has been reported by two haemophilia centres (Frankfurt and Bonn). The change from a pure to an FVIII/vWF concentrate during ITI increased the success rate significantly.

In conclusion, these clinical results and *in vitro* studies underline the importance of vWF and therefore the type of concentrate used for ITI, in particular in patients with inhibitors directed against the light chain of FVIII. On the contrary, in patients with inhibitors against the heavy chain of FVIII there seems to be no significant difference between different types of concentrates. Patients who develop FVIII inhibitors during replacement therapy with recombinant products often present inhibitory activity against the heavy chain of FVIII. However, in these patients inhibitor elimination is possible using pure FVIII concentrate. This implies that the mechanisms to induce immune tolerance in individuals with haemophilia who also have inhibitors are of multifactorial origin. However, the use of FVIII/vWF concentrates show advantages, at least in a specific cohort of inhibitor patients.

We now perform *in vitro* inhibitor testing of the patient's plasma against different types and batches of concentrates and epitope mapping is performed at our centre *before* starting ITI. This further test should help to achieve the highest success rate of complete immune tolerance in the shortest time, therefore providing maximal benefit for the patient's future course of the disease.

References

1 Ehrenforth S, Kreuz W, Scharrer I *et al.* Incidence of development of factor VIII and factor IX inhibitors in haemophiliacs. *Lancet* 1992; **339**: 594–8.
2 Kreuz W, Escuriola Ettingshausen C, Martinez Saguer I, Güngör T, Kornhuber B. Epidemiology of inhibitors in haemophilia A. *Vox Sang* 1996; **70** (Suppl. 1): 2–8.
3 Lusher J, Arkin S, Hurst D *et al.* Recombinant F VIII (Kogenate™) treatment in previously untreated patients (PUPs) with haemophilia A: update of safety, efficacy and inhibitor development after seven study years. *Thromb Haemost Suppl* 1997; ISTH Florence Abstract 663.
4 Gruppo R, Bray GL, Schroth P, Perry M, Gomperts ED. Safety and immunogenicity of recombinant factor VIII (Recombinate™) in previously untreated patients (PUPs): a 6.5 year update. *Thromb Haemost Suppl* 1997; ISTH Florence Abstract 664.
5 Courter SG, Bedrosian CL. Clinical evaluation of B-domain deleted recombinant FVIII in previously untreated patients. *Semin Hematol* 2001; **38** (Suppl.): 52–9.
6 Kreuz W. GTH–PUP study on inhibitor development: preliminary results. Presented at the Inhibitor Workshop, Bonn 2001.
7 Kreuz W, Ehrenforth S, Funk M *et al.* Immune tolerance therapy in pediatric haemophiliacs with factor VIII inhibitors: 14 years follow-up. *Haemophilia* 1995; **1**: 24–32.
8 Brackmann HH. Induced immunotolerance in factor VIII inhibitor patients. *Prog Clin Biol Res* 1984; **150**: 181–95.
9 Oldenburg J, Schwaab R, Brackmann HH. Induction of immune tolerance in haemophilia A inhibitor patients by the 'Bonn Protocol': predictive parameter for therapy duration and outcome. *Vox Sang* 1999; **77**: 49–54.
10 Mauser–Bunschoten EP, Nieuwenhuis KH, Rosendaal G, van den Berg M. Low-dose immune tolerance induction in haemophilia A patients with inhibitors. *Blood* 1995; **86**: 983–8.
11 Ewing NP, Sander NL, Dietrich SL, Kasper CK. Induction of immune tolerance to factor VIII in haemophiliacs with inhibitors. *J Am Med Assoc* 1988; **259**: 65–8.
12 Freiburghaus C, Berntorp E, Ekman M *et al.* Tolerance induction using the Malmö treatment model 1982–95. *Haemophilia* 1999; **5**: 32–9.
13 Brackmann HH, Oldenburg J, Schwab R. Immune tolerance for the treatment of FVIII inhibitors: twenty years 'Bonn protocol'. *Vox Sang* 1996; **70**: 30–5.
14 DiMichele DM, Kroner BL. Analysis of the North American Immune Tolerance Registry (NAITR) 1993–97: current practice implications. ISTH Factor VIII/IX Subcommittee Members. *Vox Sang* 1999; **77**: 28–30.
15 Kreuz W, Escuriola Ettingshausen C, Auerswald G *et al.* Immune tolerance induction (ITI) in Haemophilia A patients with inhibitors: the choice of concentrate affecting success. *Haematologica* 2001: **86**: 16–20.
16 Kreuz W, Joseph–Steiner J, Mentzer D *et al.* Successful immune tolerance therapy of FVIII-inhibitors in children after changing from high to intermediate purity FVIII concentrate: 40th Annual Meeting GTH, Interlaken 1996. *Ann Hematol* 1996; **72** (Suppl. 1): 339.
17 Brackmann HH, Gormsen J. Massive factor VIII infusion in haemophiliac with factor–çIII inhibitor, high responder. *Lancet* 1977; **29**: 933.
18 Brackmann HH, Lenk H, Scharrer I, Auerswald G, Kreuz W. German recommendations for immune tolerance therapy in type A haemophiliacs with antibodies. *Haemophilia* 1999; **5**: 203–6.
19 Berntorp E. Immune tolerance induction: recombinant vs. human-derived product. *Haemophilia* 2001; **7**: 109–13.
20 Batlle J, Lopez MF, Brackmann HH *et al.* Induction of immune tolerance with recombinant factor VIII in haemophilia A patients with inhibitors. *Haemophilia* 1999; **5**: 431–5.
21 Rothschild C, Laurian Y, Satre EP *et al.* French previously untreated patients with severe haemophilia A after exposure to recombinant factor VIII: incidence of inhibitor and evaluation of immune tolerance. *Thromb Haemost* 1998; **80**: 779–83.
22 Rocino A, Papa ML, Salerno E *et al.* Immune tolerance induction in haemophilia A patients with high-responding inhibitors to factor VIII: experience at a single institution. *Haemophilia* 2001; **7**: 33–8.
23 Unuvar A, Warrier I, Lusher JM. Immune tolerance induction in the treatment of pediatric haemophilia A patients with factor VIII inhibitors. *Haemophilia* 2000; **6**: 150–7.
24 Prescott R, Nakai H, Saenko EL *et al.* and the Recombinate™ and Kogenate™ Study groups. The inhibitor antibody response is more complex in haemophilia A patients than in most nonhaemophiliacs with factor VIII autoantibodies. *Blood* 1997; **89**: 3663–71.
25 Berntorp E, Ekman M, Gunnarsson M, Nilsson IM. Variation in factor VIII inhibitor reactivity with different commercial factor VIII preparations. *Haemophilia* 1996; **2**: 95–9.
26 Suzuki T, Arai M, Amano K, Kagawa K, Fukutake K. Factor VIII inhibitor antibodies with C2 domain specificity are less inhibitory to factor VIII complexed with von Willebrand factor. *Thromb Haemost* 1996; **76**: 749–54.

CHAPTER 10

Immune tolerance: The North American Immune Tolerance Registry

D.M. DiMichele and B.L. Kroner

Introduction

The cumulative prevalence of inhibitor development in individuals with severe or moderately severe haemophilia A has been estimated to be as high as 33% [1]. Inhibitor incidence has not been as well documented among individuals with haemophilia B, but it appears to be less frequent at 1–6% [2]. Treatment options for haemophilia patients with inhibitors are limited in number, useful in the acute management of bleeding, but less than satisfactory in providing a good long-term outcome for this population. Consequently, morbidity and mortality from the haemorrhagic complications of haemophilia remain high, although not clearly defined [3]. Inhibitor eradication with immune tolerance therapy (ITT) currently remains the best long-term option, particularly for high titre antibodies. Since the first observation of its efficacy in haemophilia A by Professor Brackmann [4], small groups of patients have been treated with multiple high- and low-dose factor regimens, with and without concomitant immune modulation therapy [5–16]. Success rates in the permanent eradication of inhibitors have been reported to be between 62 and 90% [5–16]. However, despite similar overall success, the time to induction of immune tolerance has differed significantly among regimens [5–16]. Given the low incidence of factor IX inhibitors, past experience with immune tolerance in haemophilia B has been limited to a few small series [5,14,17].

Because of the intensity, long-term expense and associated complications of the ITT performed in predominantly paediatric populations, the optimization of this treatment must be a priority. To this end, a further understanding of current ITT practices, outcomes and outcome predictors is essential. The initial report from the International Immune Tolerance Registry provided the first such aggregate data for haemophilia A [18]. It identified several good outcome predictors including:
1 dosing regimens of ≥100 U/kg/day;
2 initiation of ITT at the time when inhibitors had declined to ≤10 Bethesda Units (BU); and
3 a ≤2-year interval between inhibitor diagnosis and initiation of ITT.

A subsequent multivariate analysis of the international data set corroborated the initial significant outcome predictors of pre-induction titres ≤10 BU and factor VIII dose [19,20]. However, the high success rate was more clearly associated with a dosing regimen of ≥200 U/kg/day. Age, peak historical titre and interval between diagnosis and therapy were significant outcome predictors only in a univariate analysis [19]. Nonetheless, trends are toward a higher success rate with lower historical peak titres and inhibitor to ITT intervals of ≤5 years.

The North American Immune Tolerance Registry (NAITR) was initiated in 1992 as a project of the ISTH Factor VIII/IX Subcommittee with the goal of further determining immune tolerance practices in Canada and the United States with respect to the identification of:
1 therapeutic regimens in use;
2 therapeutic outcomes;
3 potential predictors of successful outcome;
4 complications of therapy.

This registry differs from the International Data Collection in that haemophilia B inhibitor patients were included; potential outcome predictors such as race and peak inhibitor titre on ITT were studied; and complications secondary to clotting factor administration and frequent venous access were recorded. The data presented in this chapter were collected between March 1993 and December 1999.

Data collection

Data collection forms were initially sent in March 1993 to 168 haemophilia treatment centres in the USA and Canada. Data updates were requested from participating haemophilia treatment centres from 1994 to 1999 inclusive. The collection tool was intended to study the following parameters:
1 the frequency of past and current use of ITT among haemophilia A and B patients with inhibitors;
2 the therapeutic regimens in use including treatment product purity, dosing and duration of therapy; and
3 the clinical outcome (success or failure) relative to the following aspects of treatment:
  (a) duration of ITT;
  (b) interval between inhibitor diagnosis and initiation of ITT;
  (c) age at ITT induction;

(d) HIV status;
(e) race; and
(f) inhibitor titres including:
   (i) historical pre-ITT titre;
   (ii) titre immediately prior to ITT induction (pre-induction titre); and
   (iii) peak titre on ITT.

Data were also collected on product and dosing regimens used for maintenance of tolerance following successful ITT. Lastly, the data instrument collected all ITT-associated central venous access and therapeutic complications.

Definition of variables

Inhibitor titres were expressed in Bethesda Units (BU). For the purposes of this registry, inhibitors were characterized only as 'high or low responding', with 'high responders' defined by a historical peak inhibitor titre prior to ITT of ≥5 BU. The peak historical inhibitor titre was defined as the highest recorded Bethesda titre prior to the initiation of ITT, as distinguished from the peak anamnestic response achieved by the patient while on ITT. The pre-ITT induction inhibitor titre identified the last BU measurement recorded prior to the initiation of ITT. This included titres achieved by both a spontaneous decline in antibody titre in the absence of re-exposure to factor VIII, as well as by therapeutic intervention with immunoadsorption or plasmapheresis.

ITT was primarily described by dosing regimen. The dosing regimen was defined as a daily dose in U/kg of body weight. For the purpose of analysis, variable dosing schedules (including continuous delivery, single and multiple daily infusions and alternating daily and weekly therapy) were standardized by dividing the total weekly dose of clotting factor by seven to derive a daily dose calculation. Registry participants were then assigned to the following four daily dosing regimen categories designed to most accurately reflect North American treatment practices:

1 ≥200 U/kg/day;
2 100–199 U/kg/day;
3 50–99 U/kg/day; and
4 50 U/kg/day.

Immune modulation was used to describe any concomitant ancillary treatment used for ITT and included:

1 immunoadsorption;
2 plasmapheresis;
3 immunosuppressive agents such as cyclophosphamide, immuran and cyclosporin A;
4 intravenous immunoglobulin therapy; and
5 corticosteroids.

Because of the large number of such regimens used, immunomodulation therapy could not be subcategorized in a meaningful way.

Respondents were asked to identify the clotting factor product used both at the time of inhibitor development as well as for ITT. For factor VIII or IX concentrates, brand names were requested where possible. The clotting factor concentrates were then classified as follows:

1 intermediate purity (IP) = specific activity (SA) 1–10 U/mg protein for factor VIII and <50 U/mg protein for factor IX;
2 high purity (HP) = SA 50–100 U/mg protein for factor VIII and >160 U/mg protein for factor IX;
3 monoclonal; and
4 recombinant.

ITT outcome was categorized as a success, a failure or indeterminate because of ongoing therapy. The NAITR permitted the criteria for success and failure to be determined by the treating physicians, but these criteria were all recorded to document the current standards of ITT practices in North America. Maintenance therapy was defined as any clotting factor regimen administered subsequent to the patient achieving the treating physician's criteria for successful immune tolerance.

Complications during the course of treatment were categorized as either:

1 adverse events directly related to the therapeutic regimen; or
2 events related to repeated venous access, whether by peripheral or central catheter or by venepuncture.

Results

Inhibitor and ITT demographics

A total of 68 centres (40% of those polled) submitted data to the NAITR. The data represented 5000 (43%) North American individuals with haemophilia A and 1325 (39%) with haemophilia B followed at the recognized haemophilia centres. 43% (2489) of the estimated severe haemophilia A patients and 35% (473) of the estimated severe haemophilia B patients were included in the survey (data supplied by the National Hemophilia Foundation (USA) and the Canadian Hemophilia Society).

A past or current history of an inhibitor was reported in 486/5000 haemophilia A patients (9.7%) and 32/1325 haemophilia B patients (2.4%). Among severe patients, the figures were 444/2489 (17.8%) with factor VIII- and 30/473 (6.3%) with factor IX-deficient participants. Importantly, 91% (444/486) of the factor VIII antibodies and 94% (30/32) of the factor IX inhibitors occurred in patients with severe disease. 67% (326/486) of the inhibitors in haemophilia A patients and 59% (19/32) of those in haemophilia B patients were high responding by registry definition.

Among haemophilia A patients with inhibitors, 39% (189/486) had undergone or were still undergoing ITT. Among high-responding inhibitor (HR) patients, that figure was 41% (134/326). For haemophilia B, comparable figures were 53% (17/32) and 74% (14/19), respectively. Additional data were submitted on 171 of the 189 ITT courses for haemophilia A reported and on all 17 courses of ITT undergone by haemophilia B subjects. As of December 1999, 96% (164) of haemophilia A ITT and 94% (16) of haemophilia B ITT was reported to

Table 10.1 Demographics of haemophilia A and B registry subjects who completed ITT*.

	Haemophilia A ($n = 164$)	Haemophilia B ($n = 16$)
Severe haemophilia	150 (93%)	14 (88%)
High-responding inhibitors	128 (78%)	13 (81%)
Family history of inhibitor	31 (22%)	5 (31%)
Race		
Caucasian	100 (63%)	8 (50%)
African-American	21 (13%)	4 (25%)
Latino	24 (15%)	2 (13%)
Other	15 (9%)	2 (12%)
HIV positive status	28 (17%)	0 (0%)
Mean age at (years)	9.3	5.6
ITT initiation (range)	(0.1–64)	(0.8–19.6)

ITT, immune tolerance therapy.
*Denominators vary for each variable; (%) represents the percentage of the total number of patients in the haemophilia A or B cohort.

be complete. 79% of the ITT was initiated in or after 1990. Table 10.1 depicts the demographic characteristics of both the haemophilia A and B populations that completed tolerance by the time of the last analysis.

Therapeutic regimens in use

Interval between inhibitor diagnosis and ITT

The 164 haemophilia A subjects who completed ITT had tolerance initiated after a mean interval of 55.8 months (range 0–256) following the inhibitor diagnosis. The 16 individuals with haemophilia B who completed tolerance initiated therapy at a mean of 44.1 months (range 0–227) following identification of the inhibitor.

Product purity

Among the 171 haemophilia A patients receiving ITT, 47% developed their inhibitors on very high purity products (monoclonal and recombinant). However, 75% of these subjects received monoclonal or recombinant factor VIII as part of their tolerizing regimen. Similarly, while 47% of inhibitors in haemophilia B developed on high purity/monoclonal factor IX concentrates, these products were used in 14/17 (82%) of the ITT courses.

Dosing regimens/adjunctive therapy

With respect to dosing regimen, data were available for 164/171 of the completed and ongoing ITT courses in haemophilia A patients. 52% of subjects received factor VIII doses of ≥100 U/kg/day. Only 14% of this cohort received ≥200 U/kg/day. Alternate-day dosing regimens (representing the majority, 91%, of the <50 U/kg/day group), were used in 21% of completed and ongoing ITT. Adjunctive immune modulation (IM) therapy was used in 65 (40%) of all completed and ongoing ITT in this population. IM was added to 58% of factor VIII dosing regimens using ≥100 U/kg/day and to 22% of regimens using <100 U/kg/day. Only four subjects received the complete or modified Malmö regimen, including immunoadsorption or plasmapheresis, as a part of their tolerance protocol.

Among the 17 reported courses of ITT in haemophilia B patients, the median daily dose of factor IX was 100 U/kg/day with a range of 25–200 U/kg/day. Daily dosing regimens were or are being used in 15 (88%) courses. IM was added to eight courses (47%) and plasmapheresis was used in two courses.

Duration of therapy

Haemophilia A patients who completed ITT received a mean of 16.8 months (range 0–89 months) of therapy. Those with haemophilia B were treated for a shorter time (11.6 months) before outcome was determined ($P = 0.05$).

Outcome data

Of the 164 registry subjects with haemophilia A who completed ITT, 115 (70%) successfully achieved tolerance. Among HR haemophilia A subjects, 81/128 (61%) were successfully tolerized. Of the 16 completed ITT courses in individuals with haemophilia B, only 5 (31%) were successful. The success rate remained constant during the 10-year study period.

The definition of ITT success for both haemophilia A and B included a negative inhibitor titre (≤1 BU) in 104 (87%) cases; normal factor VIII recovery (≥70% of predicted) in 69 (56%) cases; normal factor survival (≥12 h) in 34 (28%) cases; and conversion from high- to low-responder status in 12 (10%) patients. Only 30 (25%) of the patients were determined to be successfully tolerized on the basis of a negative inhibitor titre, as well as a normal factor recovery and survival.

Several reasons for ITT failure were recorded for each subject. The three most common causes for failure were:
1 a late rise in inhibitor titre after an initial decline while on ITT;
2 patient/family desire to terminate ITT; and
3 central venous access device complications.

Other reasons cited included adverse reaction to therapy; ineffectiveness of therapy; loss of medical insurance; patient relocation or death from unrelated causes; poor patient compliance; and enrolment in a bypassing agent study.

Outcome predictors: haemophilia A

Several parameters were examined as possible ITT outcome predictors in HR haemophilia A patients (Table 10.2). Although trends toward both a younger mean age and a shorter

	Success mean (range)	Failure mean (range)	P-value
Age at ITT (years)	9.5 (0.6–64.1)	10.6 (0.9–48.6)	0.6
Months between inhibitor diagnosis and ITT	55 (0–226)	65 (0–226)	0.4
Duration of ITT (months)	16.3 (0–84)	19.6 (0–89)	0.4
ITT dose (U/kg/day)	81 (11–200)	109 (11–303)	0.01
Historical peak titre (BU)	130 (5–4833)	571 (6–9999.5)	0.05
Pre-induction titre (BU)	19.5 (0–230)	61.3 (0–523)	0.005
Peak titre on ITT (BU)	154 (0–1770)	619 (34–3500)	0.001

Table 10.2 Characteristics and predictors of outcome in 128 HR haemophilia A patients who completed ITT.

BU, Bethesda units; ITT, immune tolerance therapy.

mean interval between inhibitor diagnosis and ITT initiation were noted in the successful group (9.5 years; 55 months) when compared to those who failed tolerance (10.6 years; 65 months), those differences were not statistically significant in this data set ($P = 0.6$ and 0.4, respectively). Success was predicted with only borderline significance by an interval between diagnosis and therapy of ≤5 years for both haemophilia A ($P = 0.08$) and HR haemophilia A ($P = 0.17$). Positive HIV serology (19% vs. 20%) and mean duration of ITT (15.6 vs. 18.8 months) were also not statistically different between the success and failure groups.

The higher incidence of inhibitors in African Americans has been previously reported [21,22]. Consequently, a potential differential response by race to immune tolerance was examined. However, in this case the rate of successful immune tolerance induction in a small cohort of African Americans and Latin Americans with haemophilia A was similar to a larger group of patients of other races (14/21 (67%) and 11/19 (58%) vs. 99/139 (71%), $P = 0.67$ and 0.3, respectively).

Factor VIII product purity had no impact on ITT success in haemophilia A patients. The success rate among those patients tolerized with intermediate or high purity products was 67.5% and not statistically different from either the success rate among patients tolerized with monoclonal or recombinant factor (71%), or the overall success rate in the haemophilia A population as a whole (70%).

Other potentially important parameters examined included the interval between the last exposure to human or porcine factor VIII and ITT induction; the frequency of adverse events on therapy; and the frequency of central access complications during tolerance induction. None were significant outcome predictors in this registry.

We examined the influence of inhibitor titre on ITT outcome in HR haemophilia A patients, specifically the historical peak titre prior to ITT, the peak inhibitor titre on immune tolerance and the pre-ITT induction titre. All three parameters were found to be significant predictors of outcome in univariate analyses (Table 10.2). However, in a multivariate analysis, both the peak titre on ITT ($P = 0.005$) and the pre-induction titre ($P = 0.001$) were the most significant predictors of success.

Specifically, when the pre-ITT titre was <10 BU, the ITT success rate was 83%, compared to 40% when the titre was ≥10 BU ($P = 0.001$). Furthermore, when the pre-ITT titre was <10 BU the mean time to successful tolerance was significantly shorter (14.4 months, compared to 21.7 months) when ITT was begun at titres ≥10 BU ($P = 0.02$). It is of note that among 22 HR haemophilia A patients with pre-ITT titres of <2 BU, all achieved ITT success within a mean treatment time of 7.1 months.

In dramatic contrast to the data generated by the international registry, the impact of dosing regimen on ITT success in the HR haemophilia A cohort in the North American study demonstrated no trend toward higher ITT success rates with increasing daily doses of factor VIII. In fact, there was a significant inverse association between total daily dose (U/kg/day) and success: (72% in the <50 cohort; 65% in the 50–99 group; 68% in the 100–199 cohort; and 38% in the 200 group, $P = 0.014$). The use of adjunctive immune modulation had no impact on the rate of successful ITT within each dose category. Furthermore, when the time to successful ITT was examined there was no statistically significant difference among the dosing categories, with or without immune modulation. However, in HR haemophilia A individuals who received at least 50 U/kg/day, tolerance was achieved significantly faster (mean of 15.9 months) when compared with those who infused with less than 50 U/kg/day (mean of 23.6 months) ($P = 0.04$).

Outcome predictors: haemophilia B

Potentially important outcome parameters are described in Table 10.3 for the 16 completed ITT courses administered to inhibitor patients with haemophilia B. Because of the small number of patients, only trends could be observed in variables such as pre-induction and historical peak titres, both of which were statistically significant predictors of success in haemophilia A. Of note is that 5/11 failures had a positive family history of an inhibitor compared to none of the 5 successful outcomes. Furthermore, many of the patients whose courses failed had access complications (9/11) and adverse reactions (8/11) compared to few such events among the successes.

Table 10.3 Characterization of severe haemophilia B patients who completed ITT.

	Success (n = 5)	Failure (n = 11)
Family history inhibitor	0/5	5/11
African-American race	2/5	2/11
Age at ITT induction (years)*	3.7 (2.4–4.8)	4.6 (0.8–19.6)
Interval between diagnosis/ITT (months)*	12 (0–31)	47 (0–227)
Duration of ITT (months)*	12 (5–18)	7 (2–32)
High responder (≥5 BU)	4/5	9/11
Historical peak titre (BU)*	13 (2.4–112)	50 (10–650)
Pre-induction titre (BU)*	5 (3–24)	10.5 (1–19)
Peak titre on ITT (BU)*	20 (3–38)	39.5 (6–59)
ITT dose (U/kg/day)*	100 (43–200)	75 (25–200)
Access complications	0/5	9/11
Adverse reactions	2/5	8/11
Allergic reactions	1/5	4/11

Results reported as median (range). BU, Bethesda units; ITT, immune tolerance therapy.

Maintenance of tolerance

Haemophilia A

Ninety-six (83%) of the 115 individuals in the haemophilia A cohort who successfully achieved immune tolerance were maintained on a regular regimen of factor VIII infusion for a variable period of time post-ITT induction. The median dose used for maintenance therapy was 150 U/kg/week (range 17–980) administered in variable dosing regimens. At the time of the last update in 1999, 64 (67%) subjects remained tolerant while continuing to receive regular doses of factor VIII for a median observation period of 17 months (range 0–129 months). Of the 31 patients whose maintenance therapy had been stopped, the status of 1 patient was unknown; 21 (68%) remained tolerant over a median observation period of 19 months (range 1–54 months); and 9 had relapsed. Among the relapses, 3 occurred at between 1 and 8 months following the cessation of maintenance therapy and 6 were noted while the patients were still on maintenance therapy for a median time of 11 months (range 2–61 months). Due to the questionnaire design, no outcome data was obtained on the 19/115 patients who were never placed on a factor VIII regimen after successful tolerance was induced.

The definition of tolerance was reviewed for those 9 subjects who relapsed on maintenance. Tolerance had been defined by a normal recovery and survival in only 1/9 patients and by a normal recovery alone in 4/9 patients. Three relapses had a negative Bethesda titre as the sole determinant of successful tolerance, and one success had been initially defined as a conversion from high- to low-responder status. The most stringent definition of success (normal recovery and survival) was underrepresented (11% vs. 30%) in the relapse cohort when compared to the entire group of haemophilia A patients undergoing ITT (Table 10.4). However, given the small number of relapse patients, no significant inverse correlation between the definition of success and the maintenance of successful tolerance could be definitively established.

Haemophilia B

Of the 16 haemophilia B subjects in the cohort who completed tolerance, 5 had a successful outcome using factor IX therapeutic dosing regimens of 43 and 200 U/kg/day. This included one patient with an inhibitor-associated allergic phenotype. Once tolerance was induced, all 4 individuals were placed on maintenance regimens of 15–100 U/kg/day of factor IX. At the time of the last data analysis in 1999, maintenance was ongoing for all subjects and 5/5 remained tolerant.

Adverse reactions on ITT

Haemophilia A

Among the 171 completed and ongoing ITT subjects, 10 (6%) therapeutic courses were complicated by 14 adverse reactions to treatment. Allergic reactions accounted for 6 (4.3%) events and included 4 with respiratory symptoms and 2 episodes of cardiovascular collapse. The other adverse events included 3 episodes of bleeding and 1 of each of DIC, hair loss, gastrointestinal toxicity, headache and hypertension. With respect to viral transmission, one seroconversion to HIV and no

Table 10.4 Maintenance of tolerance—comparison of successful haemophilia A (HA) ITT and relapse cohorts.

Definition	HA ITT Successes* ($n = 102$)	HA/Relapse ($n = 9$)
≤1 BU alone	30 (29%)	3 (33%)
≤1 BU/rec ≥70% expected	32 (31%)	4 (44%)
≤1 BU/normal rec/T$^{1}/_{2}$	31 (30%)	1 (11%)
Conversion to low responder	9 (9%)	1 (11%)

* Information not available on 13/115 patients. BU, Bethesda units. ITT, immune tolerance therapy.
T$^{1}/_{2}$, survival; rec, recovery.

seroconversions to hepatitis A, B or C was reported. An analysis of adverse event frequency relative to ITT dose was possible in 159 haemophilia A subjects. It is of note that none of the 33 individuals receiving <50 U/kg/day (mostly alternate-day regimens) experienced adverse reactions to therapy compared to 8/126 subjects receiving ≥50 U/kg/day (mostly daily regimens) ($P = 0.14$).

Haemophilia B

Among the 17 completed and ongoing ITT courses, 14 adverse events were reported in 10 subjects undergoing 11 (65%) courses of ITT, a frequency 10 times higher than that for individuals with haemophilia A. Allergic reactions accounted for 11 (79%) adverse events and occurred in the 10 subjects who had previously exhibited allergic reactions to factor IX replacement therapy subsequent to the development of an inhibitor. This phenomenon has been previously described [23,24]. Three of these reactions were severe and caused premature cessation of ITT. Overall, allergic reactions represented the major reason for failure in 5/11 unsuccessful courses of ITT in haemophilia B.

The registry noted an unexpected and important association between allergic reactions to factor IX and the development of nephrotic syndrome in 3 haemophilia B patients undergoing ITT. In all 3 subjects, symptoms of periorbital oedema, proteinuria and hypoalbuminaemia developed 7–9 months into ITT. All received 100 U/kg/day of either prothrombin complex concentrate or monoclonal factor IX, and had exhibited a decline in inhibitor titre prior to the diagnosis of nephrosis. Two of these patients had previously been reported by Ewenstein et al. [25]. There were no reported viral seroconversions. When adverse reactions were examined relative to dose, adverse events were not noted to occur any less frequently with doses of <50 U/kg/day (4/4) than with doses of ≥50 U/kg/day (7/13) ($P = 0.09$).

Central venous access catheter complications

There were 129 access complications reported with the use of 136 central access devices (CAD), including both external catheters and venous ports inserted for ITT. Thirteen complications were reported among 18 indwelling peripheral intravenous catheters. Infections accounted for ≥60% of the complications occurring in each type of catheter. The 79 CAD infections led to 55 hospitalizations and necessitated catheter removal in 18 cases. CAD-associated bleeding was the second most common complication (18/129, 14%), with packed red cell transfusions required in two episodes. However, among external CAD, accidental loss was reported as the second most common complication (14/88 events, 16%). CAD-associated thrombosis necessitated the removal of the access device in 5/9 (56%) events. As might be expected, access complications in subjects with haemophilia A and B were significantly less frequent in subjects receiving <50 U/kg/day or primarily alternate-day regimens (5/37), when compared to largely daily regimens of ≥50 U/kg/day (62/140) ($P = 0.0006$).

Conclusion

In summary, the NAITR accomplished its goal of characterizing ITT practices, outcomes, outcome predictors and complications in a large North American population of haemophilia A and B inhibitor patients. Importantly, several significant successful outcome predictors were identified. For high-titre haemophilia A patients, the importance of historical and preinduction inhibitor titres in predicting success corroborated the data generated by the International Immune Tolerance Registry. Peak titre on immune tolerance, a parameter not examined by the international study, was also identified by the NAITR as an important outcome predictor. With respect to immune tolerance in haemophilia A patients, the importance of the dosing regimen in predicting success remains the topic of greatest debate. The potential role in this controversy of observed differences in immune tolerance dosing practices between the two registries has been previously discussed in detail [26].

Given the low incidence of inhibitors and immune tolerance in haemophilia B, outcome predictors could not be conclusively identified. However, the North American study did demonstrate the poor outcome and potential nephrotic syndrome complication of immune tolerance therapy in the factor IX inhibitor patient with an allergic phenotype.

Ultimately, this registry—a retrospective data-gathering tool—was limited in its capacity to provide definitive answers to the many unresolved issues surrounding the practice of immune tolerance. Nonetheless, it played an important role in developing the hypotheses that must now be tested in the logical next step—a newly initiated international prospective randomized study of ITT.

Acknowledgements

The authors wish to acknowledge the contributions of the

North American Immune Tolerance Study Group to these data as well as the support of the International Society of Thrombosis and Haemostasis Factor VIII/IX Subcommittee in the development of this registry.

References

1 Ehrenforth S, Kreutz W, Scharrer I *et al*. Incidence of development of factor VIII and factor IX inhibitors in haemophiliacs. *Lancet* 1992; **339**: 594–8.
2 Briet E, Reisner HM, Roberts HR. Inhibitors in Christmas disease. *Prog Clin Biol Res* 1984; **150**: 123–9.
3 Aledort LM, Cohen M, Hilgartner M, Lipton R. Treatment of hemophiliacs with inhibitors: cost and effect on blood resources. *Prog Clin Biol Res* 1984; **150**: 353–65.
4 Brackmann HH, Gormsen J. Massive factor VIII infusion in a haemophiliac with factor VIII inhibitor, high responder (letter). *Lancet* 1977; **2** (8044): 933.
5 Brackmann HH. Induced immunotolerance in factor VIII inhibitor patients. *Prog Clin Biol Res* 1984; **150**: 181–95.
6 Aznar JA, Jorquera JI, Peiro A *et al*. The importance of corticoids added to continued treatment with factor VIII concentrates in the suppression of inhibitors in hemophilia A. *Thromb Haemost* 1984; **51**: 217–21.
7 Van Leeuwen EF, Mauser-Bunschoten EP, van Dijken PJ *et al*. Disappearance of factor VIII:C antibodies in patients with haemophilia upon frequent administration of factor VIII in intermediate or low dose. *Br J Haematol* 1986; **64**: 291–7.
8 Ewing NP, Sanders NL, Dietrich SL, Kasper CK. Induction of immune tolerance to factor VIII in hemophiliacs with inhibitors. *JAMA* 1988; **259**: 65–8.
9 Nilsson IM, Berntorp E, Zettervall O. Induction of immune tolerance in patients with hemophilia and antibodies to factor VIII by combined treatment with intravenous IgG, cyclophosphamide and factor VIII. *N Engl J Med* 1988; **318**: 947–50.
10 Gruppo RA, Valdez LP, Stout RD. Induction of immune tolerance in patients with hemophilia A and inhibitors. *Am J Pediatr Hematol Oncol* 1992; **14**: 82–7.
11 Mauser-Bunschoten EP, Nieuwenhuis HK, Roosendaal G, van den Berg HM. Low dose immune tolerance induction in hemophilia A patients with inhibitors. *Blood* 1995; **86** (3): 983–8.
12 Kreuz W, Ehrenforth S, Funk M *et al*. Immune tolerance therapy in paediatric hemophiliacs with factor VIII inhibitors: 14 years follow up. *Haemophilia* 1995; **1**: 24–32.
13 Brackmann HH, Oldenburg J, Schwaab R. Immune tolerance for the treatment of factor VIII inhibitors—twenty years' 'Bonn Protocol'. *Vox Sang* 1996; **70**: 30–5.
14 Freiburghaus C, Berntorp E, Ekman M *et al*. Immunoadsorption for removal of inhibitors: update on treatments in Malmo-Lund between 1980 and 1995. *Haemophilia* 1998; **4**: 16–20.
15 Freiburghaus C, Berntorp E, Ekman M *et al*. Tolerance induction using the Malmo treatment model 1982–95. *Haemophilia* 1999; **5**: 32–9.
16 Oldenburg J, Schwaab R, Brackmann HH. Induction of immune tolerance in haemophilia A inhibitor patients by the 'Bonn Protocol': predictive parameter for therapy duration and outcome. *Vox Sang* 1999; **77** (1): 49–54.
17 Nilsson IM, Berntorp E, Zetterval O. Induction of split tolerance and clinical cure in high-responding hemophiliacs with factor IX antibodies. *Pro Natl Acad Sci USA* 1986; **83**: 9169–73.
18 Mariani G, Ghirardini A, Bellocco R. Immune tolerance in hemophilia. Principal results from the international registry. *Thromb Haemost* 1994; **72** (1): 155–8.
19 Mariani G, Kroner B. International immune tolerance registry, 1997 update. *Vox Sang* 1999; **77** (1).
20 Kroner B. Comparison of the international immune tolerance registry and the north american immune tolerance registry. *Vox Sang* 1999; **77** (1): 33–7.
21 Gill FM. The natural history of factor VIII inhibitors in patients with hemophilia A. *Prog Clin Biol Res* 1984; **150**: 19–29.
22 Addiego JE, Jr, Kasper C, Abildgaard *et al*. Increased frequency of inhibitors in African American hemophilia A patients. *Blood Supplement* 1994; **1**: 239a.
23 Warrier I, Ewenstein BM, Koerper M *et al*. Factor IX inhibitors and anaphylaxis in hemophilia B. *Am J Pediatr Hematol Oncol* 1997; **19** (1): 23–7.
24 Warrier I. Factor IX antibody and immune tolerance. *Vox Sang* 1999; **77** (1): 70–1.
25 Ewenstein BM, Takemoto C, Warrier I *et al*. Nephrotic syndrome as a complication of immune tolerance in hemophilia B (letter). *Blood* 1997; **89** (3): 1115–16.
26 DiMichele DM, Kroner BL and the North American Immune Tolerance Study Group. The North American Immune Tolerance Registry: practices, outcomes, outcome predictors. *Thromb Haemost* 2002; **87**: 52–7.

PART 4
Medical management of bleeding episodes

CHAPTER 11

The treatment of bleeding episodes in children

J.M. Tusell

Importance of early detection of inhibitors for optimum treatment

It is well known that inhibitors appear during the initial phases of haemophilia treatment with factor concentrates. In order to ensure optimum comprehensive treatment of paediatric patients with inhibitors, it is essential that the inhibitors are detected as soon as they appear. An early diagnosis, through a systematic and rigorous search for inhibitors against factors VIII or IX in children with haemophilia undergoing replacement therapy, is essential for three fundamental reasons. First, it prevents the risks associated with ineffective treatment of the coagulopathy; these risks include the possibility of developing serious arthropathies (caused by persistent joint bleeds), ineffective coverage during surgery, and more serious risks such as inadequate treatment of intracranial haemorrhages. Secondly, and more importantly, the ideal treatment for patients with inhibitors is immune tolerance induction (ITI) and this, as is suggested by most of the registries and cohort studies [1–5], is more easily and quickly achieved if carried out at an early age, with a minimum of stimulations with factors VIII or IX between the diagnosis of inhibitors and the start of ITI. Finally, there is a financial reason; inadequate treatment of a patient with high-response inhibitors using ineffective doses of factors VIII or IX is not only dangerous but is also a wasteful use of already costly resources. Furthermore, the cost of comprehensive ITI treatment is much lower in children with lower body weights than older children and adults [6,7].

Characteristics of bleeding episodes in children with inhibitors

The frequency and location of bleeding episodes in children with haemophilia who also have inhibitors does not vary greatly from those in severe haemophilia patients without inhibitors [8]. The higher morbidity and increase in factor consumption when inhibitors are present are mostly brought about by the limited biological and clinical responses to treatment and complications during bleeding episodes, which significantly increase the risk of developing target joints and even pseudotumours. This greater tendency to develop chronic arthropathy leads, in turn, to an increase in bleeds, because of the presence of inflammation and sinovial hypertrophy as well as muscular weakness and joint instability, which develop as a further consequence in a number of cases. The most frequent type of bleeds are similar to those in patients without inhibitors: a predominance of joint bleeds (62%) over bleeds affecting tissues and other parts, and a clear association with previous injury (84%) [9].

Aims of treatment

By far the most important aim in starting a treatment programme in paediatric patients with inhibitors is to achieve immune tolerance to the infused factor in as short a time as possible. However, before this is achieved, or during the ITI itself, a rigorous and effective effort needs to be made to minimize the effects of possible bleeding episodes. Furthermore, statistics show that in a certain percentage of patients (20%) immune tolerance will not be achieved. In these cases it is vital for the patient to receive optimum treatment in order to keep the devastating effects of bleeding episodes to a minimum.

General medical treatment for stopping bleeds—therapeutic option in paediatric patients

The specific aim of treatment during bleeds is to stop the bleeding. In patients without inhibitors, this is easily performed by administering factors VIII or IX. In patients with low titre inhibitors, particularly those with a low anamestic response, it is achieved by increasing the administered doses. However, in high titre and high-response patients other therapeutic options are required, given that simple replacement therapy is completely inefficacious.

Porcine factor VIII

Factor VIII antibodies are, to a certain degree, species-specific, which enables porcine-origin factor VIII to be used in some patients with inhibitors in order to increase factor VIII levels without it being recognized and neutralized by human factor VIII antibodies. As this product is of animal origin, there is no risk of contamination by human viruses.

Older preparations of this product caused a large number of adverse reactions. However, since 1980 a purer preparation has been available, which has been widely used in Europe and the USA with few adverse effects and a high level of efficacy [10–12]. Candidates for this treatment are patients whose antibodies have been shown by pretreatment tests to have no, or very little, cross-reactivity to porcine factor VIII. However, patients with high titre inhibitors also tend to have a degree of cross-reactivity; thus patients with a titre of >5 Bethesda units (BU) are unlikely to respond to treatment using porcine factor VIII and, indeed, the antibody titre may show a tendency to increase. Furthermore, even patients who do not initially show antibody cross-reactivity tend to develop antiporcine factor VIII antibodies as the number of administrations increases [10]. Although this is a safe product that may be used in paediatric patients and even in home treatment, administration must be gradual, with suitable precautions, and regular level checks must be carried out so that the dose may be adjusted [12,13].

As with other products used for treating patients with inhibitors, this product is very expensive. Continuous infusion, which has been used successfully, given that this is a stable product [14,15], may allow consumption to be reduced, thereby reducing the costs of certain treatments and coverage in certain surgical procedures.

Bypass agents

Activated or non-activated prothrombin complex concentrates

The aim of these products is to bypass the blockage in the coagulation cascade caused by the lack of factors VIII or IX and, at the same time, avoid the influence of inhibitor presence. For over 25 years activated (aPCC) or non-activated prothrombin complex concentrates (PCC) have been used in paediatric patients for this purpose, with efficacy levels varying between 50 and 60%, depending on the study [16,17], although in a wide retrospective study carried out in France, efficacy was judged as excellent in 81.3% of cases [18]. Activated complexes are generally preferred, as they are perceived to be more effective than non-activated ones. The element responsible for the haemostatic effect, attributed to activated factors VII, IX or X, has not been well established. No laboratory test is available to measure its biological efficacy, and its clinical efficacy is not the same as that obtained using factor VIII or IX concentrates in children with haemophilia without inhibitors. Repeated doses may lead to thrombosis, disseminated intravascular coagulation and myocardial infarction, even in young patients [19,20]. Because of these complications, some authors have suggested that patients should not receive more than four successive doses of PCCs during the treatment of an individual bleeding episode [13].

Activated recombinant factor VII

Activated recombinant factor VII (rFVIIa) is one of the components of PCC to which the bypass haemostatic effect in patients with haemophilia has been attributed. Furthermore, one of its characteristics is that it is the only activated factor not, in itself, enzymatically active. It only takes effect when and where tissue factor is present. Hedner and Kisiel [21] demonstrated the efficacy of activated plasma-derived factor VII in treating bleeding episodes in haemophilia patients with inhibitors. The development by Novo Nordisk (Copenhagen) of a recombinant origin activated factor VII concentrate has provided treatment centres with an important new product for treating patients with inhibitors. Its use has spread rapidly, including in children, and represents an important new step in the treatment of bleeding episodes, with efficacy percentages around 60–90% [22]. As this is a recombinant origin product and therefore has greater viral safety, it is particularly important in the treatment of children who have never been treated with plasma-derived products. No specific pharmacokinetic studies in children have been carried out, but existing studies suggest greater clearance in paediatric patients than in adults, where a relatively short half-life (around 2.5 h) has been observed [23]. Furthermore, the specific dose for each bleed is not well established as there is a wide margin, ranging from an initial dose of 90 µg/kg (the most common) to 300 µg/kg. Because of its short half-life, the intervals for administration are from 2 to 3 h, which in paediatric patients gives rise to the problem of venous access. Another significant problem is the expense of the product, which has led a number of authors to propose administration by continuous infusion, thereby making savings in product use and venous access. However, there are no specific protocols for such treatment, nor is the ideal constant plasma level of factor VIIa required for effective treatment precisely known. Where this form of treatment is particularly recommended is in the treatment of severe life-threatening bleeding episodes or in coverage for major surgery.

Paediatric treatment scheme according to the individual patient's condition
(Table 11.1)

Low-response patients with an inhibitor titre <5 BU

In these cases, all that is required is to increase the dose sufficiently to neutralize the inhibitors and reach minimum haemostatic levels. It must be ensured that the inhibitors are completely neutralized and haemostatic levels of 30–100 U/mL (depending on the severity of the bleed) are obtained. In calculating the neutralizing dose, the inhibitor titre must be considered in conjunction with the patient's plasma volume in proportion to weight, assuming that the extravascular space contains the same inhibitor titre as the plasma. The formula for calculating the dose is as follows [24]:

Table 11.1 Scheme of treatment options for severe bleeding episodes in children with inhibitors against factors VIII/IX.

Type of patient	Inhibitor titre during haemorrhage	Treatment in ITI candidates*	Treatment in non-ITI candidates
Low response	<5 BU	rFVIIa	FVIII/FIX porcine FVIII aPCC rFVIIa
High response	<5 BU	rFVIIa	FVIII/FIX until anamnestic response porcine FVIII aPCC rFVIIa
High response	>5 BU	rFVIIa	porcine FVIII aPCC rFVIIa plasmapheresis + FVIII/FIX

*In patients undergoing ITI treatment, if the inhibitor titre is <5 BU then treatment may be carried out using high doses of factor, if >5 BU then bypass agents must be used.
aPCC, activated prothrombin complex concentrate; BU, Bethesda units; ITI, immune tolerance induction; rFVIIa, activated recominant factor VII.

$$2 \times BW [0.8 \times (100 - Ht)] \times Inh = \text{units of factor VIII},$$

where BW is body weight expressed in kilograms, Ht is the haematocrit expressed as a percentage, and Inh is the inhibitor titre expressed in BU.

In these conditions, treatment does not differ from that given to patients without inhibitors and requires periodic checks for plasma clearance of the administered factor, as this tends to be higher when inhibitors are present. Some low titre inhibitors with no anamnestic response are transitory and may disappear after a certain period of treatment, both in prophylactic treatment regimens and episodic treatment. If the patient is a candidate for ITI treatment, stimulations with factors VIII and IX must be avoided; rVIIa must therefore be used for treating episodic bleeds because aPCCs contain small quantities of factors VIII and IX.

High-response patients with current titre <5 BU

In this case treatment may be with factor VIII/IX concentrates until an antibody anamnestic response is produced. Another option is the direct use of bypass agents. If the patient is a candidate for ITI, then treatment with rVIIa is required.

High-response patients with titre >5 BU

For these patients and for even lower inhibitor titres that cannot be neutralized and overcome through high doses, the use of bypass agents is required. In special cases the anamnestic response may be eliminated by using high doses of factor or immunosuppression. This method has been advocated by Scandinavian research groups, who have had considerable success in combining factors VIII or IX with cyclophosphamide [25–27].

Other strategies involve carrying out plasmapheresis to eliminate antibodies, combined with the simultaneous administration of factor VIII and plasma [28] or, alternatively, the extracorporeal immune adsorption of the factor VIII antibody via a protein A sepharose column or by immune adsorption of the antibodies against factor IX [29–31]. These procedures involve a certain degree of sophistication and require time and specialized personnel, as well as extremely good venous access, which in children represents a further difficulty. Therefore they should only be used in special cases, such as life-threatening situations or, possibly, to achieve a low titre inhibitor before starting ITI with high doses of factor.

For patients undergoing ITI treatment, if the inhibitor titre at the onset of a bleeding episode is <5 BU then treatment may be carried out using high doses of factor; otherwise bypass agents are required.

Treatment with factor IX

Anaphylactic reactions

In recent years, this serious problem has been detected in the use of factor IX concentrates in severe haemophilia B patients with inhibitors, most of whom are of paediatric age. A large number of anaphylactic reactions have been detected, some of them accompanied by shock and often leading to the development of nephrotic syndrome, which ceases once treatment is interrupted. This complication may affect 50% of children with haemophilia B and inhibitors and is related to the type of mutation responsible for the coagulopathy, rather than the type of product infused. Large deletions or large alterations of the factor IX gene seem to be most associated with these reactions [32].

The existence of such a serious complication, along with the fact that ITI is rarely achieved in haemophilia B patients with inhibitors, has led to the questioning of the validity of ITI treatment in these patients. One clear counterindication is the use of factor IX in patients who have shown reactions of this type. For such patients, treatment of bleeding episodes must be based on bypass agents and, more specifically, rFVIIa, as aPCCs contain small quantities of factors VIII and IX. The existence of this complication requires rigorous monitoring of children with haemophilia B when starting treatment, including strict monitoring of inhibitor development. Furthermore, as a precautionary measure, it is advisable to administer the first 10–20 factor IX infusions under hospital care, especially with patients who have a high-risk mutation [33].

Treatment of specific bleeds in paediatric patients

Importance of early treatment

Home treatment

It is a general rule in haemophilia, but especially in paediatric patients, that treatment administered as rapidly as possible after the onset of a bleeding episode produces better results. Home treatment has enabled this rule to be put into practice. In the case of patients with inhibitors, the availability of rVIIa, a safe product with no adverse effects or anamnestic response, has enabled early home treatment programmes to be carried out. These programmes have achieved total efficacy levels of 79–92%, and partial efficacy of 11%, after an average of only 2–2.2 infusions [34,35], thus demonstrating that treatment administered in this way achieves optimum results with a minimum of factor consumption. In some cases spectacular results have been achieved in paediatric patients with difficulties in receiving treatment (such as venous access, or living long distances from treatment centres) by home visits from nurses specializing in this form of treatment, in order to provide early factor administration [36].

Activated recombinant factor VII vs. activated prothrombin complex concentrates

Although the published and referenced studies in this field suggest a slightly greater efficacy on the part of rVIIa concentrates, no results from comparative studies are currently available for a definitive decision to be reached on the ideal option. Arguments in favour of using rVIIa for treating children with haemophilia and inhibitors are its non-plasma origin; the absence of factor VIII or IX traces that could cause an anamnestic response; the absence of adverse effects; and the ability to measure levels of the factor. However, the short half-life; the need for repeated doses (damaging to venous access); a single market supplier; and the high costs involved are all negative arguments, which mean that aPCCs are still an option in the treatment of these patients [37]. A randomized crossed multicentre study in the use of these two bypass agents at the Malmö centre (Sweden) is now underway and may provide an answer to the controversial question of the efficacy of these products in the treatment of acute episodes in patients with inhibitors [38].

Preventative treatment of bleeds

The introduction of prophylactic treatment programmes for children with haemophilia has changed the prognosis of the disease, and allows children to live almost normal lives. Because of the characteristics of the treatments currently available, prophylaxis is not initially an option in children with haemophilia and inhibitors. However, it should be stressed that a number of experiences have produced positive results that suggest this treatment must be considered, particularly in patients with high titre inhibitors and high rates of bleeding episodes, for whom ITI programmes have failed and who are therefore likely to develop serious arthropathies. These experiences, all of which are extremely expensive, have been carried out using aPCCs, although there are also references, in some cases, to using daily doses of rVIIa concentrates [39–41].

Conclusions

The treatment of bleeding episodes in children with inhibitors is based on early treatment carried out at the patient's home. The choice of product depends on each patient's circumstances, but basically consists in factor VIII/IX concentrates when dealing with low titre inhibitors and bypass agents (preferably rVIIa) when dealing with high titres. It must be borne in mind that the main aim when treating children with haemophilia and inhibitors is to provide optimum treatment through ITI.

References

1 DiMichele DM, Kroner BL and the ISTH Factor VIII/IX Subcommittee Members. Analysis of the North American Immune Tolerance Registry (NAITR) 1993–97: current practice implications. *Vox Sang* 1999; **77** (Suppl. 1): 31–2.

2 Ehrenforth S, Kreuz W, Funk M, Auerswald G, Scharrer I. Variables that might affect the outcome of immune tolerance therapy in haemophiliacs with factor VIII inhibitors. *Thromb Haemost* 1994; **72**: 782–6.

3 Haya S, Lopez MF, Aznar JA, Batlle J and the Spanish Immune Tolerance Group. Immune tolerance treatment in haemophilia patients with inhibitors: the Spanish Registry. *Haemophilia* 2001; **7**: 154–9.

4 Lenk H. The German National Immune Tolerance Registry: 1997 update. *Vox Sang* 1999; **77** (Suppl. 1): 28–30.

5 Mariani G, Ghiardini A, Belloco R. Immune tolerance in haemophilia: principal results from the International Registry. *Thromb Haemost* 1994; **72**: 155–8.

6 Goudemand J. Treatment of patients with inhibitors: cost issues. *Haemophilia* 1999; **5**: 397–401.
7 Colowick AB, Bohn RL, Avorn J, Ewenstein BM. Immune tolerance induction in haemophilia patients with inhibitors: costly can be cheaper. *Blood* 2000; **96**: 1698–702.
8 Abilgaard CF. Management of inhibitors in hemophilia. In: Hilgartner MW, ed. *Hemophilia in the Child and Adult*. New York: Masson Publishing, 1982: 167–80.
9 Janco RL, Maclean WE, Perrin JM, Gortmaker SL. A prospective study of bleeding in boys with haemophilia. *Haemophilia* 1996; **2**: 202–6.
10 Brettler DB, Forsberg AD, Levine PH et al. The use of porcine factor VIII concentrate (Hyate:C) in the treatment of patients with inhibitor antibodies to factor VIII: a multicenter US experience. *Arch Intern Med* 1998; **149**: 1381–5.
11 Kernoff PBA. Porcine factor VIII: preparation and use in treatment of inhibitor patients. *Prog Clin Biol Res* 1984; **150**: 207–24.
12 Hay CRM, Laurian Y, Verroust F, Preston FE, Kernoff PBA. Induction of immune tolerance in patients with haemophilia A and inhibitors treated with porcine factor VIII by home therapy. *Blood* 1990; **76**: 882–6.
13 Lusher JM. Inhibitor antibodies to factor VIII and factor IX: management. *Semin Thromb Hemost* 2000; **26**: 178–88.
14 Martinowitz U, Schulman S. Continous infusion of coagulation products. *Int J Pediatr Hematol Oncol* 1994; **1**: 47–8.
15 Woloschik DMN, Rubinger M, Israels JJ, Houston S, Schwetz N. In vitro stability of porcine factor VIII (Hyate:C®). *Haemophilia* 1997; **3**: 21–3.
16 Sjamsoedin LJ, Heijnen L, Mauser-Bunschoten EP et al. The effect of activated prothrombin-complex concentrate (FEIBA) on joint and muscle bleeding in patients with hemophilia A and antibodies to factor VIII. *N Engl J Med* 1981; **305**: 717–21.
17 Lusher JM, Shapiro SS, Palascak JE et al. Efficacy of prothrombin-complex concentrates in haemophiliacs with antibodies for factor VIII: a multicenter trial. *N Engl J Med* 1980; **303**: 421–5.
18 Negrier C, Goudemand J, Sultan Y et al. and the members of the French FEIBA Study Group. Multicentre retrospective study on the utilization of FEIBA in France in patients with factor VIII and factor IX inhibitors. *Thromb Haemost* 1977; **77**: 1113–19.
19 Lusher JM. Thrombogenicity associated with factor IX complex concentrates. *Semin Hematol* 1991; **28** (Suppl. 6): 3–4.
20 Chavin SI, Siegel DM, Rocco TA Jr, Olson JP. Acute myocardial infarction during management with an activated prothrombin complex concentrate in a person with factor VIII deficiency and factor VIII inhibitor. *Am J Med* 1988; **85**: 245–9.
21 Hedner U, Kisiel W. Use of human factor VIIa in the treatment of two hemophilia A patients with high-titer inhibitors. *J Clin Invest* 1983; **71**: 1836–41.
22 Lusher JM, Roberts HR, Davignon G et al. and the rVIIa Study Group. A randomized double-blind comparison of two dosage levels of recombinant factor VIIa in the treatment of joint, muscle and mucocutaneous haemorrhages in persons within haemophilia A and B, with and without inhibitors. *Haemophilia* 1998; **4**: 790–8.
23 Lindley CM, Sawyer WT, Macik G et al. Pharmacokinetics and pharmacodinamics of recombinant factor VIIa. *Clin Pharmacol Ther* 1994; **55**: 638–48.
24 van Leuven EF, Mauser-Bunschoten EP, van Dijken PJ et al. Disappearance of factor VIII:C antibodies in patients with haemophilia A upon frequent administration of factor VIII in intermediate or low dose. *Br J Haematol* 1986; **64**: 291–7.

25 Nilsson IM, Hedner U. Immunosuppressive treatment in haemophiliacs with inhibitors to factor VIII and IX. *Scand J Haematol* 1976; **16**: 369–82.
26 Nilsson IM, Berntorp E, Zetterwall O. Induction of immune tolerance in patients with haemophilia and antibodies to factor VIII by combined treatment with intravenous IgG, cyclophosphamide and factor VIII. *N Engl J Med* 1988; **328**: 947–50.
27 Green D. Suppression of an antibody to factor VIII by a combination of factor VIII and cyclophosphamide. *Blood* 1971; **37**: 381–7.
28 Colvin BT. Role of plasma-exchange in the management of patients with factor VIII inhibitors. *La Ricerca Clin Lab* 1983; **13**: 85–93.
29 Révész T, Mátyus J, Goldschmidt B, Harsanyi V. Control of life-threatening bleeding by combined plasmapheresis and immunesuppressive treatment in a haemophilic with inhibitors. *Arch Dis Child* 1980; **55**: 641–3.
30 Nilsson IM, Jonsson S, Sundqvist S-B, Ahlberg A, Bergentz SE. A procedure for removing high titer antibodies by extracorporeal protein-A-sepharose adsorption in haemophilia: substitution therapy and surgery in a patient with haemophilia B and antibodies. *Blood* 1981; **58**: 38–43.
31 Theodorsson B, Hedner U, Nilsson IM, Kisiel W. A technique for specific removal of factor IX antibodies from human plasma: partial characterization of the alloantibodies. *Blood* 1983; **61**: 973–81.
32 Warrier I, Lusher JM. Development of anaphylactic shock in hemophilia B patients with inhibitors. *Blood Coag Fibrinolysis* 1988; **9**: S125–8.
33 Warrier I, Ewenstein BM, Koerper MA et al. Factor IX inhibitors and anaphylaxis in hemophilia B. *J Pediatr Hematol Oncol* 1997; **19**: 23–7.
34 Santagostino E, Gringeri A, Mannucci PM. Home treatment with recombinant activated factor VII in patients with factor VIII inhibitors: the advantages of early intervention. *Br J Haematol* 1999; **104**: 22–6.
35 Key NS, Aledort LM, Beardsley D et al. Home treatment of mild to moderate bleeding episodes using recombinant factor VIIa (Novoseven) in haemophiliacs with inhibitors. *Thromb Haemost* 1998; **80**: 912–18.
36 Cooper T, Wilson F, Pace B, Yang Y-M. Successful treatment of recurrent bleeding event using recombinant factor VIIa (rVIIa) provided by home nursing services in a non-compliant haemophilia patient with inhibitors. *Thromb Haemost* 2001; July Suppl: Abstract CD3374.
37 Teitel JM. Recombinant factor VIIa versus aPCC in haemophiliacs with inhibitors: treatment and cost considerations. *Haemophilia* 1999; **5** (Suppl. 3): 43–9.
38 Berntop E, Astermark J. Comparison of the efficacy of bypassing agents in patients with inhibitors. *Thromb Haemost* 2001; July Suppl: Abstract SY2411.
39 Kreuz W, Escuriola-Ettinghausen C, Martinez I et al. Effective and safe long-term prophylaxis in patients with high-titer inhibitors using factor VIII inhibitor bypass activity (FFIBA). *Thromb Haemost* 2001; July Suppl: Abstract P2542.
40 Centra M, Mancuso G, Morfini M et al. APCC (FEIBA) home treatment prophylaxis: an opportunity for inhibitor hemophilia A patients. *Thromb Haemost* 2001; July Suppl: Abstract P2546.
41 Saxon BR, Jory CB. Effective prophylaxis with daily rFVIIa (NovoSeven) in a child with high titer inhibitors and a target joint. *Haemophilia* 2000; **6**: 299.

General medical management of bleeding episodes: haemarthroses, muscle haematomas, mucocutaneous bleeding and haematuria

M. Quintana-Molina, V. Jimenez-Yuste, A. Villar-Camacho and F. Hernandez-Navarro

The development of inhibitors against factor VIII and/or factor IX is one of the most important complications of substitutive therapy for haemophilia, and a considerable problem in the treatment of these patients. The incidence of inhibitors in patients with haemophilia A is 21–33%, but less frequent in those with haemophilia B [1–3]. The incidence of inhibitors is more frequent during the 5 years after the first administration of the deficient factor [4].

The appearance of inhibitors reduces the average life of factor VIII/IX concentrates, which makes the patient become refractory to factors VIII/IX. The main consequence is a reduction in the patient's quality of life or even their life expectancy because, clinically, the alternative treatments used are not as effective as the factor concentrates used in patients without inhibitors.

The occurrence of inhibitors is a result of the interaction of different factors: specific mutations in factors VIII or IX and the immune response [5–7].

There are many factors responsible for the appearance of inhibitors [8]: duration of exposure, cumulative doses, exposure patterns, use of different products, product changes and immunogenicity of the products.

Most patients with haemophilia are routinely screened once a year for the development of inhibitors. Clinically, if the patient who had an excellent response in the past to the factor and/or the cryoprecipitate with a haemorrhage resolution presents a poor response, the appearance of an inhibitor is to be expected. Studies of mixes with normal plasma can show an activated partial prothrombin time (APPT) initial correction, but after 1 h incubation the APPT will be extended. A study of the recovery and the recovery and half-life can be carried out; these will appear considerably reduced in the case of an inhibitor.

The quantification of an inhibitor is carried out by the Bethesda test [9]. Several plasma dilutions of patient's plasma are added to constant amounts of factor VIII and incubated for 2 h at 37°C. One Bethesda unit (BU) of inhibitor is defined as the quantity of antibody needed to destroy 50% of factor VIII in normal pooled plasma.

Low-response inhibitors can be transitory, disappearing spontaneously. Those with a high response usually present with high titres, developing an anamnestic response when factor concentrate is given. In haemophilia patients with a low-responding inhibitor (<5 BU), the treatment can be with deficient factor concentrates. The dose is increased to neutralize the inhibitor; any type of concentrate can be used.

In patients with high-responding inhibitors, the main short-term objective is to control the bleeding events and in the long term to eradicate the inhibitor. Several therapeutic strategies can be used:
1 Replacement therapy:
- human factor VIII or IX concentrates;
- porcine factor VIII concentrates;
- bypassing concentrates which avoid the inhibitor effect; and
- recombinant factor VIIa concentrates.
2 Therapy to decrease the level of the antibody.
3 Therapy to decrease or prevent the development of inhibitors.
4 Immunodepletion of plasma in patients.

Replacement therapy

Human factor VIII or IX concentrates

The effect depends on the inhibitor titre and on the patient's immunological response. This is the treatment of choice for patients who have a low response to the antigenic stimulus. It can also be used when the inhibitor titre is <10 BU, taking into account that a higher dose than in conventional haemophilia treatments will be used. First, a dose to neutralize the inhibitor (1 U concentrated factor/kg/inhibitor unit), then an appropriate amount of factor VIII, depending on the clinical situation.

Use of factor VIII concentrates in patients with inhibitors who have a normal anamnestic response will be followed by an anamnestic rise of the inhibitor, reaching high levels. For this reason, factor VIII concentrate is only used in these patients for serious bleeding problems or for surgery, because the effect only covers a short period (5 or 6 days) [10].

Porcine factor VIII concentrates

It is known that the inhibitor against the human factor VIII is neutralizing. Factor VIII of other animal species is more allergenic. Porcine factor VIII concentrates have been mainly used for therapeutic purposes. Their usage was abandoned for many years because of allergic reactions and thrombopenias. Lately, the availability of a porcine factor VIII, obtained by chromatography of an ionic refill with polyelectrolytes, has become important in the treatment of patients with inhibitors [11]. Before it is used, a patient's cross-reactivity should be determined. There is usually a cross-reaction when the level of inhibitor is high, so porcine factor VIII will not be effective [12].

The recommended therapeutic regimen (Kernoff et al. [11]) is an initial dose of 10–100 U/kg, depending on the severity of the bleeding and the titre of the inhibitor. The subsequent doses are related to the response to factor VIII or the desired post infusion outcome. Its use is recommended in patients who have an inhibitor titre of between 10 and 15 BU. It must be noted that porcine factor VIII is only approved for use in inpatient treatment in the USA and in the UK for home treatments [13,14].

Prothrombin complex concentrates

The therapeutic success of prothrombin complex concentrates was initially reported in 1969. At first non-activated prothrombin complex concentrates (PCC) were used, then different procedures to obtain concentrate and/or to prevent activation were developed.

Activated prothrombin complex concentrates (aPCC) were prepared. There are two on the market: Feiba™ (Baxter Hyland, Glendale, CA, USA) (factor eight inhibitor bypassing activity) and Autoplex™ (anti inhibitor coagulant complex) (Immuno, Austria). In the Haemophilia Centre at La Paz University Hospital, Madrid, following an original therapeutic plan (Table 12.1), 78% success rates were obtained in mild to moderate musculoskeletal bleeding (Table 12.2).

Although aPCCs are not as efficient as factor VIII/IX concentrates in haemophilic patients without inhibitors, it has become one of the most popular therapies in many treatment centres for haemarthrosis and haematomas and also for elective surgery.

Table 12.3 shows the therapeutic protocol at the Haemophilia Centre in La Paz Hospital, Madrid, for the type of inhibitor response, the bleeding severity and the patient's

Table 12.1 Therapeutic protocol with activated prothrombin complex concentrates (aPCC) depending on the severity of bleeding in the Haemophilia Centre at La Paz Hospital, Madrid.

Light bleeds	50–100 U/kg in a single dose
Moderate to severe bleeds	100 U/kg initial dose
	100 U/kg after 6 h
	Then 50 U/kg every 12 h
	After 24 h check continuity depending on the results
Very severe bleeds	100–200 U/kg every 6–8 h

Table 12.2 Results obtained with aPCC with the original therapeutic protocol on the treatment of haemarthroses and haematomas in patients with haemophilia at the Haemophilia Centre at La Paz Hospital, Madrid.

Therapeutic protocol	
First injection (0 h)	100 U/kg
Second injection (6 h)	100 U/kg
Third injection (12 h)	50 U/kg
Fourth injection (24 h)	50 U/kg
After 24 h the treatment is revised depending on results	
Results	
Excellent	Spontaneous pain disappears by 6 h
Good	Pain disappears between 6 and 24 h, and mobility is recovered at 24–48 h
Doubtfully efficient	Pain disappears between 24 and 49 h. Mobility is recovered after 48 h
Inefficient	Lack of response

Results were excellent in 78% of cases for the treatment of haemarthrosis and haematomas.

Table 12.3 Therapeutic protocol for the treatment of patients with inhibitors at the Haemophilia Centre at La Paz Hospital, Madrid.

Haemorrhage	Light	Moderate	Severe
Inhibitor (low response)	FVIII:C	FVIII:C	FVIII:C
Inhibitor (high response)	aPCC	aPCC	Plasmapheresis + FVIII:C or aPCC

inhibitor titre. The main aim is to maintain the inhibitor titre at the lowest possible level, avoiding the use of factor VIII concentrate, reserving its use for life-threatening bleeding episodes.

It is important to point out that there is no clinical evidence of hypercoagulability with the use of aPCC. There are cases of an anamnestic response, limited to some patients. The mechanism of action is not very clear.

Activated recombinant factor VII concentrates

Factor VII makes a complex with tissue factor in order to activate factors IX or X. Because the tissue factor is only exposed at the site of injury, factor VIIa will only create a complex in these places, bringing about efficient haemostasis. The risk of systemic activation and DIC (disseminated intravascular coagulation) is avoided.

Recently it has been demonstrated that the therapeutic effect of factor VIIa [18,19] can be independent from the tissue factor. Monocytes can support the activation of factor X through factor VIIa without the tissue's activation to high doses of factor VIIa [20]. Clinically, highly purified factor VIIa plasma has been very efficient in the treatment of patients with inhibitors.

Most recently, factor VIIa has been used in the treatment of bleeding in patients with inhibitors in surgery [18]. Lately, activated recombinant FVII (rFVIIa) is being very successfully used for the home treatment of patients with inhibitors [21].

The Haemophilia Centre at La Paz University Hospital, Madrid, uses the following therapeutic protocol:
- initial dose of 120 µg/kg;
- repeat the dose after 2 and 4 h; and
- reinforcement dose after 6 h to avoid bleeding.

Other doses are administered depending on the clinical evaluation. rFVIIa is given simultaneously with tranexamic acid in a dosage of 500–1000 mg intravenously every 8 h. Another route of administration of rFVIIa is by continuous infusion [22].

Therapies to reduce the level of circulating inhibitor

1 Intensive plasmapheresis: repeating plasmapheresis can reduce the inhibitor titres to such a level that can be neutralized with factor VIII/IX concentrates.
2 Extracorporeal immune adsorption: the antibody can be removed by extracorporeal adsorption to staphylococcal protein A, although it can absorb only antibodies IgG1 and IgG3, and not IgG4.

Table 12.4 Immune tolerance with high doses of factor (Bonn).

Phase I
100 U FVIII/IX/kg every 12 h
50 U Feiba/kg every 12 h
Until the inhibitor <1 BU

Phase II
100 U FVIII/IX/kg every 12 h
Until the inhibitor disappears

Phase III
100 U FVIII/IX/kg every 24 h
Until the average half-life of the FVIII/IX is normal

Phase IV
Treatment with FVIII/IX in case of haemorrhage

Therapies to reduce or prevent the development of inhibitors

Methods are based on obtaining immune tolerance to factor VIII/IX. The inefficacy of immunosuppressive drugs to suppress the cell producer of the antibodies has been proved.

The methods mainly used are:
1 immune tolerance with high doses; and
2 immune tolerance with low doses.

The mechanisms by which the prolonged administration of factor VIII/IX might induce immune tolerance are not fully understood. When high doses of antigen are used, it is assumed that this can overcome the antibody produced by the specific B lymphocyte clone. When low doses of antigen are used, the immune complex formed by factor VIII and the antibody will produce an intensive suppressor effect and maintain the permanent exposition of the antigen, and immune tolerance will be induced. This effect can be achieved by the use of immunosuppressor drugs or by the union of antigen and immunoglobulins.

Immune tolerance with high doses of factor [23] was achieved by the Brackmann protocol (Table 12.4). Although efficient, this protocol is very expensive, complicated and demanding on the patient. Other authors have tried to obtain the same results with inferior quantities of concentrates with or without the addition of steroids.

Immune tolerance with low doses of factor has also been used. Patients with haemophilia and inhibitors have been treated in the same way as patients without antibodies, although with replacement therapy with higher doses than the norm [24]. This study has developed a protocol with good results in patients with moderate inhibitor titre or those in whom the development of inhibitors was recent. The therapy comprises injecting 25 U/kg/day until there is a reduction of the inhibitor titre. The doses are then reduced, depending on the rate of recovery of factor VIII, to a dose of 10 U/kg twice or three times a week. The protocol is suspended if after two weeks of treatment the inhibitor titre rises.

Table 13.1 Therapeutic options for bleeding in patients with inhibitors.

Product	Contents	Approved by FDA for this indication	Efficacy in haemorrhages	Side-effects
PCC	FII, FVII, FIX, FX, varying concentrations	No	+	Thrombotic, potential viral transmission
aPCC	As above + FIIa, FVIIa, FIXa, FXa	Yes	++	Thrombotic, potential viral transmission
Porcine FVIII	Porcine FVIII, 20–35 IU/mL	No	− → +++, depending on inhibitor titre	Allergy, thrombocytopenia, potential viral transmission
rFVIIa	Recombinant human FVIIa, 0.6 mg/mL = 30 KIU/mL	Yes	++ → +++, depending on delay	Thrombotic

aPCC, activated prothrombin complex concentrates; FDA, Food and Drug Adminstration (USA); PCC, prothrombin complex concentrates; rFVIIa, activated recombinant factor VII.

to porcine FVIII is invariably less than 100% and in 85% of the patients so low (<5 Bethesda units, BU) that it would most likely provide effective haemostasis [11]. In a worldwide survey of treatment of hospitalized patients with porcine FVIII (Hyate:C™, Speywood Pharmaceuticals Ltd, Wrexham, UK; currently Beaufour Ipsen, Berks, UK) for 491 haemorrhagic episodes, it gave a fair–excellent response in over 90% of cases [12], with insufficient effect mainly being associated with high antiporcine inhibitor titres. Unfortunately, during the past few years porcine FVIII concentrate has been virtually unavailable because of problems with porcine parvovirus and restrictions on importation from the UK related to foot-and-mouth disease.

The most recent addition to the therapeutic armament for inhibitor patients is rFVIIa (approved as NovoSeven™, Novo Nordisk, Bagsvaerd, Denmark), which was used for the first time in patients in 1988 [13]. In the following compassionate use programme, the investigators assessed this treatment as excellent or effective in 79% of haemarthroses and 65% of muscle bleeds [14]. However, with home treatment the first dose can be injected earlier, and when this was performed within 8 h of the onset of symptoms, treatment with rFVIIa was excellent or effective for peripheral intramuscular haemorrhages in 92% [15]. The mean duration from onset was actually only 1.2 h in this study, which is ideal but may not be attained in an unselected cohort of patients with varying compliance. These results have been corroborated in another study (Nosepac) with home treatment, where global efficacy was assessed as excellent or efficient in 90% [16].

None of the randomized or larger cohort studies on these factor concentrates have included patients with limb- or life-threatening bleeds. Therefore, our concept of efficacy in such situations is based on a mixture of extrapolations from the above-mentioned studies on joint bleeds and less severe muscle bleeds together with case reports and small series of serious haemorrhages. Whereas haemarthroses typically are treated sufficiently with one or two doses of almost any factor concentrate, life- or limb-threatening bleeds will, in the majority of cases, require repeated dosage over many days. In haemophilia A patients without an inhibitor, the dose of FVIII for a joint or muscle bleed is usually between 15 and 30 IU/kg, whereas for life- or limb-threatening bleeds it is increased to about 50 IU/kg. Similarly, the dose of aPCC can vary for the different types of haemorrhages from 25 to 100 IU/kg. On the other hand, the recommended dose of rFVIIa is standardized to 90 µg/kg for all kinds of events, although both higher and lower doses have been tried [17,18]. Finally, for porcine FVIII the dose is mainly dependent on the titre of the cross-reacting inhibitor, from which it can be mathematically derived [12], but also on the type of haemorrhage.

Side-effects of treatment

An adverse complication of PCC and aPCC, particularly in cases of repeated injection of high doses, is hyperactivation of coagulation, which can lead to disseminated intravascular coagulation (DIC) [9,19] and/or myocardial infarction [20], which have occurred in young patients without predisposing pathology in the blood vessel walls. Thrombotic complications have also been described in a few patients treated with rFVIIa [21]. The first case was in a patient with a massive thigh muscle haematoma, necrosis and infection and previous unresponsiveness to PCC and porcine FVIII [22]. rFVIIa was initially effective but the patient developed DIC and fatal haemorrhage. Older patients with acquired haemophilia and atherosclerosis are probably at higher risk for thrombotic complications.

Porcine FVIII has previously been associated with allergic reactions and frequent thrombocytopenia but, with improved purification of the product, a survey demonstrated that adverse reactions occurred in only 2.3% of 491 hospital-treated episodes and in 0.64% of regularly treated patients [23]. A decrease in the platelet count by at least $20 \times 10^9/L$ was observed after 61% of infusions, but this was not associated with an increased bleeding tendency [23].

In the choice of therapeutic agent for a major bleed in a patient with an inhibitor, the expected efficacy must be weighed against the risk of serious side-effects.

Intracranial bleeds

Previous experiences

In general, intracranial haemorrhage is associated with a poor prognosis, with fatal outcome in four of seven patients in one study [24]. In other studies a slightly lower mortality rate of 20–35% was reported, but 50% of the surviving patients had residual neurological deficits [25–27]. Obviously, the best alternative is to prevent this kind of bleeding by prophylactic replacement therapy early after a head trauma, which is the typical aetiology in very young patients but is only identified in 50% of older patients with haemophilia [25]. Neurosurgical intervention has been recommended in cases where there is lack of improvement in the condition within a few hours after replacement therapy [24]. However, this may not be a suitable alternative for patients with an inhibitor.

It is difficult to compare the efficacy of the different therapeutic agents for intracranial haemorrhage in inhibitor patients, because of the lack of specific studies and the paucity of data from larger studies, including all kinds of bleeds. The aPCC-product Feiba™ was used for a total of four bleeds in the central nervous system in two large studies [10,28] and the treatment was always effective. The problem is that high doses have to be given frequently over many days, and the anticipated thrombotic side-effects, described above, have been reported in some of the cases with intracranial bleeds treated with Feiba™ [29,30].

When porcine FVIII was used for central nervous system bleeds one out of four patients did not respond because of a high titre inhibitor [31].

Recombinant activated factor VII

In view of the partial efficacy and side-effects with these therapies for central nervous system bleeds, it has been a natural development to evaluate rFVIIa in this respect. One of the first cases was a patient with haemophilia A and a high titre inhibitor with an intracranial haemorrhage, who received rFVIIa at the regular dose of 90 μg/kg every 3 h for 2 days, followed by every 4 h for the next 6 days with full recovery and no side-effects [32]. The first reported series consisted of five intracranial haemorrhages in two children with haemophilia A and an inhibitor [33]. The rFVIIa regimen always consisted of an initial bolus of 90–95 μg/kg and subsequently 60 μg/kg every 3 h for the first event and 90 μg/kg every 2 h for the other events, and slowly tapered down. The efficacy for the central nervous system bleeds was excellent but, during four of these treatment courses, bleeding started from other sites (nose, joints, around central line). Addition of tranexamic acid stopped two of those bleeds. One of the patients developed transient seizures during the treatment course for his first event of subdural haematoma. The second patient also had seizures during his second episode of intracerebral haemorrhage and 8 days later ataxia and upward-gaze nystagmus were observed but subsided during the following 6 weeks without any abnormalities seen on computed tomography or magnetic resonance imaging of the brain and brainstem. The fibrinogen level decreased by 20–40% during these treatments, and a similar reduction of plasminogen activity was observed during two of the treatment courses.

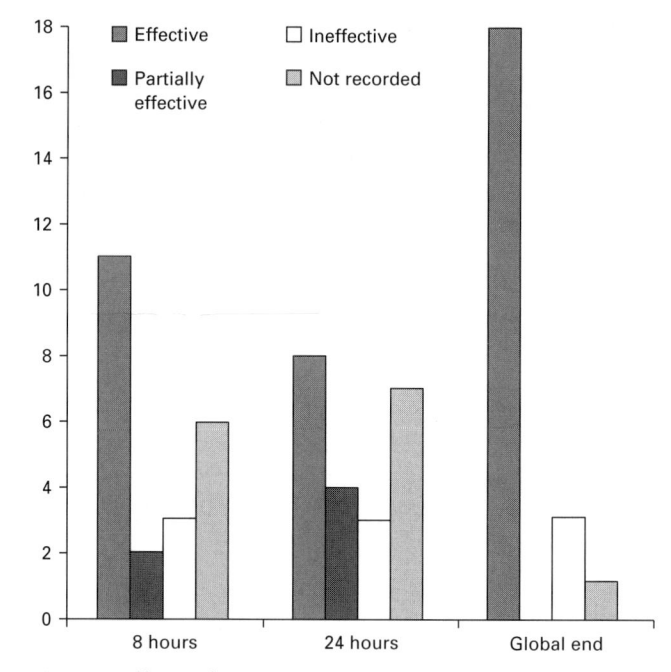

Fig. 13.1 Efficacy of rFVIIa in 22 central nervous system haemorrhages in 16 haemophilia A patients. (Adapted from Rice & Savidge [34].)

Subsequently, a series of 22 central nervous system bleeds treated with rFVIIa in two of the compassionate use studies has been presented [34]. The series included 16 patients with haemophilia A and two with haemophilia B, both groups with inhibitors. The bleeds were subdural (almost half), subarachnoid, intracerebral or spinal extradural. In at least six cases other therapies had been tried initially. The doses varied in size (mean 80–100 μg/kg) and number (2–332!). The efficacy in the 22 haemorrhages in the 16 haemophilia A patients is shown in Fig. 13.1; overall it was 82%. For three bleeds among the patients with haemophilia the treatment was assessed as ineffective and one of those—a patient with haemophilia A—died.

The indications are that treatment with rFVIIa is the best option for inhibitor patients with central nervous system bleeds, in view of the low mortality and low number and manageable nature of the side-effects. It should be realized that in a compassionate use programme some of the patients are in a very poor condition after failure of conventional therapy and, in a comparison of rFVIIa used as first-line therapy vs. salvage therapy within the compassionate use programme, efficacy was definitely worse among the latter [35].

Internal bleeds

Previous experiences

In analogy with the experience from intracranial bleeds there are very limited data on the efficacy of older therapies for other types of internal haemorrhage. There may be a bias for publishing case reports with a successful outcome, but those products, including PCC [36], can definitely be life-saving to a certain extent. On the other hand, the second patient who received rFVIIa can represent the large number of such patients in the compassionate use programme with rFVIIa, where this was used as salvage therapy. This patient with haemophilia A had a high titre inhibitor against FVIII and developed a large retropharyngeal haemorrhage, precipitated by an upper respiratory tract infection [18]. Despite treatment with seemingly adequate doses of aPCC (100 IU/kg Autoplex™ every 8 h) breathing became increasingly difficult because of tracheal compression by the haematoma and neck pain increased. Asphyxia was imminent when treatment with rFVIIa at a dosage of 60 µg/kg every 3–4 h and then every 6 h resulted in improvement after 12 h and eventual complete recovery.

Recombinant activated factor VII

In a report of the compassionate use programme using rFVIIa, the results of treatment of a relatively large number of internal bleeds are included [37]. Although two of the 43 patients had congenital factor VII deficiency, their data cannot be distinguished from the rest. An overview of the material and treatment is shown in Table 13.2. In many cases the treating physician considered the therapy as life-saving. Two patients had fatal outcomes which, according to the investigator, were unrelated to treatment with rFVIIa as one was caused by congestive heart failure in an 89-year-old man and the other by lack of availability of sufficient rFVIIa and continued haemorrhage with complicating renal failure. Thromboembolism or DIC were not seen in any of the patients in this report.

Continuous infusion with activated recombinant factor VII

So far none of the manufacturers of any factor concentrate has included continuous infusion (CI) as a mode of administration in the application for approval. In spite of this, CI has become routine at many haemophilia treatment centres when high doses of the missing factor are needed for several days, especially after surgery. For patients with inhibitors the situation is slightly different, however, as virtually all products except possibly PCC have been tested in CI. For rFVIIa this is of special interest, because of the short half-life of only 2.7 h [38]. Normally, rFVIIa has to be given as frequent bolus injections every 2–3 h with meticulous timing. By using CI with a reliable infusion pump, convenience can be improved and total doses and costs potentially reduced by eliminating the peak concentrations, as previously reported and reviewed [39–41]. There have also been several reports on failures with CI of rFVIIa, many of these for obvious reasons but occasionally without, and it seems that some centres have been more successful than others with this mode of administration.

Within the framework of the rFVIIa-CI Group we have now updated the data up to 31 August 2001 through reports from some of the participants, supplemented with a literature search on MEDLINE and abstracts from major conferences on haemophilia or thrombosis and haemostasis. We have only included data on life- or limb-threatening bleeds in patients with haemophilia (A or B) with inhibitors, or acquired haemophilia with an inhibitor against FVIII. The type of data collected corresponds to previous surveys [40,41].

We have so far only registered a single case treated with rFVIIa in CI for a central nervous system bleed [42]. There have been 23 other internal haemorrhages and major muscle bleeds, which are summarized in Table 13.3. These occurred in 23 patients: 18 with haemophilia A and inhibitor; four with haemophilia B and inhibitor; and one with acquired haemophilia A. One of the iliopsoas bleeds was bilateral and massive, requiring the longest treatment duration of 17 days [43]. Overall, 20 of 23 treatment episodes were evaluated as corresponding to excellent–effective; two had a partial response and in one case the treatment had no effect and, because there was no cross-reactivity, treatment was switched to porcine FVIII with a successful outcome [44]. Although the treatment for a forearm compartment syndrome was effective, the patient developed gross haematuria, which continued for a few days and then ceased while still on CI with rFVIIa.

Thus, the overall results from CI appear similar to those with

Table 13.2 Treatment of internal haemorrhage with rFVIIa. (Adapted from Lusher [37].)

Site of bleeding	Patients	Bleedings	Doses of rFVIIa/episode, median (range)	Excellent–effective response in global evaluation (%)
Gastrointestinal tract	12	18	30 (1–265)	83
Iliopsoas muscle	12	12	76 (15–284)	83
Other intra-abdominal or retroperitoneal	13	13	43 (24–875)	62
Renal tract	4	4	55 (9–875)	50
Other internal	4	4	38 (6–77)	75

Table 13.3 Treatment of internal haemorrhage or limb-threatening haemorrhage with rFVIIa in continuous infusion.

Site of bleeding	Bleedings	Plasma level of FVII (actual or target), IU/mL	Duration of therapy, median (range)	Excellent or effective response	Complication
Gastrointestinal tract	4	10–15	5.5 (1–10)	4/4	
Iliopsoas muscle	4	10–25	13.5 (10–17)	4/4	
Other intra-abdominal or retroperitoneal	4	10–15	4 (3–4)	1/4	2 assessed as partially effective, 1 ineffective
Renal	1	≥10	13	1/1	
Compartment (forearm, calf or massive thigh)	8	10–30	6.5 (1–12)	8/8	One developed gross haematuria day 7
Other (neck, tongue)	2	10–12	6 (6)	2/2	

bolus injections with rFVIIa for this type of haemorrhage. However, we would like to repeat previous words of caution, that CI should be used only by centres with experience thereof or in close cooperation with such expertise. There are data in favour of using a fibrinolytic inhibitor such as tranexamic acid concomitantly with rFVIIa [40]. In case of extensive trauma there is most likely a benefit of giving a few bolus doses of rFVIIa before switching to CI to ensure the required bursts of thrombin generation. A pharmacokinetic evaluation is not possible prior to treatment in these emergent cases, but the factor VII level in plasma should instead be checked early and repeatedly together with the prothrombin time. The infusion pump and the venous access must be reliable. There must be no further dilution of rFVIIa other than with the water provided by the manufacturer for reconstitution, otherwise the product will lose activity. Likewise, heparin should not be added to the infusion container. To avoid local thrombophlebitis at the infusion site a parallel infusion with saline at a rather high rate is the best alternative. Finally, the pump also needs surveillance and somebody to react appropriately when its alarm goes off.

Conclusions

Over the past three decades the treatment for life- or limb-threatening bleeds in patients with inhibitors has evolved from a total lack of treatment via partially effective PCC and probably more effective aPCC; however, both are associated with hyperactivation of the systemic coagulation at repeated and high dosage. Therapy has now moved to more selective treatment with rFVIIa which usually provides excellent haemostatic efficacy, especially when used early as the first-line therapy and this should therefore be the treatment of choice for life- or limb-threatening bleeds.

Appendix

The members of the rFVIIa-CI Group are as follows. *Australia* J. Lloyd, Institute of Medical and Veterinary Science, Adelaide; J. McPherson, University of Newcastle, Newcastle. *Belgium* K. Peerlinck, University of Leuven, Leuven. *Canada* B. Ritchie University of Alberta, Edmonton. *Finland* A. Mäkipernaa, Tampere University Hospital, Tampere. *France* R. D'Oiron, Hôpital Bicêtre, Le Kremlin-Bicêtre; M-E Briquel, Centre Hospitalier Universitaire de Nancy, Nancy; J. Goudemand, Hôpital Huriez, Lille; C. Négrier, Edouard Herriot Hospital, Lyon. *Germany* H.J. Siemens, Med. Universität Lübeck, Lübeck; C. Escuriola, Dr von Haunersches Kinderspital, Munich; R. Grossmann, University of Würtzburg, Würtzburg. *Israel* U. Martinowitz and G. Kenet, Sheba Medical Centre, Tel Hashomer. *Italy* E. Satagostino, A. Bianchi Bonomi Hemophilia and Thrombosis Centre, Milan; M. Morfini, Ospedale di Careggi, Florence; V. De Mitrio, University of Bari School of Medicine, Bari; A. Rocino, Ospedale Nuovo Pellegrini, Naples; F. Baudo, Niguarda Hospital, Milan; G. Tagariello, Ospedale di Castelfranco Veneto. *Slovakia* A.A. Botarova, University Hospital, Bratislava. *Spain* J. Tusell, Hospital Vall d'Hebron, Barcelona; J.I. Lorenzo, Hospital La Fé, Valencia. *Sweden* P. Petrini and S. Schulman, Karolinska Hospital; Stockholm. *The Netherlands* E. Mauser-Bunschoten, Academisch Ziekenhuis Utrecht, Utrecht. *UK* J. Pasi, Royal Free Hospital, London. *USA* J. Lusher, Children's Hospital of Michigan, Detroit.

References

1 Rosendaal FR, Varekamp I, Smit C et al. Mortality and causes of death in Dutch haemophiliacs, 1973–86. *Br J Haematol* 1989; **71**: 71–6.
2 Triemstra M, Rosendaal FR, Smit C, Van der Ploeg HM, Briet E. Mortality in patients with hemophilia: changes a Dutch population from 1986 to 1992 and 1973 to 1986. *Ann Intern Med* 1995; **123**: 823–7.
3 Larsson SA, Wiechel B. Deaths in Swedish hemophiliacs, 1957–1980. *Acta Med Scand* 1983; **214**: 199–206.
4 Kerr CB. Intracranial haemorrhage in hemophilia. *J Neurol Neurosurg Psychiatry* 1964; **27**: 166–73.
5 Silverstein A. Intracranial bleeding in hemophilia. *Arch Neurol* 1960; **3**: 141–57.
6 Lusher JM, Shapiro SS, Palascak JE et al. and the Hemophilia

Study Group. Efficacy of prothrombin complex concentrates in hemophiliacs with antibodies to factor VIII: a multicenter therapeutic trial. *N Engl J Med* 1980; **303**: 421–5.
7. Sjamsoedin LJM, Heijnen L, Mauser-Bunschoten EP *et al*. The effect of activated prothrombin-complex concentrate (FEIBA) on joint and muscle bleeding in patients with hemophilia A and antibodies to factor VIII. *N Engl J Med* 1981; **305**: 717–21.
8. Lusher JM, Blatt PM, Penner JA *et al*. Autoplex versus Proplex: a controlled double-blind study of effectiveness in acute hemarthroses in hemophiliacs with inhibitors to factor VIII. *Blood* 1983; **62**: 1135–8.
9. Négrier C, Goudemand J, Sultan Y *et al*. Multicenter retrospective study on the utilization of FEIBA in France in patients with factor VIII and factor IX inhibitors. *Thromb Haemost* 1997; **77**: 1113–19.
10. Hilgartner MW, Knatterud GL and the Feiba Study Group. The use of factor eight inhibitor by-passing activity (FEIBA Immuno) product for the treatment of bleeding episodes in hemophiliacs with inhibitors. *Blood* 1983; **61**: 36–40.
11. Lloyd JV, Street AM, Berry E *et al*. Cross-reactivity to porcine factor VIII of factor VIII inhibitors in patients with haemophilia in Australia and New Zealand. *Aust N Z J Med* 1997; **27**: 658–64.
12. Hay C, Lozier JN. Porcine factor VIII therapy in patients with factor VIII inhibitors. In: Aledort LM, Hoyer LW, Lusher JM, eds. *Inhibitors to Coagulation Factors*. New York: Plenum Press, 1995: 143–51.
13. Hedner U, Glazer S, Pingel K *et al*. Successful use of recombinant factor VIIa in patient with severe haemophilia A during synovectomy [Letter]. *Lancet* 1988; **2**: 1193.
14. Mølskov Bech R. Recombinant factor VIIa in joint and muscle bleeding episodes. *Haemostasis* 1996; **26** (Suppl. 1): 135–8.
15. Lusher JM. Early treatment with recombinant factor VIIa results in greater efficacy with less product. *Eur J Haematol* 1998; **81** (Suppl. 63): 7–10.
16. Laurian Y, Goudemand J, Négrier C *et al*. Use of recombinant factor VIIa as first line therapy for bleeding episodes in hemophiliacs with factor VIII or IX inhibitors (Nosepac study). XVIth Congress of the International Society on Thrombosis and Haemostasis, Florence, 6–12 June 1997.
17. Kenet G, Lubetsky A, Gitel S *et al*. Treatment of bleeding episodes in patients with hemophilia and an inhibitor: comparison of two treatment protocols with recombinant activated factor VII. *Blood Coag Fibrinolysis* 2000; **11** (Suppl. 1): S35–8.
18. Macik BG, Hohneker J, Roberts HR, Griffin AM. Use of recombinant activated factor VII for treatment of a retropharyngeal hemorrhage in a hemophilic patient with a high titer inhibitor. *Am J Hematol* 1989; **32**: 232–4.
19. Kasper CK. Thromboembolic complications. *Thromb Diath Haemorrh* 1975; **33**: 640–4.
20. Gruppo RA, Bove KE. Fatal myocardial necrosis associated with prothrombin-complex-concentrate therapy in haemophilia A. *N Engl J Med* 1982; **309**: 242–3.
21. Green D. Complications associated with the treatment of haemophiliacs with inhibitors. *Haemophilia* 1999; **5** (Suppl. 3): 11–17.
22. Stein SF, Duncan A, Cutler D, Glazer S. Disseminated intravascular coagulation (DIC) in a hemophiliac treated with recombinant factor VIIa. *Blood* 1990; **76**: 438A.
23. Hay CRM, Lozier JN, Lee CA *et al*. Safety profile of porcine factor VIII and its use as hospital and home-therapy for patients with haemophilia-A and inhibitors: the results of an international survey. *Thromb Haemost* 1996; **75**: 25–9.
24. Martinowitz U, Heim M, Tadmor R *et al*. Intracranial hemorrhage in patients with hemophilia. *Neurosurgery* 1986; **18**: 538–41.
25. Bray GL, Luban NL. Hemophilia presenting with intracranial hemorrhage: an approach to the infant with intracranial bleeding and coagulopathy. *Am J Dis Child* 1987; **141**: 1215–17.
26. Eyster ME, Gill FM, Blatt PM *et al*. for the Hemophilia Study Group. Central nervous system bleeding in hemophiliacs. *Blood* 1978; **51**: 1179–88.
27. Yoffe G, Buchanan GR. Intracranial hemorrhage in newborn and young infants with hemophilia. *J Pediatr* 1988; **113**: 333–6.
28. Hilgartner M, Aledort L, Andes A, Gill J and members of the FEIBA Study Group. Efficacy and safety of vapor-heated anti-inhibitor coagulant complex in hemophilia patients. *Transfusion* 1990; **30**: 626–30.
29. Fukui H, Fujimura Y, Taahashy Y, Mikami S, Yoshioka A. Laboratory evidence of DIC under FEIBA treatment of a hemophilic patient with intracranial bleeding and high titre factor VIII inhibitor. *Thromb Res* 1981; **22**: 177–84.
30. Dalsgaard-Nielsen J. Successful treatment with FEIBA of a spontaneous intracranial bleeding in a hemophilic patient with factor VIII inhibitor. 16th International Congress of the World Federation of Haemophilia, Rio de Janeiro, 1984.
31. Brettler DB, Forsberg AD, Levine PH *et al*. The use of porcine factor VIII concentrate (Hyate C) in the treatment of patients with inhibitor antibodies to factor VIII: a multicenter US experience. *Arch Intern Med* 1989; **149**: 1381–5.
32. Stigendal L, Tengborn L. Intracerebral bleed and antibodies to factor VIII in a patient with haemophilia A: experience with recombinant factor VIIa (rFVIIa). Proceedings from the xx International Congress of the World Federation of Hemophilia 1992; **62**: Abstract 116.
33. Schmidt ML, Gamerman S, Smith HE, Scott JP, DiMichele M. Recombinant activated factor VII (rFVIIa) therapy for intracranial hemorrhage in hemophilia a patients with inhibitors. *Am J Hematol* 1994; **47**: 36–40.
34. Rice KM, Savidge GF. NovoSeven™ (recombinant factor VIIa) in central nervous system bleeds. *Haemostasis* 1996; **26** (Suppl. 1): 131–4.
35. Hay CRM, Negrier C, Ludlam CA. The treatment of bleeding in acquired haemophilia with recombinant factor VIIa: a multicentre study. *Thromb Haemost* 1997; **78**: 1463–7.
36. Kagawa H, Yasuzawa M, Ozaki Y *et al*. The effect of non-activated prothrombin complex concentrate (Proplex™ ST) on intraperitoneal hematoma in a hemophilia A patient with a factor VIII inhibitor. *Intern Med* 1996; **35**: 319–22.
37. Lusher JM and the NovoSeven™ Compassionate Use Investigators. Recombinant factor VIIa (Novoseven™) in the treatment of internal bleeding in patients with factor VIII and IX inhibitors. *Proceedings of New Aspects of Haemophilia Treatment*, 3rd Symposium, Copenhagen 1995.
38. Lindley CM, Sawyer WT, Macik BG *et al*. Pharmacokinetics and pharmacodynamics of recombinant factor VIIa. *Clin Pharmacol Ther* 1994; **55**: 638–48.
39. Schulman S, Bech Jensen M, Varon D *et al*. Feasibility of using recombinant factor VIIa in continuous infusion. *Thromb Haemost* 1996; **75**: 432–6.
40. Schulman S for the rFVIIa-CI Group. Safety, efficacy and lessons from continuous infusion with rFVIIa. *Haemophilia* 1998; **4**: 564–7.
41. Schulman S, d'Oiron R, Martinowitz U *et al*. Experiences with continuous infusion of recombinant activated factor VII. *Blood Coag Fibrinolysis* 1998; **9** (Suppl. 1): S97–101.
42. McPherson J, Sutcharitchan P, Lloyd J *et al*. Experience with continuous infusion of recombinant activated factor VII in the Asia-Pacific region. *Blood Coag Fibrinolysis* 2000; **11** (Suppl. 1): S31–4.

PART 5

Inhibitors in haemophilia B
Inhibitors in mild and moderate haemophilia A
Acquired inhibitors to FVIII

Factor IX inhibitors and anaphylaxis

I. Warrier

Anaphylaxis occurring at the time of alloantibody development is a unique problem in haemophilia B [1,2]. Alloimmunization against factor VIII (FVIII) or factor IX (FIX) is a devastating complication of factor replacement therapy in haemophilia [3,4]. Fortunately, inhibitor development is less common in haemophilia B than in haemophilia A [5] with most studies showing an inhibitor incidence of 1–3% in haemophilia B and 15–30% in haemophilia A [3,5]. The concurrent complication of anaphylaxis characterizes FIX inhibitor development while anaphylaxis is not a common problem in haemophilia A patients with inhibitors. Thus, there are clear differences between haemophilia A and B with regard to incidence of inhibitors and concurrent complications.

Incidence of factor IX inhibitors

Factor IX inhibitors occur 10 times less often in haemophilia B when compared to FVIII inhibitors in haemophilia A. Several hypotheses have been postulated to explain this difference in inhibitor incidence [6] (Table 14.1).

Smaller size of factor IX

Factor IX protein is fivefold smaller than factor VIII protein, thus has less epitopes against which antibodies can be formed. Although this may be true, currently no information is available regarding epitopes or their affinity.

Fewer patients with severe haemophilia B

The 1970 survey of US Hemophilia Treatment Centers showed that only 44% of haemophilia B patients had severe disease, while 60% of haemophilia A patients had severe disease [7]. More recent surveys by Katz in 1993 [8] and Warrier in 1997 [9] showed that only 30–37% had severe haemophilia B. Thus, at any time, there are fewer severe haemophilia B patients at risk for the development of inhibitors.

Table 14.1 Postulated reasons for low incidence of factor IX inhibitors.

Smaller size of factor IX
Fewer patients with severe haemophilia B
Genetic predisposition
Structural analogy to other vitamin K-dependent factors
Higher non-plasma of factor IX

Genetic predisposition

Genetic predisposition and presence of certain mutations determine the prevalence of inhibitors in haemophilia. Although 50% of all inhibitors occur in severe haemophilia B with total FIX gene deletions, these deletions account for only 1–3% of severe haemophilia B [7,10]. Severe haemophilia B patients with total absence of FIX antigen, as seen in complete FIX gene deletion, have the greatest risk for inhibitor formation. Green et al. [11] have demonstrated a 3% risk of inhibitors in haemophilia B if the mutation is unknown, but a near-zero risk for patients with single amino acid substitutions. Patients with frameshift, stop codons or splice-site mutations have a 20% risk for inhibitors. Thus it appears that the nature of the FIX mutation is an important predisposing factor and mutations leading to loss of coding information are more frequently associated with inhibitor development.

Structural analogy to other vitamin K-dependent factors

There is considerable conservation of amino acid sequence among the vitamin K-dependent factors, which include FIX. Thus, in the absence of FIX, presence of other vitamin K-dependent factors II, VII, and X may confer some tolerance to FIX infused.

Higher non-plasma level of factor IX

Factor IX protein content is much higher (5 µg/mL) when compared to FVIII protein (100 ng/mL). Although patients with severe haemophilia B may synthesize only very small amounts

Table 14.2 Characteristics of haemophilia B patients with inhibitors and allergy to factor IX.

Reactions occur generally early in life—median age 19.5 months (range 7–156 months)
Median number of exposure days 11 days (range 2–180 days)
Peak inhibitor titre—median 30 BU (range 1–950 BU)
All ethnic and racial groups are affected
Various products including FFP, PCC, aPCC, monoclonal and recombinant FIX have been used prior to inhibitor development and anaphylaxis
Complete gene deletions, stop codon abnormalities and missing exons are the most commonly observed genotypes

---

aPCC, activated prothrombin complex concentrates; BU, Bethesda units; FFP, fresh frozen plasma; PCC, prothrombin complex concentrates.

of FIX protein, that small amount of antigen is adequate for induction of tolerance.

International registry of factor IX inhibitors

Alloimmunization in haemophilia B occurs only after exposure to FIX-containing products. Similar to the FVIII inhibitors in haemophilia A, FIX antibodies in haemophilia B patients are detected either by routine surveillance or when the patients do not have the expected response to treatment. More recently, anaphylaxis occurring simultaneously with inhibitor development has identified a subset of patients with FIX inhibitors in haemophilia B [1].

Under the direction of the International Society on Thrombosis and Haemostasis (ISTH), a registry for FIX inhibitors has been developed. Currently there are known to be 46 haemophilia B patients worldwide who have both FIX inhibitors and allergy to FIX: 24 of these patients are from the USA and the other 22 are from Canada, Europe, Japan and Australia. In the majority of these patients there was a temporal relationship between inhibitor development and anaphylaxis. Other common features include:
1 high titre inhibitors;
2 allergy to alternate FIX products; and
3 genotyping demonstrating either deletion or stop codon abnormalities of the FIX gene.

The information collected on 30 of the 46 patients with FIX inhibitors and anaphylaxis has revealed the characteristics shown in Table 14.2.

Characteristics of FIX antibodies

Antibodies to FIX are not well characterized. From what is known, they are predominantly IgG4 subclass and do not activate complement. They have equal affinity to both heavy and light chains of FIX. An interesting finding reported from Japan was the presence of IgG1 antibody in addition to IgG4 at the time of anaphylaxis in FIX inhibitor patients [12]. IgG1 antibody was found to bind strongly to heavy chain of FIX suggesting that heavy chain of FIX may be involved with anaphylaxis. Some other studies have demonstrated that these reactions may be mediated by immunoglobulins of the IgE class by skin and radioallergosorbent test (RAST) testing [13].

Anaphylaxis with inhibitor development

Anaphylaxis at the time of inhibitor development in haemophilia B is a relatively newly observed phenomenon. The reasons for this complication are not clear at present. One of the reasons may be a better recognition of a previously existing complication. A second reason may relate to the availability of highly purified FIX products, which evoke a more vigorous immune response than other forms of treatment (fresh frozen plasma, FFP or prothrombin complex concentrates, PCC). Similar reactions have also been reported with recently licensed recombinant FIX (J. Gill, personal communication; FIX Registry, personal communication).

The exact pathogenesis of the anaphylactic reactions occurring at inhibitor development is largely unknown. However, the nature of gene mutations appear to be a contributing factor. Large gene deletions account for only 1–3% of haemophilia B patients, but constitute 50% of the mutations observed in patients with inhibitors. It has been suggested that complete gene deletions confer the greatest risk for anaphylaxis, with a minimum risk of 26% [14]. It has been stated that the complete absence of endogenous FIX protein results in inhibitor formation after exposure to exogenous FIX in replacement therapy, while deletion of neighbouring immunomodulatory genes may contribute to the development of anaphylaxis. Complete FIX gene deletions often extend a megabase or more beyond the FIX gene and thus codeletion of neighbouring genes is an attractive hypothesis to consider. However, why some unrelated patients have anaphylactic reactions while other related patients do not, is not known. It is possible that a subtle interplay of factors such as mutation type, environmental exposure—including type and frequency of FIX use—and genetic factors may contribute to the development of anaphylaxis in patients with FIX inhibitors.

Management

Management of haemophilia B patients with inhibitors and anaphylaxis is complicated and challenging and should only be attempted at a well-equipped haemophilia treatment centre with experienced staff. Treatment in these patients has two major goals:
1 to control acute bleeding episodes most effectively in an efficient manner; and
2 to eradicate the inhibitor by immune tolerance induction (ITI).

Treatment of an acute bleeding episode

Acute bleeding episodes in haemophilia B patients with inhibitors and allergy to FIX should be treated in a well-equipped haemophilia treatment centre with experienced staff. Anaphylaxis with the development of a FIX inhibitor makes it impossible for these patients to receive any FIX-containing products unless the patient is desensitized thoroughly. Desensitization has been attempted in many of these patients with no consistent response. Although a few patients are able to tolerate FIX after desensitization, the majority required premedication with antihistamines and steroids. Thus, FIX-containing PCC and activated PCC (aPCC) does not appear to be the treatment of choice in haemophilia B patients with inhibitors and allergy to FIX. For those patients with allergy, recombinant FVIIa (rFVIIa) appears to be the most appropriate treatment. The recommended dose of rFVIIa is 90–120 μg/kg every 2 h until bleeding stops. Intervals between doses can be increased once the bleeding has been controlled. Several of the patients in the registry have received rFVIIa for treatment of acute bleeding episodes with excellent results [1]. No serious adverse events with rFVIIa have been reported in haemophilia B patients with inhibitors [15]. Disadvantages of rFVIIa include:

1 the need for multiple doses of rFVIIa for treatment of a bleeding episode;
2 the requirement for an easy venous access (central venous line) and/or hospitalization (especially in infants and children);
3 no suitable easy laboratory test for monitoring response; and
4 last, but not least—the high cost of multiple doses of rFVIIa.

Eradication of the inhibitor by immune tolerance induction

In general, the experience with immune tolerance induction (ITI) in haemophilia B patients with inhibitor is limited. Thrombogenicity of PCC, the lack of availability of high purity FIX products until recently, the small number of patients and the lack of a standard method for ITI are some of the reasons for this limited experience. In addition, ITI has had minimal success in haemophilia B patients with inhibitor; in those with concurrent allergy, the success rate is even lower [1].

The only reported success with ITI in haemophilia B patients is from the late Professor Nilsson's group [16]. Utilizing the Malmö treatment protocol incorporating the removal of antibody by extracorporeal adsorption to protein A followed by cyclophosphamide IV or PO (orally), factor concentrate is given daily to maintain the patient's factor concentration at a level of 40–100 IU/dL for 2–3 weeks. Intravenous gammaglobulin is administered for 5 days starting on day 4 of ITI. Once the inhibitor is undetectable, patients are started on a prophylactic regimen with FIX infusions twice per week.

Seven haemophilia B patients with inhibitors underwent ITI on the Malmö protocol. None had anaphylaxis or allergic manifestations at the time of inhibitor development. Four out of seven haemophilia B patients with inhibitors responded well with no detectable antibody after 2–4 weeks of treatment. Two patients required two treatment courses to eradicate the inhibitors completely and the seventh patient failed ITI.

Disadvantages of ITI in haemophilia B patients with inhibitor and anaphylaxis include:
1 the need for desensitization to FIX prior to starting ITI; and
2 the development of nephrotic syndrome [17–21].

Nephrotic syndrome

This unique problem has been reported mainly in haemophilia B patients with inhibitors and allergy to FIX. We are aware of 12 cases of nephrotic syndrome on ITI: four are from the USA and the other eight are from Europe. Ten of 12 cases had allergy to FIX in addition to inhibitors. In each case, nephrotic syndrome typically developed after 8–9 months of ITI. Two of the 12 cases had a renal biopsy performed which showed membranous glomerulonephritis. Immunohistochemical stains on renal biopsy specimen from one patient showed no FIX-containing deposits.

Clinical features common to the patients with nephrotic syndrome on ITI include:

1 young age (<12 years);
2 anaphylaxis at the time of inhibitor development;
3 exposure to frequent and high doses of FIX (100–325 μ/kg/day);
4 FIX deficiency either because of deletion or stop codon abnormality;
5 poor response to steroids and cyclophosphamide;
6 resolution with decrease or discontinuation of FIX.

Despite the vastly larger clinical experience with ITI in haemophilia A patients with inhibitors, nephrotic syndrome has never been reported in haemophilia A patients undergoing ITI. Although there are no clear reasons to explain this difference, several hypotheses can be considered. One hypothesis relates to the smaller molecular size of FIX. Unlike higher molecular weight FVIII complexed with von Willebrand factor that stays confined to intravascular spaces, FIX with a smaller molecular size of 55 000 daltons distributes to extravascular spaces more readily. This unique feature of extravascular dissemination of FIX may be a contributing factor for the development of an immune complex-mediated reaction, leading to membranous glomerulonephritis (MGN).

A second hypothesis relates to the fact that haemophilia B patients are exposed to much larger amounts of exogenous protein when treated. This is because of the higher plasma concentration of FIX protein (5 μg/mL) when compared to FVIII (100 ng/mL) [7]. This protein load becomes even higher in patients on ITI and that in some way could influence the quantity or tissue distribution of immune complexes formed with FIX inhibitor.

A third hypothesis relates to the genetic predisposition of the patient for inhibitor development with concurrent anaphylaxis. Lack of tolerance because of total absence of any FIX gene product has been commonly cited as one of the reasons for inhibitor development. Whether there is also a deletion of immunomodulatory genes adjacent to FIX is not known at present.

The occurrence of 12 cases of nephrotic syndrome in haemophilia B patients with inhibitor on ITI, and biopsy-proven membranous nephropathy in two patients, suggests a direct causal relationship between ITI and nephrotic syndrome. The hypothesis of causal relationship is further strengthened by the observation that cessation of exposure to the offending agent alone resulted in resolution of the nephropathy.

Conclusions and practical suggestions

Management of bleeding episodes in haemophilia B patients with inhibitors with a history of anaphylaxis is extremely difficult and should only be attempted in a well-equipped haemophilia treatment centre with experienced staff. Now that rFVIIa is approved for use in patients with inhibitors most minor bleeding episodes can be managed on an outpatient basis with 1–2 doses of rFVIIa. Those who require more than 1–2 doses of factor per treatment episode (young patients) may require an easy reliable venous access (central venous line). It appears that rFVIIa is the most appropriate treatment for this group of patients. As the success rate with ITI is poor with high rates of complication (nephrotic syndrome) it is advisable not to start ITI especially when rFVIIa is readily available. However, if a decision is made to start a patient on ITI, routine urinalysis must be included on follow-up and the shortest possible method for inhibitor elimination must be considered.

In summary, anaphylaxis concurrent with inhibitor development is a unique problem in haemophilia B. This complication occurs in early life and is seen in patients with either FIX gene deletions or stop codon abnormalities. Response to ITI is poor in this group and nephrotic syndrome is seen as a complication of ITI. rFVIIa is the most likely treatment of choice in these patients although it has several disadvantages.

Although the reasons for the recently described problems in FIX inhibitor patients are not very clear, haematologists can use certain guidelines when treating young patients with severe haemophilia B. Guidelines include the following.

1 Identify the children at risk by obtaining molecular diagnosis of severe haemophilia B at the time of initial presentation. Those with large deletions or stop codon abnormalities can then be monitored closely during the early period of treatment at a medical facility equipped to handle life-threatening emergencies.

2 Discuss these unique problems with the family at the time of initial presentation and counsel as needed.

3 Consider ITI only when there are no other treatment options and shorten the period of ITI as much as possible.

4 Treat the bleeding episode in patients with inhibitors with rFVIIa early in order to avoid repeated doses as well as to prevent a target joint.

Further studies are required to determine the characteristics of FIX inhibitors and the mechanism of anaphylaxis at inhibitor development. An international registry established under the guidance of ISTH must be utilized to collect clinical data regarding FIX inhibitor patients, experience with ITI and complications. This will help gain further knowledge with regard to prevention of treatment-related problems in haemophilia B.

References

1 Warrier I, Ewenstein BM, Koerper MA *et al*. Factor IX inhibitors and anaphylaxis in hemophilia B. *J Pediatr Hematol Oncol* 1997; **19**: 23–7.

2 Warrier I, Lusher JM. Development of anaphylactic shock in hemophilia B patients with inhibitors. *Blood Coag Fibrinolysis* 1998; **9**: 125–8.

3 Lusher JM, Arkin S, Abildgaard CF, Schwartz RS. Recombinant FVIII in previously untreated subjects with hemophilia A: a 3–5 year observational study of safety, efficacy and inhibitor development. *N Engl J Med* 1993; **328**: 453–9.

4 Ehrenforth S, Kreuz W, Scharrer I *et al*. Incidence of development of factor VIII and factor IX in inhibitors in hemophiliacs. *Lancet* 1992; **339**: 594–8.

5 Hoyer LW. The incidence of factor VIII inhibitors in patients with severe hemophilia A. In: Aledort L, Hoyer L, Reisner H, Lusher JM White GC, eds. *Inhibitors to Coagulation Factors*. New York: Plenum Press, 1995: 35–45.

6 Briët E. Factor IX inhibitors in hemophilia B patients: their incidence and prospects for development with high purity factor IX products. *Blood Coag Fibrinolysis* 1991; **2** (Suppl.): 47–50.

7 High KA. Factor IX: molecular structure, epitopes and mutations associated with inhibitor formation. In: Aledort L, Hoyer L, Reisner HM, Lusher JM, White GC, eds. *Inhibitors to Coagulation Factors*. New York: Plenum Press, 1995: 79–86.

8 Katz J. Prevalance of FIX inhibitors among patients with hemophilia B: results of a large North American survey. *Hemophilia* 1996; **2**: 28–31.

9 Warrier I. Immune tolerance therapy in hemophilia B: possibilities and problems. *Hemophilia Monitor* 2000; **8**: 3–6.

10 Ljung RC. Gene mutations and inhibitor formation in patients with hemophilia B. *Acta Hematol* 1995; **94** (Suppl. 1): 49–52.

11 Green PM, Montandon AJ, Bentley DR, Giannelli F. Genetics and molecular biology of hemophilias A and B. *Blood Coag Fibrinolysis* 1991; **2**: 539.

12 Sawamoto Y, Midori S, Masakuni Y *et al*. Measurement of antifactor IX subclass in hemophilia B patients who developed inhibitors with episodes of allergic reactions to factor IX concentrates. *Thromb Res* 1996; **83**: 279–86.

13 Dioun AF, Ewenstein BM, Geha RS *et al*. IgE-mediated allergy and sensitization to factor IX In hemophilia B. *J Allergy Clin Immunol* 1998; **102**: 113–17.

14 Thorland EC, Drost JB, Lusher JM *et al*. Anaphylactic response to factor IX replacement therapy in hemophilia B patients: complete gene deletions confer the highest risk. *Haemophilia* 1999; **5**: 101–5.

15 Warrier I. Management of hemophilia B patients with inhibitors and anaphylaxis. *Hemophilia* 1998; **4**: 574–6.

16 Nilsson IM, Berntorp E, Freiburghaus C. Treatment of patients with factor VIII and IX inhibitors. *Thromb Hemost* 1993; **70**: 56–9.

17 Ewenstein BM, Takemoto C, Warrier I et al. Nephrotic syndrome as a complication of immune tolerance treatment in hemophilia B. *Blood* 1997; **89**: 115–16.

18 Lenk H, Bierback U, Schille R. Inhibitor to FIX in hemophilia B and nephrotic syndrome in the course of immune tolerance treatment [Abstract]. *Haemophilia* 1996; **2** (Suppl.): 104.

19 Pollman H. A haemophilia B patient with complete FIX gene deletion and high titer antibody against FIX [Abstract]. *Haemophilia* 1996; **2** (Suppl.): 104.

20 Tengborn L, Hansson S, Fasth A et al. Anaphylactoid reactions and nephrotic syndrome: a considerable risk during FIX treatment in patients with hemophilia B and inhibitors—a report on the outcome of two brothers. *Haemophilia* 1998; **4**: 854–9.

21 Perez R, Martinez ML, Sosa R. Nephrotic syndrome in A 12-year-old hemophilia B patient in the course of immune tolerance [Abstract]. *Haemophilia* 2000; **6**: 311.

CHAPTER 15

Inhibitors in mild and moderate haemophilia A

C.R.M. Hay and C.A. Lee

Introduction

Although inhibitors occur in patients with mild and moderate haemophilia A they are believed to be uncommon. However, there have been few large studies to estimate prevalence and it is possible that with increasing diagnosis and the widespread availability of treatment this is a complication that develops more frequently than previously thought.

Incidence

Using the UK Haemophilia Centre Doctors' Organisation (UKHCDO) inhibitor register, Hay et al. reported that 57 new inhibitors were reported between January 1990 and January 1997, of which 16 (28%) arose in patients with mild or moderate haemophilia [1]. At this time the registry included 1939 patients with severe, 1615 with moderate and 1546 with mild haemophilia. Thus, the annual incidence of inhibitors in the UK was 3.5 per 1000 patients registered with severe haemophilia and 0.84 per 1000 patients registered with mild and moderately severe haemophilia [2].

Clinical presentation

The presence of an inhibitor is usually suggested by a change in a patient's bleeding pattern with a reduced recovery and half-life. Cross-reactivity of the inhibitor with the patient's endogenous factor VIII (FVIII) usually reduces the circulating FVIII concentration to a very low level. Although many patients develop the classical bleeding pattern of severe haemophilia [3], about two-thirds develop a bleeding pattern similar to that of acquired haemophilia [4]. Those with an acquired bleeding pattern—the majority in the series reported by Hay et al.—may have a severe, life-threatening bleeding tendency in which large ecchymoses, muscle bleeds, gastrointestinal and urogenital bleeding are commonly observed. Such bleeding resulted in the deaths of two patients: one from uncontrollable gastrointestinal haemorrhage and one from retroperitoneal haemorrhage [5] (Table 15.1).

The inhibitor usually arises after a period of intensive treatment for operative intervention or trauma. Although these inhibitors may occur relatively early in a patient's treatment history after a median 5.5 (range 1–107) treatment episodes they tend to occur later in life than those observed in severe haemophilia at a median age of 33 (range 7–71) years [5]. This reflects the infrequency with which such patients require treatment. Thus, in the patient reported by Bovill et al. an inhibitor developed following bilateral knee replacement which was covered with cryoprecipitate—the clinical presentation was a haemarthrosis in the knee whilst the limb was still in splints on the tenth postoperative day [3].

Inhibitor characteristics

Both type I and type II inhibitor reaction kinetics have been described.

Thus, a patient reported by Kesteven et al. with mild haemophilia, who had previously been treated with cryoprecipitate, received 23 400 units of intermediate FVIII following a road traffic accident and developed an inhibitor which produced an anamnestic response at the time of a retroperitoneal haematoma 8 months later [6]. It was possible to demonstrate that the inhibitor plasma collected at the time of plasmapheresis rapidly neutralized the post 1-deamino-8-D-arginine vasopressin (DDAVP) sample, indicating its property as an autoantibody. However, there was no anamnestic response to DDAVP although the patient had demonstrated anamnesis to exogenous FVIII. It was suggested that the patient's endogenous FVIII was less immunogenic than the exogenous material. Following the publication of this case report Lowe and Forbes also described a patient who failed to develop anamnesis with DDAVP [7].

In contrast Capel et al. reported a boy with a basal FVIII between 14 and 26% who developed an inhibitor following an increase in treatment for haematomas in a toe, thigh and calf muscle over a short period. This showed type I kinetics [8].

Natural history

Although the majority of reported cases are in the second, third or even later decades of life there are rare reports in children [4,9,10]. This may be due to limited exposure to FVIII in

Table 15.1 Treatment, inhibitor titre, change in basal VIIIC and bleeding pattern.

Patient number	Age (years)	Pre-inhibitor bleeding episodes	Basal VIIIC (IU/mL)	Post-inhibitor VIIIC (IU/mL)	Presenting anti-human/porcine titre BU or OU/mL	Post-inhibitor treatment*	Bleeding pattern†
1	18	8	0.05	0.01	7.3/17.4	H, F	c
2	7	107	0.05	0.01	0.5/0.0	H	b
3	53	6	0.08	0.01	52.0/47.0 OU	H, P, F	c
4	35	2	0.04	0.01	162.0/21 OU	H, F	c
5	26	–	0.08	0.00	108.0/61.0	H, P, F, I[U]	c
6	67	7	0.06	0.01	See text	H, I[S]	See text
7	48	63	0.03	0.01	3.6/0.5	None	c
8	30	–	0.14	0.01	14.0/–	H, F, D	c
9	71	2	0.06	0.00	1.3/0.0	F	c
10	33	5	0.08	0.01	23.0/5.0	H, P, I[S]	c
11	33	6	0.08	0.00	8.0/1.0	H, P, I[P]	c
12	44	–	0.04	0.00	10.0/1.0	H, P	a
13	60	2	0.10	0.10	4.0/1.0	H	a
14	68	8	0.12	0.00	See text	H, P, F, I[P]	b
15	49	4	0.11	0.01	67.0/1.8	H	c
16	61	5	0.36	0.01	2.0/0.0	D	c
17	60	30	0.03	0.00	568/16	None	c
18	19	40	0.11	0.01	3.3/1.9	H	b
19	8	1	0.06	0.00	10.0/–	H, D, I[P]	c
20	23	2	0.09	0.01	28.0/7.0	H, P, D, F	b
21	30	–	0.12	0.01	14.0/–	H, D	b
22	28	10	0.08	0.00	80.0/–	F, D	c
23	16	4	0.05	0.05	11.2/1.1	H, F, D	a
24	70	4	0.12	0.01	12.0/0.9	H, P, F, I[U]	c
25	19	40	0.23	0.00	22.0/–	H, F, D, I[P]	c
26	19	3	0.23	0.05	9.0/–	F	c

*H, human VIIIC; P, porcine VIIIC; F, Feiba; D, DDAVP; I[s], successful immune tolerance induction (ITI); I[p], partial response to ITI; I[u], unsuccessful ITI.
†a, unchanged; b, classical severe haemophilia phenotype; c, similar to acquired haemophilia.
BU, Bethesda units; OU, Oxford units.

children with mild haemophilia therefore requiring more time for inhibitors to develop. One child reported by Puetz et al. had an anamnestic response following DDAVP and it is suggested that he developed an autoantibody that cross-reacted with 'wild-type' FVIII rather than the conventional explanation that an antibody develops against 'wild-type' FVIII which cross-reacts with endogenous FVIII.

The question of why patients with mild to moderate haemophilia develop inhibitors is not easy to answer. Certain kindreds have a predisposition for inhibitor development. Thus, Hay et al. found that a family history of inhibitor development was present in 19 patients from 11 kindreds (Table 15.2) [5]. These families included 49 individuals with haemophilia A who had been treated with blood products at some time and 20 (41%) had a history of inhibitor formation. The series included four brother pairs with moderate haemophilia and inhibitors including one pair of monozygotic twins. This high prevalence can only partly be explained by selection bias. The prevalence amongst the first and second-degree relatives of patients is far greater than would have been expected by chance.

Genetic predisposition

Most patients with mild to moderate haemophilia have missense mutations and since there is circulating FVIII the occurrence of inhibitors is not easily explained. It has been suggested that the mutation may give rise to conformational change in the FVIII molecule. This would result in the patient's FVIII molecule being antigenically distinct from 'wild-type' and thus infused FVIII would be recognized as non-self and result in inhibitor production. In the study of Hay et al. nine patients carried missense mutations which introduced a new Cys residue (Tyr2105Cys, Trp2229Cys, and Arg593Cys) which can affect the formation of disulphide bridges leading to stable abnormal conformations [5].

Hay et al. showed that six of the patients with a missense mutation introducing a new Cys residue were from three unrelated families (Fig. 15.1). It is possible that these families share a common ancestor. It does seem from this study that the identification of this mutation may identify individuals who are particularly at risk. However, not all individuals carrying the mutation Arg593Cys develop antibodies and Knobe et al. have

Table 15.2 Natural history, genotype and family history.

Patient number	Elimination/follow-up time (months)	Genotype: Nuc. no.	aa-change	Family members with haemophilia with or without inhibitors
1	12/133			No family history of haemophilia A
2	9/61	[2044]G-T	Val[663]-Phe	No family history of haemophilia A
3	8/22			No family history of haemophilia A
4	Persists/99	[6744]G-T	Trp[2229]-Cys	No family history of haemophilia A
5	26/38	[6371]A-G	Tyr[2105]-Cys	No family history of haemophilia A
6	0.5/6			No family history of haemophilia A
7	Persists/22	[6853]T-A	Phe[2260]-Ile	Uncle and grandfather lacking inhibitors
8	Persists/31			Brother with an inhibitor. No other relatives
9	4/25			No family history of haemophilia A
10	13/37	[6744]G-T	Trp[2229]-Cys	Grandfather and cousin lacking inhibitors
11	Persists/119	[6744]G-T	Trp[2229]-Cys	Patients 11 and 12 are brothers and also
12	46/130	[6744]G-T	Trp[2229]-Cys	the nephews of patients 13 and 14
13	2/3	[6744]G-T	Trp[2229]-Cys	The kindred includes a further 8 haemophilic
14	1/1	[6744]G-T	Trp[2229]-Cys	relatives lacking inhibitors (Fig. 15.1)
15	12/14	[6082]G-A	Gly[2009]-Arg	Cousin with haemophilia lacking an inhibitor
16	9/15	[6599]G-T	Glu[2181]-Asp	Brother with haemophilia lacking an inhibitor
17	Persists/17	[6545]G-A	Arg[2163]-His	Uncle with haemophilia lacking an inhibitor
18	1/156			No family history of haemophilia A
19	6/190			7 cousins, one with an inhibitor
20	Persists/205	[6507]G-A	Arg[2150]-His	Patients 20 and 21 are brothers with one
21	Persists/202	[6507]G-A	Arg[2150]-His	haemophilic uncle lacking an inhibitor
22	12/204			Brother lacking inhibitor
23	Persists/433			Brother lacking inhibitor
24	Persists/17			Five second-degree relatives lacking inhibitors
25	Persists/196	[1834]C-T	Arg[593]-Cys	Patients 25 and 26 are monozygotic twins
26	12/184	[1834]C-T	Arg[593]-Cys	lacking other haemophilic relatives

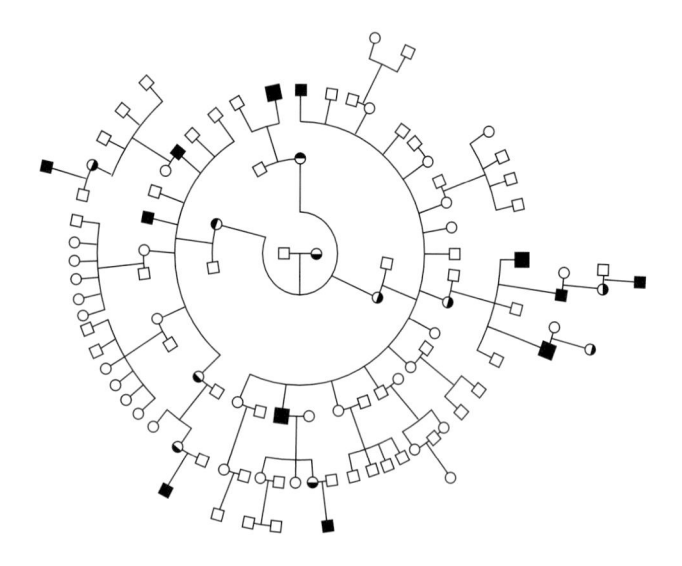

- ● Carrier
- ○ Unaffected female
- ■ Affected male
- □ Unaffected male
- ■ Affected male with inhibitor

postulated an alternative mechanism of action [11]. It has been shown that the conformation of the A2 domain of the recombinant Arg593Cys mutant behaves in a similar way to 'wild-type'. This would therefore mean that it is unlikely that the Arg to Cys mutation induces abnormal disulphide bonds or damaged 3-dimensional folding. A recent study by Roelse et al. revealed intracellular accumulation of the Arg593Cys mutant in transfected cells [12]. Knobe et al. suggest that the local conformational changes induced by the mutation impede intracellular contacts during the processing of FVIII. Once in the bloodstream the mutant protein might be slightly less stable but functions like wild FVIII. However, the mutant protein must behave differently locally because of changes displayed on loop 593. Clearly there must be other factors involved because not all individuals with this mutation develop an inhibitor. It would seem that an intensive antigenic challenge such as a period of intensive treatment is a particular triggering factor.

Fig. 15.1 (left) Family tree showing six generations of an extensive kindred with moderate-severity haemophilia caused by a [6744]G-T mutation resulting in a Trp[2229]-Cys amino acid substitution. 12 family members have haemophilia A and four have inhibitors. (From Hay et al. [5].)

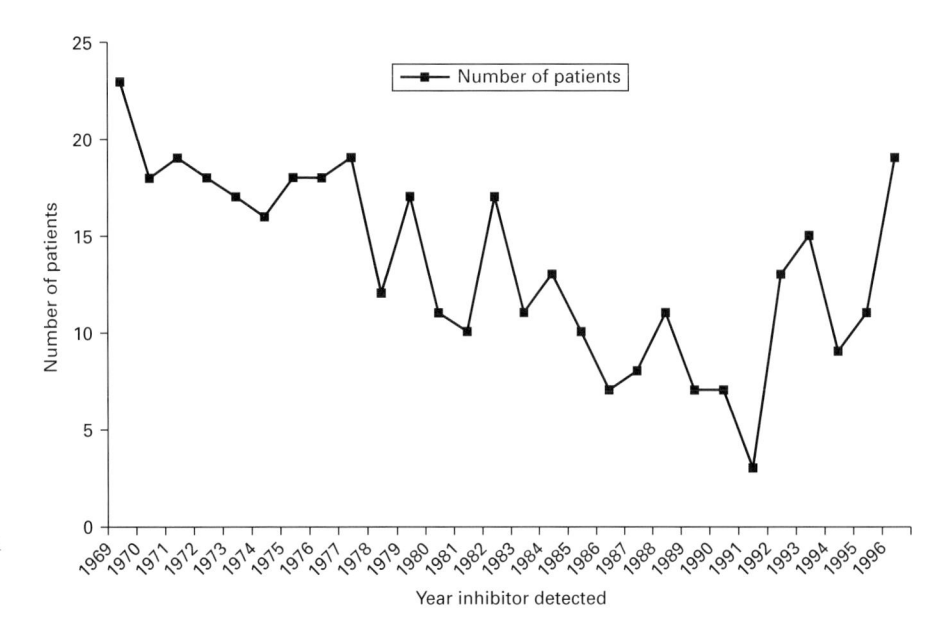

Fig. 15.2 The number of new inhibitors (FVIII antibodies) detected in haemophilia A patients in 1969–96 in the UK. (From Rizza et al. [1].)

Van den Brink et al. conducted an elegant longitudinal analysis of the inhibitor specificity in a patient with mild haemophilia with the mutation Arg593Cys [13]. They showed in immunoprecipitation studies the presence of antibodies directed toward the light chain and A2 domain of FVIII. Almost complete neutralization could be achieved by the addition of the A2 domain. However inhibitor binding was unaffected by Arg593Cys substitution in the recombinant A2 fragment. Evaluation of the inhibitor at different time points showed the initial development of a low titre antibody directed to the A2 domain and FVIII light chain. A second period of FVIII therapy resulted in a dramatic rise of inhibitor titre. These workers hypothesize that in treatment of patients with the Arg593Cys mutation using 'wild-type' FVIII, subsequent processing of the normal FVIII antigen peptides containing amino acid Arg593 may evoke T-helper cell activation. This may result in loss of tolerance to the patient's Arg593Cys FVIII, which coincides with the formation of B-cell clones expressing antibodies directed toward the major inhibitor epitopes on FVIII.

Hay et al. [5] found that substitutions within 100 bases of the C1/C2 junction were found in nine families. Three apparently unrelated families have the same Trp2229Cys mutation and are thought to share the same common ancestor. This uncommon mutation results in the substitution of a non-polar by an uncharged polar amino acid and may therefore give rise to a conformational change in the C2 domain. A pair of brothers, one of whom was patient 17 reported in this series, have mutations Arg2163His and Phe2101Cys both in the C1 domain. Arg2163His was first reported in relation to their case history [14]. These brothers had two mutations, described in [14]; they were also mentoned in [5] but only one mutation was discussed. Santagostino et al. also included two brothers in the series who have the mutation Arg2150His (patients 20 and 21 in Fig. 15.2) in the C1 domain [15].

Type and delivery of treatment as a risk factor

It is possible that both the type of clotting factor concentrate and the mode of delivery influence the development of an inhibitor in the individual patient. Most inhibitors develop after a period of intensive replacement therapy such as for cover of surgery—two-thirds of the series reported by Hay et al. [5].

It is unlikely that neo-antigenicity is involved as described in the reports from Belgium [16] and The Netherlands [17] because inhibitors have been reported in patients receiving a variety of low or intermediate products from many different manufacturers and none received concentrates associated with a high rate of inhibitor development [5]. However, in one patient reported by Yee et al. three different types of concentrate, recombinant, intermediate purity and monoclonally purified concentrates, were used over a period of 6 months in intensive bursts of treatment [3]. In a second patient who had previously only received treatment with intermediate purity concentrate, treatment with a monoclonal product by continuous infusion was followed by a different monoclonal product as bolus therapy; within a month a further operative episode was covered by continuous infusion with the original monoclonal product. It is speculated that the change in concentrate type, particularly over a short period of time, could have played a role in inhibitor development. Furthermore, it is interesting to note that these therapeutic decisions were driven by economic issues rather than clinical; the clinical priority being avoiding unnecessary change of product (C.A. Lee, personal observation).

Baglin et al. reported two patients who developed an inhibitor following continuous infusion of FVIII concentrate to cover liver biopsy [18]. It was suggested that delivery of the concentrate by continuous infusion rather than bolus could have been a factor in the development of the inhibitor. In view of this experience Yee et al. reviewed patients in their centre who had been treated with continuous infusion [14]. From 1995 until 1999 high purity monoclonal purified plasma derived or recombinant FVIII were used in continuous infusion to cover 50 treatment episodes in 19 non-severe and 26 severe patients with haemophilia [14]. Three of 26 patients with severe haemophilia with a past history of inhibitor development received continuous infusion but the inhibitor remained undetectable. Two of 19 non-severe haemophilic patients received concentrate by continuous infusion but they had also received many changes of concentrate over a short period. Although these authors promote the use of DDAVP whenever possible they conclude that the role of continuous infusion in inhibitor development was speculative and there was as of then no evidence to support this theory.

More recently White *et al.* [19] reported two patients with mild haemophilia who were treated with recombinant factor VIII (rFVIII) by continuous infusion and developed an inhibitor. They reported that over a period of 20 years 120 patients had been treated with factor replacement and none had developed an inhibitor [19]. They point out that the recent increase in inhibitor development reported from the UKHCDO [1] may reflect improved data collection. It is also possible it results from recent changes in clinical practice, namely, the switch to higher purity or recombinant products and the use of continuous infusion [1,19] (Fig. 15.2).

The treatment of inhibitors

The frequency of bleeding increases following inhibitor development and is often extremely severe—characteristic of acquired inhibitors. Bleeding can be controlled with human and porcine FVIII, prothrombin complex concentrates, activated prothrombin complex, rVIIa and DDAVP. Bypass therapy with prothrombin complex concentrates (PCCs) or rVIIa has the advantage of avoiding anamnesis. It has also been shown that patients can successfully be treated with DDAVP. Desmopressin is an effective option for bleeding management because DDAVP releases endogenous FVIII which is less immunogenic than exogenous FVIII [5,15].

There are few published cases of patients with mild haemophilia and inhibitors undergoing attempted immune tolerance [4,20]. The Malmö regimen has been successful in two patients with a partial response in a further two patients [5,21]. The Van Creveld regimen has been used unsuccessfully in one individual and achieved a partial response in a second [5,22]. Two patients achieved a complete response and one a partial response using the standard therapy used for acquired inhibitors—high dose immunoglobulin IVIgG, steroid 1 mg/kg and cyclophosphamide 50 mg daily (C.A. Lee, unpublished observation). However, there are insufficient data to determine the optimum approach to immune tolerance for this group. Immune tolerance seems to be unusually difficult to achieve in patients with mild haemophilia perhaps because the patients are usually adult when their inhibitors arise.

Prevention of inhibitor formation

Clearly, as far as is possible, DDAVP should be the treatment of choice in order to avoid an inhibitor occurring in the first place. However, there are also prophylactic approaches which can be taken when a patient is particularly at risk. Savidge *et al.* have described an approach to surgery in a patient who demonstrated an at-risk mutation [23]. A combination of rVIIa, DDAVP and antifibrinolytics were used to cover total hip replacement. The surgery was successful and no inhibitor occurred but the cost was high at £108 000. On the other hand, if this avoided inhibitor occurrence then it certainly could be argued this is cost effective.

The brother of the second patient in the letter by Yee *et al.* [17] has not developed an inhibitor and is treated with rVIIa for any bleeding episode because he has the same Arg2163His mutation and is therefore thought to be at particular risk (T.T. Yee, unpublished observations).

Savidge *et al.* have suggested that all patients with mild haemophilia should have their genetic mutation identified in order to identify those particularly at risk [23]. It is also probably advisable to avoid large exposure to FVIII unless really necessary and certainly to avoid frequent changes of product type.

Conclusion

It is interesting to read the conclusion in a report published from Cardiff by Beck, Giddings and Bloom in 1969: 'The development of a FVIII inhibitor in a mild haemophiliac is apparently a rare occurrence, but when it does occur it is a serious event.' 'A lengthy course of treatment with AHG should therefore be controlled by FVIII assays; if resistance occurs, examination of the patient's plasma may reveal the presence of a FVIII inhibitor' [24].

Although this still holds true we now have information from genetic studies which helps identify patients at particular risk as well as the availability of better therapeutic strategies. The study of patients with mild haemophilia who have inhibitors is also revealing much about the structural functional relationships of the FVIII molecule.

References

1 Rizza CR, Spooner RJD, Giangrande PLF on behalf of the UK Haemophilia Centre Doctors' Organisation (UKHCDO). Treatment of haemophilia in the United Kingdom 1981–96. *Haemophilia* 2001; 7: 349–59.
2 United Kingdom Haemophilia Centre Directors Organisation Annual Returns for 1994. Copyright UKHCDO 1995.
3 Bovill EG, Burns SL, Golden EA. Factor VIII antibody in a patient with mild haemophilia. *Br J Haematol* 1985; **61**: 323–8.

4 Hay CR. Factor VIII inhibitors in mild and moderate–severity haemophilia A. *Haemophilia* 1998; **4**: 558–63.
5 Hay CRM, Ludlam CA, Colvin BT *et al*. Factor VIII inhibitors mild moderate severity haemophilia A. *Thromb Haemost* 1998; **79**: 762–6.
6 Kesteven PJ, Holland LJ, Lawrie AS, Savidge GF. Inhibitor to factor VIII in mild haemophilia. *Thromb Haemost* 1984; **52**: 50–2.
7 Lowe GDO, Forbes CD. Inhibitor to factor VIII in mild haemophilia. *Thromb Haemost* 1985; **53**: 159.
8 Capel P, Toppet M, van Remoortel E, Fondu P. Factor VIII inhibitor in mild haemophilia. *Br J Haematol* 1986; **62**: 786–7.
9 Karayalun G, Goldberg B, Cherrick I *et al*. Acute myocardial infarction complicates prothrombin complex concentrate therapy in an 8-year-old boy with haemophilia A and factor VIII inhibitor. *Am J Pediatr Hematol Oncol* 1993; **15**: 416–19.
10 Puetz JJ, Bouhasin JD. High-titre factor VIII inhibitor in two children with mild haemophilia A. *Haemophilia* 2001; **7**: 215–19.
11 Knobe KE, Villoutreix BO, Tengborn LI, Petrini P, Ljung RC. Factor VIII inhibitors in two families with mild haemophilia A: structural analysis of the mutations. *Haemostasis* 2000; **30**: 268–79.
12 Roelse JC, De Laaf RT, Timmermans SM, Peters M, Van Mourik JA, Voorberg J. Intracellular accumulation of factor VIII induced by missense mutations Arg593→Cys and Asn618→Ser explains cross-reacting material-reduced haemophilia A. *Br J Haematol* 2000; **108**: 241–6.
13 Van den Brink EN, Timmermans SM, Turenhout EA *et al*. Longitudinal analysis of factor VIII inhibitors in a previously untreated mild haemophilia A patient with an Arg593→Cys substitution. *Thromb Haemost* 1999; **81**: 723–6.
14 Yee TT, Lee CA. Is a change of factor VIII product a risk factor for the development of a factor VIII inhibitor? *Thromb Haemost* 1999; **81**: 852.
15 Santagostino E, Gringeri A, Tagliavacca L, Mannucci PM. Inhibitors to factor VIII in a family with mild hemophilia: molecular characterization and response to factor VIII and desmopressin. *Thromb Haemost* 1995; **74**: 619–21.
16 Peerlinck K, Armout J, Gilles JG, Saint-Remy JM, Vermylen J. A higher than expected incidence of factor VIII inhibitors in multi-transfused haemophilia A patients treated with an intermediate purity pasteurised factor VIII concentrate. *Thromb Haemost* 1993; **69**: 115–18.
17 Mauser-Bunschoten EP, Rosendaal FR, Nieuwenhuis HK *et al*. Clinical course of factor VIII inhibitors developed after exposure to a pasteurised Dutch concentrate compared to classical inhibitors in haemophilia A. *Thromb Haemost* 1994; **71**: 703–6.
18 Baglin T, Beacham E. Is a change of factor VIII product a risk factor for the development of a factor VIII inhibitor? *Thromb Haemost* 1998; **80**: 1036–7.
19 White B, Cotter M, Byrne M, O'Shea E, Smith OP. High responding factor VIII inhibitors in mild haemophilia—is there a link with recent changes in clinical practice? *Haemophilia* 2000; **6**: 113–15.
20 Thompson AR, Murphy ME, Liu M *et al*. Loss of tolerance to exogenous and endogenous factor VIII in a mild hemophilia A patient with an Arg593→Cys mutation. *Blood* 1997; **90**: 1902–10.
21 Nilsson IM, Berntorp E, Zettervall O. Induction of immune tolerance in patients with haemophilia and antibodies to factor VIII by combined treatment with intravenous IgG, cyclophosphamide and factor VIII. *N Engl J Med* 1988; **318**: 947–50.
22 Brackmann HH, Oldenberg J, Schwaab R. Immune tolerance for the treatment of factor VIII inhibitors—twenty years of the 'Bonn Protocol'. *Vox Sang* 1996; **70** (Suppl. 1): 30–5.
23 King E, McCartney A, Savidge G. Prevention of inhibitor formation in mild haemophilia in patients undergoing major surgery. *Haemostasis* 2001, in press.
24 Beck P, Giddings JC, Bloom AL. Inhibitor of factor VIII in mild haemophilia. *Br J Haematol* 1969; **17**: 283–8.

CHAPTER 16

Acquired inhibitors to factor VIII

C.M. Kessler and L. Nemes

The development of a circulating immunoglobulin, which interferes with and/or neutralizes the procoagulant function of a specific coagulation protein, may precipitate severe and often life-threatening haemorrhagic complications. Most commonly, these circulating anticoagulants are detected as oligoclonal **alloantibodies** (IgG subclasses 1 and 4) in individuals with congenital coagulopathies who have received multiple infusions of therapeutic replacement products containing an exogenous source of their deficient coagulation factor protein. Such alloantibodies directed selectively against factor VIII (FVIII) procoagulant activity have been reported to develop in 25–45% of patients with haemophilia A [1–4]. Previously non-coagulopathic, non-haemophilic individuals with no antecedent history of bleeding problems spontaneously can develop polyclonal **autoantibodies** directed specifically against FVIII. This is designated acquired haemophilia and is an extremely rare event, with an estimated incidence of 1 case per 3 million population [5,6]. This is probably a significant underestimate since there has never been a systematic attempt to determine its precise epidemiology. Furthermore, because the detection of antifactor VIII autoantibodies requires multiple rigorous and somewhat esoteric coagulation assays, which are not universally available or quality controlled in the clinical coagulation laboratories of many hospitals, the diagnosis of acquired haemophilia is often missed or misdiagnosed as a lupus anticoagulant. There is also the possibility that the incidence of acquired haemophilia is indeed rising or is increasingly recognized, due to its association with numerous malignant and autoimmune disease processes.

Interestingly, the overall median age of presentation and diagnosis for acquired haemophilia is around 65 years with a wide age distribution, characterized by a bimodal trend. This and the predominance of young females and elderly males reflect the variety of accompanying clinical conditions, which must in some way alter the host immunity to promote generation of autoimmune antibodies against FVIII, e.g. pregnancy and autoimmune collagen vascular diseases in young women and malignancies in elderly men.

Associated disease states

The several larger comprehensive surveys of acquired haemophilia have reported consistent disease associations, although the majority of elderly individuals with antifactor VIII autoantibodies have no obvious underlying disease state [7–9]. As an autoimmune epiphenomenon, the autoantibody inhibitor of acquired haemophilia could theoretically be a harbinger of otherwise occult and indolent evolving autoimmune processes and lymphoproliferative malignancies. As indicated in Table 16.1, the most common diseases associated with acquired haemophilia reside in the category of collagen vascular disorders, including systemic lupus erythematosus (SLE), rheumatoid

Table 16.1 Clinical conditions associated with development of acquired haemophilia.

Clinical condition (%)	[7] ($n = 215$)	[8] ($n = 65$)	[9] ($n = 34$)
Idiopathic	46.1	52.5	47.1
Collagen vascular and other autoimmune disorders	18.4	17.0	11.7
Autoimmune dermatoses	2.0	4.5	5.9
Medication related	3.0	5.6	2.9
Pregnancy and postpartum	11.0	7.3	8.9
Lymphoproliferative malignancies	1.5	2.3	2.9
Solid tumour malignancies	12.0	3.2	11.7
Other (multiple transfusions, asthma, hepatitis, diabetes mellitus, etc.)	1.5	11.8	8.8

Modified from Hay [18].

arthritis, and Sjögren's syndrome [10,11]. Other disease states believed to be autoimmune in nature, such as multiple sclerosis, asthma, inflammatory bowel disease and graft vs. host disease, have also been anecdotally associated with clinically significant bleeding due to neutralizing FVIII autoantibodies. Autoimmune dermatoses [8–10,12], such as pemphigus vulgaris, psoriasis and exfoliative dermatitis, and certain medications including antibiotics [8–10,12] (penicillins, sulphonamides, chloramphenicol, etc.), and anticonvulsants (diphenylhydantoin) [8–10,13], have well established associations with acquired haemophilia. Frequently, the medication-induced autoantibodies arise following hypersensitivity reactions, suggesting that alteration of the host immunological state in this context does not necessarily depend on medication-specific antibody formation. On the other hand, the FVIII inhibitor usually remits shortly after withdrawing the offending drug. Altered immune status is probably the basis for the development of autoimmune FVIII inhibitors associated with the use of biological response modifiers (interferon-alpha) [14] and fludarabine [15]. Acute viral infections have produced acquired inhibitors of coagulation in non-haemophilic children [16]. The spontaneous, although curious, appearance of FVIII autoantibody inhibitors (all low titre with <5 Bethesda units) was recently described in three patients to occur within days of simple surgery [17]. There were no other obvious risk factors. All required re-exploration to achieve haemostatic control of their bleeding complications.

The relationship between FVIII autoantibody inhibitor development in the context of solid tumour and lymphoproliferative malignancies and in the context of pregnancy and the postpartum period deserves closer attention. These very important associated clinical conditions may provide insight into the cause–effect mechanisms of acquired haemophilia.

Acquired haemophilia in the pregnant and postpartum woman

The occurrence of FVIII autoantibodies is an infrequent complication of pregnancy, but pregnancy is the underlying clinical condition in approximately 10% of the overall population with acquired haemophilia. These neutralizing antibodies induce severe haemorrhagic manifestations in women who previously had no abnormal bleeding tendencies. According to a recent review of 40 cases [19], acquired haemophilia usually arises in the postpartum period, most commonly from 1 to 4 months after delivery but may occur as late as 1 year afterwards. They were rarely detected during pregnancy or labour and when they did occur during this time, they were frequently associated with severe uterine bleeding, ultimately necessitating hysterectomy [20]. There was a 10% death rate due to severe bleeding [20]. The FVIII inhibitor also has developed in one patient after spontaneous miscarriage [19]. Multiparous and primagravid women have been affected and the sex of the child does not seem to be a risk factor. The majority of autoantibody inhibitors (over 60%) dissipate spontaneously after a mean period of 30 months. In a few cases the autoantibody inhibitor has persisted through subsequent pregnancies, only to appear in the plasmas of the second children by transplacental transfer of the IgG antibodies, which then precipitated transient but life-threatening haemorrhagic complications [21–23]. If the inhibitor has remitted, it very rarely recurs during subsequent pregnancies. There is a single case report of the spontaneous disappearance of a persistent inhibitor during a subsequent pregnancy [24].

The pathogenesis of these pregnancy-related autoantibody inhibitors directed against FVIII remains unclear. Their predominant postpartum appearance suggests that the mother was exposed to fetal FVIII, perhaps during delivery [25]; however, the absence of an anamnestic response on subsequent pregnancies is not consistent with this theory. Furthermore, if these inhibitors were strictly due to alterations of the host immune system, immunosuppressive therapies with corticosteroids and cytotoxic chemotherapeutic agents should be effective in eradicating the autoimmune phenomenon. In fact, these inhibitors are not very responsive to these therapeutic manoeuvres [26]. This is in contrast to the apparent effectiveness of these agents to induce remission of the FVIII inhibitors associated with the other aetiologies of acquired haemophilia.

Autoantibody inhibitors of FVIII in patients with cancer

The occurrence of autoimmune inhibitors directed against the FVIII procoagulant protein in cancer patients was astutely recognized in 1940 [27]. Population surveys have indicated that cancer is the predominant risk factor in 5–15% of individuals with acquired haemophilia (Table 16.1) and that this epiphenomenon of cancer is found mainly in elderly males [8,9,28–30]. Males are affected three times more commonly than females and the median age of presentation is 69 years [28]. No specific tumour type appears to predispose to the development of the inhibitor, although solid tumour malignancies are much more commonly associated with these inhibitors than lymphoproliferative malignancies. Among the solid tumours, lung and prostate predominate, but tumours of all types have been described. The autoimmune inhibitors against FVIII are usually of the IgG variety; however, rare IgA and IgM monoclonal antibodies also have been implicated in lymphoproliferative malignancies, including multiple myeloma, chronic lymphocytic leukaemia, Waldenström's macroglobulinemia and non-Hodgkin's lymphoma.

The aetiology of the autoimmune inhibitor against FVIII in the cancer patient remains to be elucidated. It is unlikely that they are coincidental concurrent disease entities; however, if they arise after treatment of the tumour has been initiated, it may be difficult to exclude the possibility that use of corticosteroids, cytotoxic agents, biological response modifiers, and/or radiation therapy could have altered host immunity and predisposed the patient to the development of autoimmune phenomena. Perhaps these autoantibody inhibitors represent an immune response to the tumour; no tumour antigen(s) has yet been described to have homology to FVIII. The inhibitor may not remit despite successful eradication of the tumour by chemotherapy, immunosuppressive agents, and radiation therapy. There is a 50–70% success rate in eradicating the inhibitor [28,29]. Conversely, the re-emergence of an autoantibody inhibitor is not a reliable surrogate marker of tumour recurrence. Lower titre inhibitors (median = 12 Bethesda units) associated with early stage tumours were more likely to disappear with successful treatment of the tumour than high titre inhibitors (median = 78 Bethesda units, $P = 0.007$); overall survival was significantly higher in patients who achieved a complete remission of their inhibitor compared with persistence of the inhibitor (75% vs. 17%, $P = 0.0006$) [29]. Treatment of acute bleeding events for tumour-induced antifactor VIII autoantibodies follows the same algorithms as for the benign aetiologies of acquired haemophilia A.

Clinical manifestations of acquired haemophilia

Acquired haemophilia is characterized by the acute onset of severe bleeding complications and diffuse bruising in individuals who previously have had no prior history of excessive bleeding. The bleeding is usually spontaneous in nature, although minimal trauma may produce extensive and disproportionately large ecchymoses. Autoimmune inhibitory antibodies directed against FVIII most commonly are associated with severe soft-tissue and mucocutaneous haemorrhage, e.g. epistaxis, gross haematuria and melaena, protracted and profuse menses, etc. Intracerebral, intramuscular and retroperitoneal bleeding are less common, but can be rapidly progressive. Haemarthroses are unusual. Approximately 87% of individuals with auto-immune FVIII antibody inhibitors present with severe haemorrhagic events [7], many of which are life-threatening. Mortality in this disease is high, with an incidence varying in large literature surveys from 7.9% [31] to 22% [7]. Morrison et al. reported a mortality rate of 14.3% [8]. This bleeding pattern is distinctly more severe and anatomically varied than that observed in congenital severe haemophilia A (FVIII activity <2% of normal) or severe haemophilia A complicated by alloantibody inhibitors against FVIII, in which intra-articular haemarthroses and intramuscular bleeds predominate. This is despite the fact that the FVIII activity levels in acquired haemophilia A are usually higher (3–25% of normal) than in severe congenital haemophilia A. The inordinate mortality rates in acquired haemophilia may reflect the consequences of invasive diagnostic and/or therapeutic procedures performed in this group of patients in an attempt to reverse their severe bleeding episodes. Alternatively, the high mortality may result from delays in diagnosis and subsequent delays in initiating appropriate and effective replacement therapy with haemostatically active products. Mortality may be influenced by the type of treatment implemented for acute bleeding problems. The administration of recombinant FVIIa [31] and porcine FVIII concentrate [8] as first-line therapies appear to reduce the mortality rates associated with use of unactivated or activated prothrombin complex concentrates [32]. No prospective randomized trials have been performed to compare the efficacy of these various replacement products in the setting of acquired haemophilia. Acquired haemophilia may remit spontaneously after months to years in about a third of patients, who have received no immunosuppressive therapy [5]. The administration of corticosteroids and cytotoxic chemotherapy hastens the disappearance of the autoantibody inhibitor in approximately 70% of patients [33]. Factor VIII autoantibodies in the context of pregnancy and the postpartum state may have a different natural history and response to immunosuppression [21–23,26].

Immunological characteristics of the autoimmune anti-factor VIII antibody inhibitor in acquired haemophilia A

Factor VIII autoantibody inhibitors are usually polyclonal IgG_4 antibodies, which are directed against specific functional epitopes on one or more domains of the FVIII protein [34]. Factor VIII antibodies of the IgG_1 subclass are less common in acquired haemophilia and rarely, IgA and IgM monoclonal antibodies inhibit FVIII activity as epiphenomena of lymphoproliferative malignancies. Alloantibody inhibitors of FVIII in severe congenital haemophilia A are typically oligoclonal IgG_1 or IgG_4, also with IgG_4 predominating. Systematic epitope mapping of inhibitory antibodies has indicated that autoantibodies to FVIII usually are directed against either the A2 or C2 domains of the FVIII molecule (62%) but typically not to both. There appears to be preferential targeting of the autoantibodies to the C2 domain on the FVIII light chain (67%) compared to the A2 domain on the heavy chain of FVIII (33%). In contrast, alloantibody inhibitors of FVIII overwhelmingly interact with both A2 and C2 domains concurrently in 85%. Occasional FVIII inhibitors may target epitopes on the A3 domain. Interestingly, both allo and autoantibody inhibitors recognize the same epitopes on each domain [35–37].

Factor VIII inhibitors neutralize procoagulant activity via several mechanisms. Inhibitors which recognize epitopes on the A2 (epitope 484–508) or A3 (epitope 1811–1818) domains may directly block high affinity binding sites involved in the interaction of FVIIIa with FIXa. C2 domain directed inhibitors prevent FVIII binding to phospholipid, a critical step in the assembly of the Xase complex for coagulation. Because there is overlap between phospholipid and von Willebrand factor binding sites within the C2 domain of FVIII, inhibitors with C2 specificity may also prevent FVIII binding to von Willebrand factor protein. This will result in increased FVIII catabolism and increases the susceptibility of the protein to proteolytic cleavage by activated protein C, factor IXa and factor Xa. Nogami et al. [38] have described an autoantibody which blocks FVIII binding to activated protein C (APC), resulting in an inhibition of APC-mediated FVIII proteolytic inactivation. The formation of circulating immune complexes of FVIII and autoantibodies may increase clearance of FVIII from the circulation. These features of autoantibody inhibitors do not explain why bleeding manifestations in acquired haemophilia are so much more severe than in severe haemophilics with or without alloantibodies. They also do not provide insight into why mucocutaneous bleeding complications predominate in acquired haemophilia vs. the intra-articular and intramuscular bleeding problems with haemophilia A. Acquired haemophilia bleeding episodes are most consistent with interference with von Willebrand factor protein and/or platelet function. However, such properties remain to be elucidated for autoantibody inhibitors.

Several unique aspects of autoantibody FVIII inhibitors (similarly for alloantibodies) deserve attention. First of all, these antibodies neutralize FVIII coagulant activity in a time and temperature dependent manner, which must be considered in the laboratory detection of the inhibitor. Secondly, when circulating immune complexes form between FVIII and autoantibodies, no complement fixation occurs and thus end organ damage is averted. Third, there is considerably less cross-reactivity between autoantibodies and heterologous sources of FVIII compared to alloantibody cross-reactivity. This justifies the use of porcine FVIII complex concentrate in the treatment and prevention of bleeding complications in acquired haemophilia. Lastly, autoantibody inhibitors follow type II inactivation kinetics so that there is incomplete neutralization of FVIII activity. Thus, there is typically residual FVIII activity in the patient plasmas with acquired haemophilia. Alloantibody inhibitors follow type I kinetics in which FVIII activity is totally neutralized [34].

Establishing the diagnosis of FVIII autoantibody inhibitors in the laboratory

Inhibitors directed against FVIII coagulant protein prolong the activated partial thromboplastin time (aPTT) assay. The prothrombin time is normal. Mixing studies with patient plasma and normal plasma will not normalize the aPTT, confirming the presence of an inhibitor rather than a clotting factor deficiency as the cause of the increased aPTT. When high titre FVIII inhibitors are present, activities of factors XII, XI and IX may be decreased; however, if the assays are repeated in increasing dilutions of patient plasma with physiological buffer, these clotting factor levels will increase into the normal range while the FVIII activity remains depressed. A key characteristic of the antifactor VIII antibody is that its inhibitory expression *in vitro* is time and temperature dependent. Thus, the mixing study aPTT result at time 0 (immediately after mixing patient plasma with normal plasma) may actually show correction. Incubation of this same mixture at 37°C for 1–2 h allows for maximum interaction between the antibody and the FVIII protein and for full inhibitory expression, resulting in a prolonged aPTT. There is no difference between allo and autoantibody inhibitors against FVIII in this regard. In contrast, the so-called lupus anticoagulant typically prolongs the aPTT in a mixing study at time 0 to the same degree as after 1–2 h of incubation. Other studies, such as the dilute Russell viper venom time, kaolin clotting time, dilute thromboplastin inhibition assay, and the platelet neutralization assay, subsequently can be performed to confirm the presence of the lupus anticoagulant. If a prolonged incubation mixing study aPTT is not performed, a FVIII inhibitor may be missed or misinterpreted as a lupus anticoagulant because there is no indication of increasing inhibitor expression. Furthermore, the lupus anticoagulant is less likely than a FVIII inhibitor to produce profuse and active bleeding complications.

The inhibitory potency of the FVIII inhibitor, whether an allo or autoantibody, is quantified with the Bethesda assay. Despite attempts to standardize this assay, there is significant interlaboratory variability of results. This is compounded in the context of autoantibody FVIII inhibitors since the interactions between the acquired antibody and FVIII follow non-linear kinetics and the potency of the inhibitor may be underestimated. From a clinical perspective, however, the assay usually can discriminate very easily between low titre (<5 Bethesda units and no anamnestic response of antibody titre after re-exposure to FVIII) and high titre (>10 Bethesda units and anamnestic responses to re-exposure to FVIII) inhibitors. These results will determine approach to treatment. By definition, one Bethesda unit is the quantity of antibody which will inactivate 50% of normal FVIII activity in a mix of normal plasma and patient plasma (or dilutions thereof) after incubation at 37°C for 1–2 h [39]. The reciprocal of the dilution to achieve this 50% residual FVIII activity arbitrarily provides the potency level of the inhibitor, e.g. if a 1/100 dilution of patient plasma is the lowest dilution which results in 50% residual FVIII activity, there are 100 Bethesda units FVIII inhibitor. The Bethesda assay can be modified to use porcine FVIII (1 U/mL) as the source of normal FVIII activity and then the degree to which the patient inhibitory antibody inactivates porcine FVIII can be determined. There is usually little cross-reactivity between the antihuman autoimmune antifactor VIII antibody and the heterologous porcine FVIII. If this can be confirmed, then the use of porcine FVIII concentrate can be administered to treat or prevent active bleeding despite the presence of very high titre inhibitor [8].

Autoantibodies differ from alloantibodies in their inactivation pattern of FVIII neutralization [34]. Most haemophilic (alloantibody) antibodies inactivate FVIII with second order kinetics, characterized by complete inactivation and linear relationship between antibody concentration and residual FVIII activity. This is designated a type I antibody reaction. In contrast, autoantibodies typically follow a type II kinetic pattern in which there is incomplete inactivation of FVIII activity. Thus, individuals with autoantibody inhibitors have residual low levels of FVIII activity in their plasma (usually 3–15% of normal). The complex type II kinetic pattern of the FVIII autoantibodies may underestimate the potency of the inhibitor [42]. The mechanistic explanation for the differences in the kinetics between allo and autoantibody FVIII inhibitors remains unclear; however, the answer may lie in the manner in which type I and type II antibodies interact with their target epitopes. Type I alloantibodies may react with FVIII epitopes in close proximity to the site which conveys procoagulant function. Type II autoimmune antibodies do not bind to FVIII with great avidity and their target epitopes may be partially blocked by von Willebrand factor–FVIII interactions. After FVIII is dissociated from von Willebrand factor, complete inactivation of FVIII activity can be achieved by the type II antibody [40]. It is unclear how these different kinetics can mediate the different types of bleeding complications observed in acquired haemophilia vs. alloantibody inhibitors in congenitally severe haemophilia A.

The treatment of acquired inhibitors to FVIII

The management of acquired haemophilia is aimed at two distinct goals: ensuring haemostasis by the treatment of the acute bleeding episodes and elimination of the FVIII autoantibody by long-term eradication therapy for consequent cure of the condition. Whereas the principles of the management of the acute bleedings are comparable in allo- and autoantibody patients, the methods used in the eradication therapy differ more fundamentally. In contrast to congenital haemophilia with alloantibody inhibitors, acquired inhibitors to FVIII represent a rather heterogeneous group of patients with different basic and accompanying diseases and prognostic factors. Therefore, the detailed assessment of the patient's individual characteristics is essential before the decision on the particular therapeutic interventions is made. The factors that usually should be taken into consideration are listed in Table 16.2.

The heterogeneous nature of acquired FVIII inhibitors, the rarity of the condition and the absence of published prospective, controlled trials have made the prescription of definite therapeutic algorithms difficult. Most haemophilia centres have only limited experience in the management of acquired inhibitors to FVIII and the therapeutic approach to these patients may vary considerably between centres.

The treatment of acute bleeding episodes

Although there are some exceptions, most patients with acquired haemophilia present with major life- or limb-threatening bleeding that must be treated aggressively. For those rare patients who have only minor (e.g. subcutaneous) haemorrhages at presentation, withholding blood-product therapy and awaiting the results of long-term immunomodulatory treatment intended to suppress inhibitor formation may be justified. The bleeding-related mortality rate approaches 15%, mainly due to early haemorrhagic complications, which fact also underscores the importance of quick and effective diagnosis and treatment [41].

The choice of blood-product therapy is typically determined by the clinical setting and the initial and peak titres of antihu-

Table 16.2 Factors influencing the decision on initial therapy in acquired haemophilia.

Clinical presentation
Location of the haemorrhage at presentation
Severity of the bleeding episode at presentation
Unstable vital signs and clinical parameters

Characteristics of the patient
Underlying disorder (e.g. idiopathic/primary vs. secondary)
Age of the patient (e.g. postpartum and autoimmune vs. idiopathic)
Accompanying diseases (e.g. comorbid states like cardiovascular diseases and diabetes)

Prognostic factors
Expected response to immunosuppression
Probability of spontaneous remission
Expected side-effects of therapy

Characteristics of the inhibitor
Initial human inhibitor titre
Initial porcine cross-reactivity
Anamnestic response to human (porcine-) factor VIII

man and/or antiporcine FVIII inhibitor antibodies. Because the measured inhibitor titre is less predictive of the effectiveness of FVIII administration in acquired than it is for the alloantibodies of congenital haemophilia, careful laboratory (APTT, FVIII:C, human and porcine inhibitor titre) and clinical (subjective and objective) monitoring [43] are necessary. FVIII autoantibodies are most commonly 'low-responder' inhibitors; an anamnestic rise of antihuman and/or porcine FVIII antibodies after human (or porcine) FVIII challenge is not characteristic of acquired haemophilia [44].

The theoretical treatment options for achieving haemostasis during acute bleeding episodes in the context of FVIII antibodies include: (1) 'overwhelming' the inhibitor; (2) 'bypassing' the inhibitor; or (3) physically removing the inhibitor from the plasma (Table 16.3). The products and the dosing schedules used in the treatment of acute bleeds in acquired haemophilia are summarized in Table 16.4.

Human plasma-derived and recombinant FVIII concentrates

The use of FVIII (human or porcine) replacement is generally preferred over the 'by-pass methods' (traditional unactivated Prothrombin Complex Concentrates, Activated Prothrombin Complex Concentrates and recombinant factor VIIa) for patients with low titre, low-responder FVIII autoantibodies. The FVIII products convey better predictability of therapeutic efficacy, ease of laboratory monitoring (these products allow for the measurement of FVIII levels), low potential for inducing anamnestic antibody responses, and minimal thrombogenicity side-effects. Therefore, the acute bleeds complicating low titre acquired autoantibody inhibitors to FVIII should initially be treated with high doses of human (or porcine) FVIII concentrates [46].

Table 16.3 Therapeutic options for the management of acute bleeding episodes in patients with autoantibody inhibitors to FVIII.

Overwhelming the inhibitor
High-dose human plasma-derived FVIII
High-dose human recombinant FVIII
Heterologous porcine FVIII (Hyate:C)
Desmopressin (1-deamino-8-D-arginine vasopressin, DDAVP, Stimate, Octostim)

Bypass of the inhibitor
Traditional Prothrombin Complex Concentrates
Activated Prothrombin Complex Concentrates (Feiba, Autoplex)
Recombinant activated factor VII (rFVIIa, NovoSeven, NiaStase)

Removal of the inhibitor
Exchange plasmapheresis
Extracorporeal protein A sepharose adsorption
Intravenous immunoglobulin

Miscellaneous methods
Platelet transfusion
Calcium gluconate infusion
Fibrin glue
Local antifibrinolytic agents

As most acquired inhibitor patients have never been exposed to blood products previously and may not require them in the future, the safety of replacement therapy is of major concern. The two most prominent safety issues involve eradication of potential blood-borne pathogens, e.g. viruses, prions, etc., and product purity. It is generally accepted that the viral safety of the double virus-inactivated high purity or monoclonal antibody purified ultra-high purity plasma-derived FVIII concentrates is excellent [47]. However, many patients and physicians prefer to use genetically engineered recombinant human FVIII concentrates, which also have shown an excellent safety profile [2,3,48] and are less prone to inadvertent viral contamination or manufacturing errors. The high-dose administration of intermediate-purity FVIII concentrates has the potential side-effect of inducing immune haemolysis, because of the anti-A isoantibody content of these products [49]. Severe haemolysis was also observed during the treatment course of patients with acquired haemophilia [50]. Therefore, the use of low or intermediate purity concentrates should be avoided, especially if high-dose or protracted FVIII administration appears to be necessary.

Recently, there has been debate about the best choice of FVIII concentrate to administer for immune tolerance induction (ITI) regimens in congenital haemophilia with FVIII alloantibody inhibitors. A few physicians believe that *in vitro* laboratory [51,52] and clinical observations [53] support the use of plasma-derived FVIII concentrates rich in von Willebrand factor protein rather than highly purified or recombinant FVIII products for treatment of acute bleeds and for ITI. Although it is difficult to extrapolate these data from alloantibody to autoantibody patients, it is still possible that plasma-derived FVIII products stabilized by the natural carrier von Willebrand factor protein may offer better outcomes in the acute and ITI treatment of acquired haemophilia as well. Randomized, controlled, prospective studies would be helpful in this regard.

The dose of human FVIII should be high enough to 'overwhelm' the inhibitor and induce measurable increases in FVIII activity, consequently promoting haemostasis. Several schemes

Table 16.4 Products used in the treatment of acute bleeding in acquired haemophilia.

Generic name	Brand name	Company	Recommended initial dose	Route	Subsequent dosing
Recombinant FVIII concentrate	Kogenate(FS)	Bayer	50–100 IU/kg	i.v.	2–3 × 50–75 IU/kg/day or continuous infusion
	Recombinate	Baxter			
	ReFacto	Wyeth			
Plasma-derived FVIII	–	–	50–100 IU/kg	i.v.	2–3 × 50–75 IU/kg/day or continuous infusion
Porcine FVIII concentrate	Hyate:C	Speywood	50–100 IU/kg	i.v.	2–3 × 50–75 IU/kg/day or continuous infusion
Desmopressin acetate	Octostim	Ferring	0.3 µg/kg	i.n., i.v./s.c.	1 × 0.3 µg/kg/day
	Stimate		300 µg/kg	i.n.	1 × 300 µg/kg/day
Activated PCC	Feiba VH	Baxter-Immuno	50–100 IU/kg	i.v.	2–3 × 50–75 IU/kg/day
	Autoplex T				
Recombinant FVIIa	NovoSeven	Novo Nordisk	70–90 µg/kg	i.v.	Repeat q2 h until cessation of bleeding, then q3–6 h
	NiaStase				
Aminocaproic acid	Amicar	Wyeth-Ayerst	4–5 g	100 p.o. or i.v. infusion	3–4 × 4–5 g/day
			mg/kg		2–3 × 100 mg/kg/day
Tranexamic acid	Cyklokapron	Pharmacia Upjohn	25 mg/kg	10 p.o. or i.v. infusion	3 × 25 mg/kg/day
			mg/kg		2–3 × 10 mg/kg/day

have been proposed for the calculation of the initial dose of FVIII: 20 IU/kg for each Bethesda Unit (BU) of inhibitor + an additional 40 IU/kg [54]; double or triple the amount of FVIII that would have been considered adequate to achieve 30–50% incremental rise in FVIII activity in a non-inhibitor patient with haemophilia A [46]; and 200–300 IU/kg bolus followed by continuous infusion of FVIII at a rate of 4–14 IU/kg/h [55]. As the recovery and circulating half-life of the infused FVIII cannot be predicted because of the variable kinetics of autoantibodies, subsequent dosing should be based on close laboratory monitoring with frequent APTT and FVIII:C measurements. It is always necessary to determine the initial response by measuring FVIII:C activity levels 15–30 min after bolus dosing. Some originally low titre inhibitors are actually high responders and during the treatment course with human FVIII may be transformed to high titre inhibitors secondary to anamnesis. Therefore, longitudinal follow up of inhibitor titres (quantities in BU) is also essential.

Porcine FVIII

The antihuman FVIII autoantibodies, in contrast to alloantibodies in congenital haemophilia complicated by inhibitors, usually demonstrate rather low cross-reactivity to porcine FVIII [56]. Therefore, generally speaking, there is more flexibility for the use of porcine FVIII (Hyate:C, Speywood) in acquired FVIII inhibitors than there is in congenital haemophilia A with an inhibitor. Porcine FVIII seems to be a very safe and effective therapeutic option even in high titre autoantibody inhibitors [8]; haemostasis may even be achieved in the absence of detectable levels of circulating FVIII, as has been demonstrated in alloantibody patients [57]. The advantages of porcine FVIII in the treatment of acquired inhibitors to FVIII include its relative viral safety compared to human plasma derived products, its predictable and monitorable FVIII incremental rise, its low level of neutralization by cross-reactive antihuman FVIII autoantibodies, and its decreased propensity to induce significant anamnestic responses (10–15% of patients). An additional mechanistic advantage contributing to the clinical effectiveness of porcine FVIII concentrate may lie in its ability to activate platelets *in vivo*.

Adverse reactions were seen much more commonly with first-generation formulations of porcine FVIII concentrates, which contained higher amounts of contaminating porcine von Willebrand factor causing *in vivo* platelet agglutination and consequent thrombocytopenia [58]. The recent formulations possess much higher specific activity, achieved by serial polyelectrolyte fractionation, and only rarely cause side-effects such as anaphylaxis, back pain, rigors, significant thrombocytopenia and resistance to treatment [41,59–62].

The indications for the use of porcine FVIII include the treatment of acute bleeding episodes, prophylaxis in the surgical setting, adjunctive therapy to extracorporeal pheresis and plasma-exchange, and possibly in immune tolerance induction protocols [63]. Before treatment, the determination of antihuman inhibitor titre and porcine cross-reactivity is essential. The initial dose of porcine FVIII can be calculated with this formula: plasma volume (ml) × antiporcine antibody titre (U/mL) (neutralizing dose) + desired FVIII:C increment × body weight (kg)/1.5 (augmenting dose) [59]. In emergency situations, when the antiporcine inhibitor quantity is not known, 50–100 IU/kg porcine FVIII concentrate should be given to patients whose antihuman antibody titre is less than 50 BU/mL. In patients with inhibitor titres between 50 and 100 BU/mL, porcine FVIII can be administered at 100–200 IU/kg [64]. After the initial bolus dose, porcine FVIII also can be administered as continuous infusion [65].

Desmopressin

Desmopressin (1-deamino-8-D-arginine vasopressin; abbreviated to DDAVP; trade names: Octostim, Stimate; Ferring, Sweden) is a non-transfusional form of treatment for inherited and acquired bleeding disorders [66]. The therapeutic mechanism of this synthetic vasopressin analogue consists of the release of FVIII/von Willebrand factor from preformed endothelial and platelet storage sites [67]. Therapeutic FVIII elevations have been reached in moderate FVIII deficiency states like von Willebrand disease and mild haemophilia A. DDAVP also has been used in healthy blood donors to increase the FVIII content of the plasma recovered during pheresis [68,69]. Desmopressin is usually administered intravenously at a dose of 0.3 µg/kg diluted in 50 mL saline and infused intravenously over 30 min.

Desmopressin has been used successfully to treat a small number of autoantibody patients, but the experience with this form of treatment is limited and anecdotal. The first report describes the application of DDAVP in a non-haemophilic patient with classical type II FVIII inhibitor prior to dental procedures [70]. Subsequently, desmopressin has repeatedly been tried in non-life-threatening haemorrhagic complications of acquired inhibitors to FVIII with various success [71–76]. In a review of the use of DDAVP in 21 previously published cases of acquired FVIII inhibitors, higher residual FVIII:C activities in the presence of lower titre inhibitors appeared to be the best predictors of clinical response [70–74,77]. The therapeutic failures occurred predominantly in high titre inhibitors [74,75]. The DDAVP-induced elevation of FVIII has been occasionally also coupled with the disappearance of the inhibitor [73,77]. Cost–benefit considerations obviously favour desmopressin vs. transfusional modalities of treatment. DDAVP also carries no risk of transmitting blood-borne infectious agents. However, the well-known antidiuretic and vasomotor side-effects of DDAVP should be considered even more carefully in patients with acquired haemophilia that predominantly belong to the older age groups, than in congenital haemophilia or von Willebrand disease. Tachyphylaxis was also observed following repeated daily treatment of non-haemophilic inhibitor patients. The problem of tachyphylaxis can be circumvented by alternating DDAVP with FVIII administration [74]. Although the low incidence of this autoimmune coagulopathy means that no prospective, randomized, controlled studies will likely be conducted to verify the effectiveness of desmopressin, the available data certainly are compelling and support a role for DDAVP in the treatment of patients with low titre acquired FVIII inhibitors (<5 Bethesda units or >5% FVIII:C) and non-life-threatening bleeding [77].

'Protected' forms of FVIII and miscellaneous methods of therapy

According to several previous *in vitro* studies, various phospholipids may protect FVIII from inactivation by inhibitors [78,79]. FVIII molecules complexed with factor IX and bound to platelet surfaces were inactivated by inhibitors at a much slower rate ($T^1/_2$: 13 min) compared to free FVIII ($T^1/_2$: one minute) [80]. These theoretical considerations provide some scientific rationale for the effectiveness of platelet transfusions to enhance haemostasis in patients with FVIII inhibitors [81]. In animal experiments phospholipid and factor Xa containing lipid vesicles also have been used with some success to bypass FVIII inhibitors and affect bleeding [82]. Although studies of the clinical efficacy of concentrates containing phospholipids have not yet been published, platelet transfusions, as a source of coagulation-active phospholipids have been used occasionally in conjunction with high-dose FVIII in the treatment of high-responder inhibitors. Another approach to improve the clinical response to FVIII concentrates in inhibitor patients is the concurrent administration of high-dose intravenous calcium infusion. The effectiveness of this treatment is thought to be mediated by altering antigen–antibody interactions. In a published report of two autoantibody patients, the calcium gluconate infusion decreased the inhibitor potency and improved the clinical response to FVIII replacement [83].

The advent of the more reliable new therapeutic modalities, particularly of porcine FVIII and recombinant activated FVII, has rendered these miscellaneous forms of therapy as just important historical steps in the evolution of effective treatment regimens for FVIII inhibitors. Nevertheless, the local or topical application of fibrin glue or antifibrinolytic agents could have a role as important adjunctive measures in the management of acquired inhibitor patients [84].

Exchange plasmapheresis and extracorporeal adsorption

In high titre inhibitors, the efficacy of human or porcine FVIII can be temporarily restored through the removal of the antibodies by extracorporeal methods. For this purpose, plasmapheresis was the only available method for many years. Exchange plasmapheresis at a rate of 40 mL plasma/kg (3–4 L in adults) can reduce inhibitor levels by half [85]. The recognition of the relatively common serious side-effects associated with this procedure (10–20%) has prompted the development of new methods for more specific depletion of the inhibitors [86]. The procedures currently available for immunoadsorption include sepharose-bound staphylococcal protein A (Immunosorba, Excorim), protein A bound to silica matrix (Prosorba, Imre), sepharose-bound polyclonal sheep antihuman antibodies (Ig-TheraSorb, TheraSorb) and experimental columns with sepharose-bound factor IX or VIII (Excorim). The introduction of immunoadsorption techniques has increased the volume of processed plasma and the efficacy of the procedure may be achieved without the need for obligatory replacement of plasma-derived products [87]. Although the overall safety profile of immunoadsorption methods is superior to traditional plasma exchange, every specific procedure has its own particular limitations, e.g. single-use design, the need for special technical equipment and expertise, allergic and circulatory side-effects due to leeching of staphylococcal proteins, complete depletion of the entire spectrum of immunoglobulins and exceedingly high cost. In the clinical setting of an acute, life-threatening bleed, this manoeuvre may be lifesaving. If the patient with an acquired FVIII autoantibody also has accompanying cardiovascular disorders and is marginally stable from a haemodynamic perspective, the employment of pheresis techniques would be relatively contraindicated.

The treatment of acquired inhibitors to FVIII with extracorporeal methods has demonstrated only temporary reductions in inhibitor titres, so FVIII replacement in the form of either human or porcine FVIII concentrates must be administered immediately after the treatment cycle to achieve haemostasis [88,89]. After the removal of the autoantibody, a consequent rebound increase of the inhibitor titre typically occurs, necessitating simultaneous immunosuppression. Durable elimination of the FVIII autoantibody is achieved within a median of 18 days [90]. Intravenous immunoglobulins are often used simultaneously with immunoadsorption to restore the lost intrinsic IgG repertoire and also to suppress the rebound formation of pathological autoantibodies by providing an exogenous source of anti-idiotypic antibodies. Because of the high cost and the relative impracticality of the extracorporeal techniques, their use is usually limited to life-threatening haemorrhage or to preparation of the high titre patient for immune tolerance induction regimens. With the introduction of new columns containing FVIII or specific (e.g. A2 and/or C2) epitope amino acid sequences of FVIII, the safety and efficacy of pheresis might be improved in the future.

Traditional and activated prothrombin complex concentrates

Although the mechanism of action for the traditional and purposely activated prothrombin complex concentrates (PCCs and aPCCs) has not been elucidated completely, their activated vitamin K-dependent factor content (VIIa, IXa, Xa) seems to play an important role with bypassing the inhibitor and promoting haemostasis. Recent preclinical observations suggest that the main active components of the anti-inhibitor complex Feiba are activated factor X and prothrombin as the two essential participants of the full prothrombinase complex [91]. PCCs and aPCCs are widely used in the management of congenital haemophilia with high titre alloantibody formation. Although in the early open label studies the haemostatic efficacy was close to 90% [92–94], the randomized controlled studies of PCCs and aPCCs vs. placebo demonstrated efficacy only around 40–60% [95–97]. Two controlled, double-blind trials compared the efficacy of the intentionally activated anti-inhibitor coagulant complexes (Feiba and Autoplex T, Baxter-Immuno, Vienna, Austria) to non-activated PCCs in joint and muscle haemorrhages of alloantibody patients. Only slight advantages for the aPCCs were demonstrated. However, it is still widely accepted that aPCCs are superior to PCCs in more serious or 'open' types of bleedings [98,99]. The recommended dose of PCCs and aPCCs is between 50 and 100 IU/kg every 8–12 h.

Activated PCCs were also used in haemostatic treatment of high titre autoantibodies to FVIII [100]. The advantages of the use of aPCCs include the ease of availability, reconstitution and administration, and reasonable efficacy. However, their potential drawbacks should also be considered seriously: (1) these products are not subjected to as rigorous viral-inactivation techniques as are most of the currently used FVIII/FIX concentrates; no documented transmission of blood-borne pathogens has been reported to date, but the theoretical possibility exists; (2) their mechanism of action is not absolutely clear and their efficacy is not predictable; (3) there is no laboratory tool to monitor the effects; (4) very large doses are required to affect haemostasis; and (5) their slight but definite risk of inducing venous thromboembolic adverse effects, such as disseminated intravascular coagulation, deep venous thrombosis and pulmonary embolism, or acute myocardial infarction [101]. In conclusion, there is no doubt that aPCCs have a place in the therapeutic armamentarium of products for the treatment of high titre inhibitors unresponsive to high doses of human or porcine FVIII [102]. Another noteworthy aspect of management is that porcine FVIII and rFVIIa given as first-line therapy rather than for salvage may be superior to aPCCs [41].

Recombinant activated factor VII (rFVIIa)

An additional therapeutic option available to bypass inhibitors to FVIII is the administration of genetically engineered recombinant activated factor VII (FVIIa). After the early encouraging experiences with plasma-derived, highly purified FVIIa [103,104], the gene of FVII was cloned and expressed in mammalian cells in the mid-1980s [105]. Subsequently, recombinant activated factor VII (rFVIIa, NovoSeven, NiaStase; Novo Nordisk, Denmark) has been widely used in the treatment of haemophilic alloantibody inhibitors with remarkable success [106]. In the therapeutic dose range FVIIa will bind to activated platelets, as well as to tissue factor (TF) bearing cells. Consequently direct activation of factor X can take place with sufficient thrombin generation and coagulation can be triggered in the absence of FVIII (or FIX) [102]. Because of the short half-life of approximately 2.9 h, rFVIIa is usually administered at 2–3 h intervals as an intravenous bolus injection of 70–90 μg/kg.

After the successful sporadic use of rFVIIa in acquired haemophilia [107–110], the multicentre clinical experience was described retrospectively in 38 patients gleaned from the NovoSeven compassionate-use program [111]. A good clinical response was noted when rFVIIa was used as first-line therapy. In the 60 bleeding episodes in which rFVIIa was administered as salvage therapy, the response was good in 75%, partial in 17% and poor in 8%. The conclusion of the analysis was that rFVIIa is a safe, useful and effective treatment for bleeding in patients with acquired haemophilia. For monitoring the effect of the rFVIIa, the measurement of the postinjection prothrombin time and FVII:C has been suggested, although adequate haemostatic levels have not been defined. There is no assay to predict the efficacy in any particular case. Although one case of fatal disseminated intravascular coagulation (DIC) has been reported in the literature [112], the overall adverse event and thromboembolic complication rate seem to be fairly low [113]. Another important advantage of rFVIIa over PCCs and aPCCs is the viral safety owing to the recombinant nature of the product. The disadvantages of rFVII include the relatively high cost, the lack of monitoring possibilities and predictive factors and the short half-life necessitating frequent dosing. As no comparative trials have been completed between porcine FVIII (or aPCC) and rFVIIa, it is not easy to determine the place of rFVIIa in the management of acquired haemophilia at the moment [114]. The most definite indication would be the life-threatening haemorrhages in high titre autoantibody patients unresponsive to other treatment modalities, but the first-line use of rFVIIa was also suggested under such clinical circumstances [111].

Eradication therapy

The primary aim of long-term management in acquired haemophilia is to eradicate the FVIII autoantibody so that further bleeding can be averted. This can be achieved through immunomodulation, employing immunosuppressant medications, administration of intravenous gammaglobulin and by immune tolerance induction regimens. Although FVIII autoantibodies may remit spontaneously [5], clinical studies indicate that early initiation of eradication treatment is advantageous [7,8]. Dramatic and life-threatening bleeding complications are experienced in 80–90% of patients at some time in the course of their disease. The 10–22% mortality rate for this autoimmune disorder is directly or indirectly attributable to the inhibitor. Most published therapeutic guidelines and algorithms recommend that immunosuppressive therapy to abolish the autoantibody be instituted as soon as the diagnosis has been established [43,115]. The controversy lies in how best to accomplish this in such an extremely heterogeneous disorder. It is possible that different strategies for long-term management may be suitable for the various subgroups of patients [116]. A conservative 'watch and wait' approach for children, peripartum and drug-induced cases in whom spontaneous remission can be reasonably expected may be more appropriate than the combined immunomodulatory therapies for idiopathic, autoimmune and malignancy-associated cases. In any case, individuals presenting with acquired haemophilia and severe haemorrhage need rapid and effective treatment.

High-dose intravenous immunoglobulin

Intravenous immunoglobulin (IVIG) provides a transition between the acute treatment of haemorrhages and eradication therapy. IVIG preparations made from the plasma of several thousands of blood donors contain a significant amount of anti-idiotypic antibodies directed against FVIII antibodies. Consequently, IVIG has a direct and immediate effect on the autoantibody quantity by complexing immunoglobulins directed against FVIII coagulant activity [117] and removing them from the plasma circulation. IVIG can often lower the inhibitor titre and even the normalization of FVIII activity has been observed in some low titre cases. High titre inhibitors usually do not remit or transform into low titre inhibitors after IVIG. Repeated administration of IVIG may also suppress antibody synthesis through modulation of the immune system [118,119]. This long-term effect on the immune system is unpredictable but probably real and potentially beneficial.

High-dose intravenous immunoglobulin (HD-IVIG) therapy is usually given according to two different dosage schedules: 1 g/kg body weight for 2 consecutive days or 0.4 g/kg for 5 consecutive days. In a prospective multicentre study evaluating the efficacy of HD-IVIG in acquired inhibitors to FVIII rather low complete (25–37%) or partial response rates were observed [120]. Complete remissions occurred almost exclusively in low titre inhibitors. The partial responses induced in high titre autoantibody patients were usually not associated with a reduced incidence of haemorrhagic complications and did not affect the response to treatment of the acute bleeding episodes. Multiple HD-IVIG courses were needed to sustain therapeutic responses. This represents an important limitation of HD-IVIG use, considering the high cost of the product. Acute oliguric renal failure has been described as a rare side-effect of HD-IVIG therapy in acquired haemophilia [121]. HD-IVIG does not appear to be an effective first-line or monotherapy in acquired haemophilia but may be a useful adjunct to plasmapheresis or immune tolerance induction [114].

Immunosuppression

The most effective responses to immunosuppression of the FVIII autoimmune inhibitor have been observed in the presence of associated autoimmune and neoplastic diseases and in idiopathic cases of the elderly. Postpartum inhibitors are considerably more resistant to immunosuppressive therapy [19], but like the drug-induced antibodies may remit spontaneously, so an expectant approach in the absence of major haemorrhages may be reasonable. In contrast to alloantibodies of congenital haemophilia, corticosteroids [122] and cytotoxic drugs, including cyclophosphamide, azathioprine and 6-mercaptopurine [123], have been successfully used to eliminate the autoantibody. In a randomized, prospective multicentre trial organized in the late 1980s, every patient with newly diagnosed acquired haemophilia received an initial 3-week course of prednisone 1 mg/kg/day. If the inhibitor persisted, they were then randomized to receive either prednisone alone 1 mg/kg/day, cyclophosphamide alone 2 mg/kg/day or the combination of both for an additional 6 weeks [124]. After 6 weeks of steroid treatment approximately 40% of inhibitors disappeared. Fifty per cent of the other two cohorts showed inhibitor suppression following cyclophosphamide alone or in combination with steroid. The main predictive factors for success were lower initial inhibitor titres and higher residual FVIII activities. Even better response rates were reported in a small series of patients who were treated initially with a combination of cyclophosphamide and prednisone [125]. The prolonged administration of this combination therapy eradicated the autoantibody in eight consecutive patients. Nevertheless, for low titre inhibitors a short course of steroid monotherapy may be adequate [126]. Another successful approach has combined immunosuppression with cyclophosphamide, vincristine and prednisone administered after an intravenous bolus dose of FVIII concentrate. Eleven of 12 complete responses were observed [123]. The toxicities of cytotoxic agents including myelosuppression, risk of infections and development of secondary malignancies should also be considered, particularly if prolonged or high-dose administration is needed. In large retrospective analyses, life-threatening infections added significantly to the morbidity and mortality of the condition [7,41].

Cyclosporine [127–129] and interferon-alpha [130] have also been occasionally used in the treatment of autoantibodies to FVIII with success. On the other hand, the therapeutic administration of interferon-alpha for other indications has been also associated with the development of autoantibodies to FVIII as an autoimmune side-effect [131,132]. Recently, newer biological response modifiers have also been developed. The calcineurin inhibitor tacrolimus (FK506) and the antilymphoproliferative agent sirolimus (rapamycin) have the potential of being more specific than the traditional immunosuppressive agents and therefore may cause fewer infectious complications [114,133]. The anti-CD20 monoclonal antibody (rituximab, Rituxan) has been tried in the treatment B-cell related lymphoproliferative disorders such as NHL, B-CLL and cold agglutinin disease [133–136].

Autoantibody formation against FVIII is a T-cell dependent process. T cells are activated when antigens interact with specific T-cell receptors and a peptide contained in the MHC complex on the antigen presenting cell (APC). T cells also require costimulation signals for B-cell activation for antibody production or for cytokine production by other T cells. This 'cross-talk' is mediated by interactions between B7 and CD40 on APCs and CD28 and CD40L on the T cells. Theoretically, interference of 'cross-talk' mechanisms and MHC-peptide presentation to the T-cell receptor offer potential therapeutic possibilities to affect immune tolerance. Such approaches have been attempted in haemophilia A mice, which were treated with human FVIII. No detectable T-cell responses or antibody formation occurred in these mice when they received infusions of anti-CD40L or B7 antibodies or CTLA-4-Ig to block CD28 interactions [137]. Thus, the anti-CD40 ligand (Biogen) could also possibly be used in autoantibody patients [114,136]. Studies in patients with allo- and autoantibody FVIII inhibitors are being planned.

Immune tolerance induction

Immune tolerance induction (ITI) is a twist on the theme of immunomodulatory treatment, in which the antigen (FVIII) is repeatedly administered in a regimen intended to 'desensitize' the immune system and eradicate the inhibitor. ITI was originally described by Brackmann and Gormsen in 1972 [138]. Subsequently, ITI has been generally accepted as the most effective form of treatment in congenital haemophilia complicated by alloantibody formation [139]. ITI regimens with human FVIII concentrates until recently were rarely implemented in adult patients with autoantibody inhibitors even though they have been used with increasing frequency for alloantibody suppression primarily in young children with congenital haemophilia [46]. Although ITI could be utilized in the management of both conditions, the actual mechanisms of the effect, dosage schedules should be fundamentally different. ITI for alloantibodies is a typical desensitizing therapy in the immunological sense, so large daily doses of FVIII are given for a prolonged period aiming at 'exhausting' the alloantibody producing clones. The duration of ITI is generally between some months and one year, and the addition of vigorous immunosuppression is of doubtful importance. On the other hand, in the ITI treatment of acquired haemophilia the FVIII administration serves to enhance the stimulation of the autoantibody-producing lymphocyte clones and is a useful adjuvant to immunosuppression. The duration of therapy is limited to some weeks. For successful ITI in autoantibody patients small, repeated FVIII doses seem to provide the adequate stimulation for the subsequent successful immunosuppression and there is no obvious need for the 'exhaustive' high-dose FVIII administration, as in the original Bonn and Malmö protocols.

The theoretical basis for the development of ITI in acquired haemophilia [140–142] was derived from the successes of plasma exchange therapy for progressive autoimmune disorders unresponsive to conventional immunosuppressive treatment. It was hypothesized that plasmapheresis induced proliferation of pathogenic clones and that partial clonal depletion could subsequently be produced by giving large doses of cytotoxic drugs during the assumed period of increased B-cell vulnerability. The stimulation induced by plasma exchange was synchronized with pulse immunosuppressive therapy [141,142]. Similarly, in ITI of autoantibody inhibitors, exogenous FVIII administration may result in additional stimulation with a corresponding increase in the susceptibility of the immunocytes to the effect of the cytotoxic drugs [143]. In other words, the repeated administration of the antigen (FVIII) causes extra stimulation of the inhibitor-producing B-cell clones, making them more susceptible to immunosuppression. In the original case report of Green [143], massive doses of FVIII were given simultaneously with 1.5 g intravenous cyclophosphamide to an acquired inhibitor patient previously unresponsive to combined immunosuppressive medication. In this report, the distinction between the acute treatment of bleeding and eradication of inhibitor as the aim of the administration of high-dose FVIII was not yet completely clear. In a later trial in 1989, Lian et al. used a combination of single high-dose FVIII bolus followed by a modified cyclophosphamide, vincristine and prednisone cytostatic protocol for the treatment of serious acute bleeding in acquired haemophilia [123]. In 1996, another successful use of the same protocol was published [144]. On the basis of this historical background, a new aggressive protocol has been developed in 1992 for the ITI treatment of acquired haemophilia patients presenting with serious bleeds [145]. This Budapest protocol consists of 3 weeks of treatment with (1) human FVIII concentrates (30 IU/kg/day for the first week, 20 IU/kg/day for the second week and 15 IU/kg/day for the third week); plus (2) intravenous cyclophosphamide (200 mg/day to a total dose of 2–3 g); plus (3) methylprednisolone (100 mg/day intravenously for the first week and then tapering the dose gradually over the next 2 weeks). After the completion of the 3 weeks of ITI treatment, no further maintenance immunosuppression is given. In contrast to the earlier reports mentioned above, in this regimen FVIII is administered daily in lower doses, simultaneously with cyclophosphamide and steroids, aiming for the rapid disappearance of the inhibitor. ITI resulted in eradication of the autoantibody in more than 90% (present-day: 22/23) of the cases. The main difference between patients treated by ITI vs. traditional immunosuppression was in the time needed for complete disappearance of the inhibitor (4.6 weeks vs. 28.3 weeks). No bleeding-related mortality occurred. ITI should be reserved for the eradication of idiopathic, autoimmune- and malignancy-associated FVIII autoantibodies in patients presenting with severe bleeding. It will be necessary to confirm these initial promising results in multicentre prospective studies. In the late 1990s three leading German haemophilia centres (Bonn, Frankfurt and Heidelberg) also adopted the concept of ITI in the management of acquired haemophilia. Brackmann et al. introduced a modified Malmö protocol consisting of long-term immunoadsorption by Ig-apheresis, plus high-dose (100–200 IU/kg/day) human FVIII, plus HD-IVIG, plus cyclophosphamide, plus steroids and rFVIIa for acute haemorrhages. In the Heidelberg modification of this regimen FVIII is given as a 200-IU/kg bolus followed by high-dose continuous infusion aiming to achieve a FVIII:C level greater than 60% of normal. Recently, an oral ITI regimen has been reported in acquired haemophilia [146]. These preliminary data indicate that ITI regimens may be applied successfully in acquired haemophilia, but further studies are warranted to establish feasibility and cost–benefit relationships [114].

References

1 Addiego J, Kasper C, Abildgaard C, Hilgartner M, Lusher J, Glader B, Aledort L. Frequency of inhibitor development in

haemophiliacs treated with a low-purity factor VIII. *Lancet* 1993; **342**: 462–4.
2. Lusher JM, Arkin S, Abildgaard CF, Schwartz RS. Recombinant factor VIII for the treatment of previously untreated patients with hemophilia. Safety, efficacy and development of inhibitors. Kogenate Previously Untreated Patient Study Group. *N Engl J Med* 1993; **328**: 453–9.
3. Bray GL, Gomperts ED, Courter S, Gruppo R, Gordon EM, Manco-Johnson M, Shapiro A *et al*. A multicenter study of recombinant factor VIII (Recombinate): Safety, efficacy and inhibitor risk in previously untreated patients with hemophilia A. The Recombinate Study Group. *Blood* 1994; **83**: 2428–35.
4. Ehrenforth S, Kreuz W, Scharrer I *et al*. Incidence of development of factor VIII and factor IX inhibitors in haemophiliacs. *Lancet* 1992; **339**: 594–8.
5. Lottenberg R, Kentro TB, Kitchens CS. Acquired hemophilia. A natural history study of 16 patients with factor VII inhibitor receiving little or no therapy. *Arch Intern Med* 1987; **147**: 1077–81.
6. Shapiro SS, Hultin M. Acquired inhibitors to the blood coagulation factors. *Sem Thromb Haemostas* 1975; **1**: 336–85.
7. Green D, Lechner K. A survey of 215 non-hemophilic patients with inhibitors to factor VIII. *Thromb Haemost* 1981; **45**: 200–3.
8. Morrison AE, Ludlam CA, Kessler CM. Use of porcine factor VIII in the treatment of patients with acquired hemophilia. *Blood* 1993; **81**: 1513–20.
9. Bossi P, Cabane J, Ninet J, Dhote R, Hanslik T, Chosidow O *et al*. Acquired hemophilia due to factor VIII inhibitors in 34 patients. *Am J Med* 1998; **105**: 400–8.
10. Margolius A, Jackson DP, Ratnoff OD. Circulating anti-coagulants. a study of 40 cases and review of the literature. *Medicine* 1961; **40**: 145–202.
11. Soriano RM, Matthews JM, Guerado-Parra E. Acquired haemophilia and rheumatoid arthritis. *Br J Rheumatol* 1987; **26**: 381–3.
12. Ishikawa O, Tamura A, Ohnishi K *et al*. Pemphigus vulgaris associated with acquired hemophilia A due to factor VIII inhibitor. *Acta Derm Venereol* 1993; **73**: 229–30.
13. O'Reilly RA, Hamilton RD. Acquired hemophilia, meningioma and diphenylhydantoin therapy. *J Neurosurg* 1980; **53**: 600–5.
14. English KE, Brien WF, Howson-Jan K, Kovacs MJ. Acquired factor VIII inhibitor in a patient with chronic myelogenous leukemia receiving interferon-alfa therapy. *Ann Pharmacother* 2000; **34**: 737–9.
15. Tiplady CW, Hamiliton PJ, Galloway MJ. Acquired haemophilia complicating the remission of a patient with non-Hodgkin's lymphoma treated by fludarabine. *Clin Lab Haematol* 2000; **22**: 163–5.
16. Brodeur GM, O'Neill PJ, Williams JA. Acquired inhibitors of coagulation in nonhemophilic children. *J Pediatr* 1980; **96**: 439–41.
17. Alumkal J, Rice L, Vempathy H, McCarthy JJ, Riggs SA. Surgery-associated factor VIII inhibitors in patients without hemophilia. *Am J Med Sci* 1999; **318**: 350–2.
18. Hay CRM. Acquired haemophilia. *Baillière's Clin Haematol* 1998; **11**: 287–303.
19. Michiels JJ. Acquired hemophilia A in women postpartum. Clinical manifestations, diagnosis, and treatment. *Clin Appl Thrombosis/Hemostasis* 2000; **6**: 82–6.
20. Michiels JJ, Bosch LJ, van der Plas PM *et al*. Factor VIII inhibitor postpartum. *Scand J Haematol* 1978; **20**: 97–107.
21. Ries M, Wolfel D, Maier-Brandt B. Severe intracranial hemorrhage in a newborn infant with transplacental transfer of an acquired factor VIII. C inhibitor. *J Pediatr* 1995; **127**: 649–50.
22. Vicente V, Alberca I, Gonzalez R, Alergre A. Normal pregnancy in a patient with a postpartum factor VIII inhibitor. *Am J Hematol* 1987; **24**: 107–9.
23. Frick PG. Hemophilia-like disease following pregnancy. *Blood* 1953; **8**: 598–9.
24. Voke J, Letssky E. Pregnancy and antibody to factor VIII. *J Clin Pathol* 1997; **30**: 928–32.
25. Coller BX, Hultin MB, Hoyer LW *et al*. Normal pregnancy in a patient with a prior postpartum factor VIII inhibitor with observations on the pathogenesis and prognosis. *Blood* 1981; **58**: 619–24.
26. Hauser I, Schneider B, Lechner K. Postpartum factor VIII inhibitors. A review of the literature with special reference to the value of steroid and immunosuppressive treatment. *Thromb Haemost* 1995; **73**: 1–5.
27. Lozner EL, Jolliffe LS, Taylor FHL. Hemorrhagic diathesis with prolonged coagulation time associated with a circulating anticoagulant. *Am J Med Sci* 1940; **199**: 318–27.
28. Hauser I, Lechner K. Solid tumors and factor VIII antibodies. *Thromb Haemost*, 1999; **82**: 1005.
29. Sallah S, Wan JY. Inhibitors against factor VIII in patients with cancer. *Cancer* 2001; **91**: 1067–74.
30. Sallah S, Nguyen NP, Abdallah JM *et al*. Acquired hemophilia in patients with hematologic malignancies. *Arch Pathol Lab Med* 2000; **124**: 730–4.
31. Hay CRM, Negrier C, Ludlam CA. The treatment of bleeding in acquired haemophilia with recombinant factor VIIa: a multicentre study. *Thromb Haemost* 1997; **78**: 1463–7.
32. Lusher JM, Shapiro SS, Palascek JE, Rao AV, Levine PH, Blatt PM and the Hemophilia Study Group. Efficacy of prothrombin complex concentrates in hemophiliacs with antibodies to factor VIII. A multicenter therapeutic trial. *N Engl J Med* 1980; **303**: 421–5.
33. Green D, Rademaker AW, Briet E. A prospective randomized trial of prednisolone and cyclophosphamide in the treatment of patients with factor VIII autoantibodies. *Thromb Haemost* 1993; **70**: 753–7.
34. Hoyer LW, Gawryl MS, de la Fuente B. Immunological characterization of factor VIII inhibitors. In: Hoyer LW, ed. *Factor VIII Inhibitors*. New York: Liss, 1995: pp. 73–85.
35. Scandella D, Mattingly M, de Graaf S, Fulcher CA. Localization of epitopes for human factor VIII inhibitor antibodies by immunoblotting and antibody neutralization. *Blood* 1989; **74**: 1618–26.
36. Scandella D, Gilbert GE, Shima M, Eagleson C, Felch M, Prescott T, Rajalakshmi KJ *et al*. Some factor VIII inhibitor antibodies recognize a common epitope corresponding to C2 domain amino acids 2248–2312 which overlap a phospholipid binding site. *Blood* 1995; **86**: 1811–19.
37. Prescott R, Nakai H, Saemko EL *et al*. The inhibitor antibody response is more complex in hemophilia A patients than in most nonhemophiliacs with factor VIII autoantibodies. *Blood* 1997; **89**: 3663–71.
38. Nogami K, Shima M, Giddings JC *et al*. Circulating factor VIII immune complexes in patients with type 2 acquired hemophilia A and protection from activated protein C-mediated proteolysis. *Blood* 2001; **97**: 669–77.
39. Kasper CK. Laboratory tests for factor VIII inhibitors, their variation, significance and interpretation. *Blood Coagul Fibrinolysis* 1991; **2** (Suppl. 1): 7–10.
40. Gawryl MS, Hoyer LW. Inactivation of factor VIII coagulant activity by two different types of human antibodies. *Blood* 1982; **60**: 1103–9.
41. Kessler CM, Ludlam CA. The treatment of acquired factor VIII inhibitors: worldwide experience with porcine factor VIII concentrate. *Semin Hematol* 1993; **30**: 22–7.

42 Green D. The management of factor VIII inhibitors in non hemophilic patients. *Prog Clin Biol Res* 1984; **150**: 337–52.
43 Hay CRM, Colvin BT, Ludlam CA, Hill FGH, Preston FE. Recommendations for the treatment of factor VIII inhibitors: from the UK Haemophilia Centre Directors' Organisation Inhibitor Working Party. *Blood Coagul Fibrinolysis* 1996; **7**: 134–8.
44 Kessler CM. An introduction to factor VIII inhibitors. The detection and quantitation. *Am J Med* 1991; **91** (Suppl. 5A) 1S–5S.
45 Kessler CM. Factor VIII inhibitors—an algorithmic approach to treatment. *Semin Hematol* 1994; **31** (Suppl. 4): 33–6.
46 Cohen AJ, Kessler CM. Acquired inhibitors. In: Lee CA, ed. Haemophilia. *Baillière's Clinical Haematology* 1996; **9**: 331–54.
47 Drohan W. Current infectious risks of plasma products. *Haemophilia* 2000; **6** (Suppl. 2): 8S–9S.
48 Lusher JM. Factor VIII inhibitors with recombinant products: prospective clinical trials. *Haematologica* 2000; **85** (Suppl. 10): 2S–6S.
49 Orringer EP, Koury MJ, Blatt PM. Hemolysis caused by factor VIII concentrates. *Arch Intern Med* 1976; **136**: 1018–20.
50 Hach-Wunderle V, Teixidor D, Zumpe P, Kühnl P, Scharrer I. Anti-A in factor VIII concentrate: a cause of severe hemolysis in a patient with acquired factor VIII C antibodies. *Infusionstherapie* 1989; **16**: 100–1.
51 Berntorp E, Ekman M, Gunnarsson M, Nilsson IM. Variation in factor VIII inhibitor reactivity with different commercial factor VIII preparations. *Haemophilia* 1996; **2**: 95–9.
52 Hodge G, Han P. Effect of factor VIII concentrate on antigen-presenting cell (APC)/T-cell interactions *in vitro*: relevance to inhibitor formation and tolerance induction. *Br J Haematol* 2000; **109**: 195–200.
53 Berntorp E. Immune tolerance induction: recombinant vs. human-derived product. *Haemophilia* 2001; **7**: 109–13.
54 Kasper CK. The therapy of factor VIII inhibitors. *Prog Hemostas Thrombosis* 1989; **9**: 57–86.
55 Blatt PM, White IIGC, McMillan CW, Roberts HR. Treatment of antifactor VIII antibodies. *Thromb Haemost* 1977; **38**: 514–23.
56 Lloyd JV, Street AM, Berry E *et al.* Cross-reactivity to porcine factor VIII of factor VIII inhibitors in patients with haemophilia in Australia and New Zealand. *Aust N Z J Med*, 1997; **27**: 658–64.
57 Gribble J, Garvey MB. Porcine factor VIII provides clinical benefit to patients with high levels of inhibitors to human and porcine factor VIII. *Haemophilia* 2000; **6**: 482–6.
58 Hay CRM, Bolton-Maggs P. Porcine factor VIIIC in the management of patients with factor VIII inhibitors. *TransfusMed Rev* 1991; **5**: 145–51.
59 Gatti L, Mannucci PM. Use of porcine factor VIII in the management of seventeen patients with factor VIII antibodies. *Thromb Haemost* 1984; **51**: 379–84.
60 Gringeri A, Santagostino E, Tradati F. Adverse effects of treatment with porcine factor VIII. *Thromb Haemost* 1991; **65**: 245–7.
61 Brettler DB, Forsberg AD, Levine PH *et al.* The use of porcine factor VIII concentrate (Hyate: C) in the treatment of patients with inhibitor antibodies to factor VIII. *Arch Intern Med* 1989; **149**: 1382–5.
62 Morrison AE, Ludlam CA. The use of porcine factor VIII in the treatment of patients with acquired hemophilia: the United Kingdom experience. *Am J Med* 1991; **91**(Suppl. 5a): 23S–26S.
63 Hay CRM. Innovative use of porcine factor VIII C for immune tolerance induction. *Am J Med* 1991; **91** (Suppl 5a): 27S–29S.
64 Kernoff PBA. Porcine factor VIII: preparation and use in treatment of inhibitor patients. *Prog Clin Biol Res* 1984; **150**: 207–24.
65 Bona RD, Riberio M, Klatsky AU, Panek S, Magnifico M, Rickles FR. Continuous infusion of porcine factor VIII for the treatment of patients with factor VIII inhibitors. *Semin Hematol* 1993; **30** (Suppl 1): 32S–35S.
66 Mannucci PM. Desmopressin: a nontransfusional form of treatment for congenital and acquired bleeding disorders. *Blood* 1988; **72**: 1449.
67 Hashemi S, Palmer DS, Aye MT, Ganz PR. Platelet-activating factor secreted by DDAVP-treated monocytes mediates von Willebrand factor release from endothelial cells. *J Cell Physiol* 1993; **154**: 496–505.
68 Mannucci PM, Rugger ZM, Pareti FI, Capitanio A. A new pharmacological approach to the management of hemophilia and von Willebrand disease. *Lancet* 1977; **i**: 869–72.
69 Mannucci PM, Federici AB. Management of inherited von Willebrand disease. In: Michiels JJ, ed. Von Willebrand factor and von Willebrand disease. *Best Practice and Research Clinical Haematology* 2001; 14455–62.
70 De la Fuente B, Panek S, Hoyer LW. The effect of 1-deamino 8 D-arginine vasopressin (DDAVP) in a nonhaemophilic patient with an acquired type II inhibitor. *Br J Haematol* 1985; **59**: 127.
71 Hasson DM, Poole AE, de la Fuente B, Hoyer LW. The dental management of patients with spontaneous acquired factor VIII inhibitors. *J Am Dent Assoc* 1986; **113**: 633.
72 Naorose-Abidi SM, Bond LR, Chitolie A, Bevan DH. Desmopressin therapy in patients with acquired factor VIII inhibitors. *Lancet* 1988; **i**: 366.
73 Chistolini A, Ghirardini A, Tirindelli MC, Moretti T, Mancine F, Di Paolantonia T, Mariani G. Inhibitor to factor VIII in a non-haemophilic patient. Evaluation of the response to DDAVP and the in vitro kinetics of factor VIII. A case report. *Nouv Rev Fr Hematol* 1987; **29**: 221.
74 Muhm M, Grois N, Kier P *et al.* 1-Deamino-8-D-arginine vasopressin in the treatment of non-haemophilic patients with acquired factor VIII inhibitor. *Haemostasis* 1990; **20**: 15.
75 Vincente V, Alberca I, Gonzales R, Lopez Borrasca A. DDAVP in a non-haemophilic patient with an acquired factor VIII inhibitor [letter]. *Br J Haematol* 1985; **60**: 585.
76 Nilsson IM, Lethagen S. Current status of DDAVP formulations and their use. *Excerpta Med* 1991; **943**: 443.
77 Mudad R, Kane WH. DDAVP in acquired hemophilia A. Case report and review of the literature. *Am J Hematol* 1993; **43**: 295–9.
78 Barrowcliffe TW, Kemball-Cook G, Gray E. Overcoming factor VIII inhibitors: a possible new approach. *Scand J Haematol* 1984; **33** (Suppl. 40): 207–12.
79 Littlewood JD, Barrowcliffe TW. The development and characterization of antibodies to human factor VIII in haemophilic dogs. *Thromb Haemost* 1987; **57**: 314–21.
80 Váradi K, Elõdi S. Protection of platelet surface bound factors IXa and VIII against specific inhibitors. *Thromb Haemost* 1982; **47**: 32–5.
81 Bloom AL, Hutton RD. Fresh-platelet transfusions in haemophilic patients with factor-VIII antibody. *Lancet* 1975; **2**: 369–70.
82 Giles AR, Mann KG, Nesheim ME. A combination of factor Xa and phosphatidylcholine-phosphatidylserine vesicles bypasses factor VIII *in vivo*. *Br J Haematol* 1988; **69**: 491–7.
83 Muhleman AF, Glueck HE, Miller MA, Coots M. Factor VIII inhibitors: in vivo decrease of inhibitory activity during calcium infusion. *Am J Hematol* 1985; **20**: 107–17.
84 Sahu S, Raipancholia R, Pardiwalla FK, Pathare AV. Hemostasis in acquired hemophilia—role of inracavitary instillation of EACA. *J Postgrad Med* 1996; **42**: 88–90.

85 Francesconi M, Korniger C, Thaler E, Niessner H, Hocker P, Lechner K. Plasmapheresis: 1st value in the management of patients with antibodies to factor VIII. *Haemostasis* 1982; **11**: 79–86.

86 Mokrzycki MH, Kaplan AA. Therapeutic plasma exchange. complications and management. *Am J Kidney Dis* 1994; **23**: 817–27.

87 Knöbl P, Derfler K. Extracorporeal immunoadsorption for the treatment of haemophilic patients with inhibitors to factor VIII or IX. *Vox Sang* 1999; **77** (Suppl. 1): 57S–64S.

88 Gjörstrup P, Berntorp E, Larsson L, Nilsson IM. Kinetic aspects of the removal of IgG and inhibitors in hemophiliacs using protein A immunoadsorption. *Vox Sang* 1991; **61**: 244–50.

89 Négrier C, Dechavanne M, Alfonsi F, Tremisi PJ. Successful treatment of acquired factor VIII antibody by extracorporeal immunoadsorption. *Acta Haematol* 1991; **85**: 107–10.

90 Jansen M, Schmaldienst S, Banyai S et al. Treatment of coagulation inhibitors with extracorporeal immunoadsorption (Ig-Therasorb). *Br J Haematol* 2001; **112**: 91–7.

91 Turecek PL, Váradi K, Gritsch H, Auer W, Pichler L, Eder G, Schwartz HP. Factor Xa and prothrombin: mechanism of action of FEIBA. *Vox Sang* 1999; **77** (Suppl. 1): 72S–79S.

92 Kurczynski EM, Penner JA. Activated prothrombin concentrate for patients with factor VIII inhibitors. *N Engl J Med* 1974; **291**: 164–7.

93 Kelly PE, Penner JA. Antihemophilic factor inhibitors. Management with prothrombin complex concentrates. *JAMA* 1976; **236**: 2061–4.

94 Penner JA, Kelly PE. Management of patients with factor VIII or IX inhibitors. *Semin Thromb Hemost* 1975; **1**: 386–99.

95 Lusher JM. Controlled clinical trials with prothrombin complex concentrates. *Prog Clin Biol Res* 1984; **150**: 277–90.

96 Sjamsoedin LJM, Heijnen L, Mauzer-Bunschoten EP. The effect of activated prothrombin complex concentrate (FEIBA) on joint and muscle bleeding in patients with hemophilia A and antibodies to factor VIII: a double-blind clinical trial. *N Engl J Med* 1981; **305**: 717–21.

97 Lusher JM, Blatt PM, Penner JA. Autoplex versus Proplex: a controlled, double-blind study of effectiveness in acute hemarthosis in hemophiliacs with inhibitors to factor VIII. *Blood* 1983; **62**: 1135–8.

98 Lusher JM. Use of factor VIII bypassing agents to control bleeding in patients with acquired factor VIII inhibitors. In: Kessler CM, ed. Acquired hemophilia. *Excerpta Medical*, 2nd edn, 1995: 113–29.

99 White IIGC. Seventeen years' experience with Autoplex/Autoplex T: evaluation of inpatients with severe haemophilia A and factor VIII inhibitors at a major haemophilia center. *Haemophilia* 2000; **6**: 508–12.

100 Lusher JM. Perspectives on the use of factor IX complex concentrates in the treatment of bleeding in persons with acquired factor VIII inhibition. *Am J Med* 1991; **91** (Suppl. 5a): 30S–34S.

101 Chavin SI, Siegel DM, Rocco TA, Olson JP. Acute myocardial infarction during treatment with an activated prothrombin complex concentrate in a patient with factor VIII deficiency and a factor VIII inhibitor. *Am J Med* 1988; **85**: 245–9.

102 Roberts HR. The use of agents that by-pass factor VIII inhibitors in patients with haemophilia. *Vox Sang* 1999; **77** (Suppl. 1): 38–41.

103 Hedner U, Kisiel W. Use of human factor VIIa in treatment of two hemophilia A patients wth high titer inhibitors. *J Clin Invest* 1983; **71**: 1836–41.

104 Hedner U, Bjoern S, Bernvil SS. Clinical experience with human plasma-derived factor VIIa in patients with hemophilia A and high titer inhibitors. *Haemostasis* 1989; **19**: 335–43.

105 Hagen FS, Gray CL, O'Hara P. Characterization of a cDNA coding for human factor VII. *Proc Natl Acad Sci USA* 1986; **86**: 1382–6.

106 Hedner U, Feldstedt M, Glaser S. Recombinant FVIIa in hemophilia treatment. In: Lusher, JM, Kessler, CM, eds. *Hemophilia and Von Willebrand's Disease in the 1990s*. Elsevier Science Publishers B.V., Amsterdam, 1991: 283–92.

107 Majundar G, Phillips JK, Lavallee H, Savidge GF. Acquired haemophilia in association with type III von Willebrand's disease: successful treatment with high purity von Willebbrand's factor and recombinant factor VIIa. *Blood Coagul Fibrinolysis* 1993; **4**: 1035–7.

108 Shafi T, Jeha MT, Black L, Al Douri M. Severe acquired haemophilia A treated with recombinant factor VIIa. *Br J Haematol* 1997; **98**: 910–12.

109 Maliekel K, Rana N, Green D. Recombinant factor VIIa in the management of a pseudotumor in acquired haemophilia. *Haemophilia* 1997; **3**: 54–8.

110 Papadaki HA, Xylouri I, Valatas W, Petinarkis J, Kontopoulou I, Eliopoulos GD. Severe acquired hemophilia A successfully treated with activated recombinant human factor VII. *Ann Hematol* 1998; **77**: 123–5.

111 Hay CRM, Negrier C, Ludlam CA. The treatment of bleeding in acquired haemophilia with recombinant factor VIIa: a multicentre study. *Thromb Haemost* 1997; **79**: 1463–7.

112 Stein SF, Duncan A, Cutter D. Disseminated intravascular coagulation (DIC) in a hemophiliac treated with recombinant factor VIIa. *Blood* 1990; **76** (Suppl. 1): 1743S.

113 Scharrer I. Recombinant factor VIIa for patients with inhibitors to factor VIII or IX or factor VII deficiency. *Haemophilia* 1999; **5**: 253–9.

114 Kessler CM. Acquired factor VIII autoantibody inhibitors. current concepts and potential therapeutic strategies for the future. *Haematologica* 2000; **85** (Suppl. 10): 57–63.

115 Rubinger M, Rivard GM, Teitel J, Walker IL. Suggestions for the management of factor VIII inhibitors. *Haemophilia* 2000; **6** (Suppl. 1): 52S–59S.

116 Morrison AE, Ludlam CA. Acquired hemophilia and its management. *Br J Haematol* 1995; **89**: 231–6.

117 Sultan Y, Maisonneuve P, Kazatchkine MD, Nydegger UE. Anti-idiotypic suppression of auto-antibodies to factor VIII (anti-haemophilic factor) by high-dose intravenous gammaglobulin. *Lancet* 1984; **ii**: 765–8.

118 Sultan Y. Acquired hemophilia and its treatment. *Blood Coagul Fibrinolysis* 1997; **8** (Suppl. 1): 15S–18S.

119 Sultan Y, Kazatchkine MD, Algiman M, Dietrich G, Nydegger U. The use of intravenous immunoglobulins in the treatment of factor VIII inhibitors. *Semin Hematol* 1994; **31** (Suppl. 4): 65S–66S.

120 Schwartz RS. Gabriel DA, Aledort LM, Green D, Kessler CM. A prospective study of acquired (autoimmune) factor VIII inhibitors with high-dose intravenous gammaglobulin. *Blood* 1995; **86**: 797–804.

121 McColl MD, Omran A, Walker ID, Lowe GD. Acute renal failure complicating high-dose intravenous immunoglobulin therapy for acquired haemophilia. *Haemophilia* 1999; **5**: 124–6.

122 Spero JA, Lewis JH, Hasiba U. Corticosteroid therapy for acquired factor VIII. C inhibitors. *Br J Haematol* 1981; **48**: 639–42.

123 Lian ELY, Larcada AF, Chiu AYZ. Combination immunosuppressive therapy after factor VIII infusion for acquired factor VIII inhibitor. *Ann Intern Med* 1989; **110**: 774–8.

124 Green D, Rademaker AW, Briet E. A prospective, randomized trial of prednisone and cyclophosphamid in the treatment of

patients with factor VIII autoantibodies. *Thromb Haemost* 1993; **70**: 753–7.
125 Shaffer LG, Phillips MD. Successful treatment of acquired hemophilia with oral immunosuppressive therapy. *Ann Intern Med* 1997; **127**: 206–9.
126 Green D. Oral immunosuppressive therapy for acquired hemophilia. *Ann Intern Med* 1998; **128**: 325.
127 Pfiegler Gy Boda Z, Hársfalvi J, Flóra-Nagy M, Sári G, Pecze K, Rák K. Cyclosporine treatment of a woman with acquired haemophilia due to factor VIII. C inhibitor. *Postgrad Med J* 1989; **65**: 400–2.
128 Hart HC, Kraaijenhagen RJ, Kreckhaert JA, Verdel G, Freen M, van de Weil A. A patient with a spontaneous factor VIII. C autoantibody: successful treatment with cyclosporine. *Transplant Proc* 1988; **20** (Suppl. 4): 323–8.
129 Shulman S, Langevitz P, Livneh A, Martinowitz U, Seligsohn U, Varon D. Cyclosporine therapy for acquired factor VIII inhibitor in a patient with systemic lupus erythematosus. *Thromb Haemost* 1996; **76**: 344–6.
130 Schwedtfeger R, Hintz G, Huhn G. Successful treatment of a patient with postpartum factor VIII inhibitors with recombinant human interferon alpha-2a. *Am J Hematol* 1991; **45**: 190–3.
131 Castenkiold EC, Colvin BT, Kelsey SM. Acquired factor VIII inhibitor associated with chronic interferon-alpha therapy in a patient with haemophilia A. *Br J Haematol* 1994; **87**: 434–6.
132 Makris M, Preston FE. Interferon-alpha treatment and formation of factor VIII antibodies. *Ann Intern Med* 1997; **126**: 829–30.
133 Maloney DG. Advances in immunotherapy of hematologic malignancies. *Curr Opin Hematol* 1998; **5**: 237–43.
134 Ahrens N, Kingreen D, Seltsam A, Salama A. Treatment of refractory autoimmune haemolytic anaemia with anti-CD20 (rituximab). *Br J Haematol* 2001; **114**: 244–5.
135 Bauduer F. Rituximab. A very efficient therapy in cold agglutinins and refractory autoimmune haemolytic anaemia associated with CD20-positive, low-grade non-Hodgkin's lymphoma. *Br J Haematol* 2001; **112**: 1085–6.
136 Glennie MJ, Johnson PWM. Clinical trials of antibody therapy. *Immunol Today* 2000; **21**: 403–10.
137 Qian J, Collins M, Sharpe AH, Hoyer LW. Prevention and treatment of factor VIII inhibitors in murine hemophilia A. *Blood* 2000; **95**: 1324–9.
138 Brackmann HH, Gormse J. Massive factor-VIII infusion in haemophiliac with factor-VIII inhibitor, high responder [letter]. *Lancet* 1977; **2**: 933.
139 Di Michele DM. Immune tolerance. A synopsis of the international experience. *Haemophilia* 1998; **4**: 568–73.
140 Schroeder JO, Euler HH, Loffler H. Synchronization of plasmapheresis and pulse cyclophosphamid in severe systemic lupus erythematosus. *Ann Intern Med* 1987; **107**: 344–6.
141 Lupus Plasmapheresis Study Group. Plasmapheresis and subsequent pulse cyclophosphamid versus pulse cyclophosphamid alone in severe lupus: design of the LPSG trial. *J Clin Apheresis* 1991; **6**: 40–7.
142 Jarreusse B, Blinchet P, Gayrand M. Synchronization of plasma exchanges and cyclophosphamide in severe systemic diseases. *Presse Med* 1993; **22**: 293–8.
143 Green D. Suppression of an antibody to factor VIII by a combination of factor VIII and cyclophosphamide. *Blood* 1971; **37**: 381–7.
144 Bucalossi A, Marotta G, De Regis F, Galieni P, Dispensa E. A case report of acquired idiopathic hemophilia successfully treated with immunosuppressive drugs and factor VIII concentrates. *Clin Appl Thromb/Hemost* 1996; **2**: 222–3.
145 Nemes L, Pitlik E. New protocol for immune tolerance induction in acquired hemophilia. *Haematologica* 2000; **85** (Suppl): 64S–68S.
146 Lindren A, Wadenvik H, Tarkowski A, Tengborn L. Does peroral administration of factor VIII induce oral tolerance in patients with acquired haemophilia A? *Thromb Haemost* 2000; **83**: 632–3.

PART 6

Musculoskeletal issues

CHAPTER 17

Pathogenesis of musculoskeletal complications of haemophilia

E.C. Rodriguez-Merchan

About 90% of bleeding complications in haemophilia occur within the joints (haemarthrosis) while 10% of such complications occur within the muscles. This chapter reviews the pathogenesis of articular bleeding episodes and haemophilic arthropathy, the pathogenesis of intramuscular haematomas and pseudotumours, as well as the pathogenesis of the neurological problems related to haemophilia.

Articular problems

Haemarthrosis is undoubtedly the most common and potentially most disabling manifestation of haemophilia, and it is the complications of such bleeding episodes that will bring the patient to the attention of the orthopaedic surgeon. Safran *et al.* [1] carried out an experimental study in order to investigate the effect of a single injection of unpreserved blood on joint stiffness and synovium and cartilage histology in the ankle joints of rabbits at 10 and 28 days following injection. Their results demonstrated that the presence of blood in an otherwise normal joint did not lead to ultimate compromise in the integrity of the cartilage or the joint function. Therapeutic aspiration of an acute post-traumatic haemarthrosis did not appear to be necessary for the prevention of long-term problems.

Roosendaal *et al.* [2] later investigated the direct effect of blood and blood components on human cartilage *in vitro*. They showed that whole blood anticoagulated with heparin, coagulated blood, mononuclear cells, red cells and plasma in this order of potency increased proteoglycan synthesis in a dose-dependent manner. The effect of the combination of mononuclear cells and red cells in concentrations equivalent to those in whole blood was significantly greater than the effects of the isolated components alone and did not differ from that of whole blood. They found that cartilage exposed to this combination for more than 4 days exhibited irreversible inhibition of proteoglycan synthesis. The effect was similar to that of whole blood and the opposite to that of the individual components or other combinations.

Haemarthrosis (synovitis and cartilaginous damage)

If a single joint becomes the site of a recurrent haemarthrosis, this is generally referred to as a target joint. When the joint fails to recover fully between bleeding episodes there is hypertrophy of the synovium and the joint becomes permanently warm and swollen. Arnold and Hilgartner [3] found that hydrolytic enzymes increase in both haemophilic synovium and the joint fluid. Acid phosphatase and cathepsin D may also have a role in maintaining chronic inflammation in the synovium. The release enzymes responsible for the breakdown of protein have a destructive effect not only on the free blood but also, as might be expected, on the synovium, the cartilage and the bone itself. The role of iron deposition in the pathophysiology of haemophilic arthropathy has not been fully established.

Stein and Duthie [4] described well defined cytoplasmic deposits of iron (siderosomes) in the synovial cells, the subsynovial tissues and in chondrocytes of the superficial layers of the articular cartilage. The haemosiderin staining of the synovium and the cartilage bears testimony to the destructive elements of the proteolytic enzymes.

Roosendaal *et al.* [5] obtained synovial specimens from patients with haemophilia and found that the haemosiderotic deposit in the tissue was often adjacent to macroscopically normal tissue in the same joint. Samples from both affected and unaffected synovial tissue were analysed histologically and biochemically in order to determine the catabolic activity. This showed that in patients with haemophilic arthropathy local synovial iron deposits were associated with increased catabolic activity. As early as 4 days after the onset of a single haemarthrosis, the synovium often shows focal areas of villus formation.

Recurrent haemarthroses stimulate the synovium, which in turn hypertrophies within the joint. This hypertrophic synovial tissue comes to occupy space and is likely to be further injured and cause new bleeding. This process initiates, and is responsible for, the chronic synovitis found within the joints of patients affected with haemophilia. Radiological evaluation of affected joints shows a pronounced increase in the density of the joint space and the recesses caused by accumulation of fluid or, alternatively, the hypertrophic synovium.

Ultrasound is a safe and reliable diagnostic technique for distinguishing synovial hypertrophy from a simple effusion or haemarthrosis. Magnetic resonance imaging may also be used as an alternative diagnostic aid in the case of synovial proliferation or joint effusion in patients with haemophilia. The hypertrophic synovium is characterized by villus formation, marked increase in vascularity and chronic inflammatory cells.

The cells in the synovium can absorb a limited amount of iron. However, once the quantity is exceeded the cells may disintegrate and release lysosomes which not only destroy the adjacent articular cartilage but result in further inflammation of the synovium. Blood from breakdown products also has an adverse effect on the chondrocytes. The abnormal synovium will act in a similar manner to that of arteriovenous fistula, producing an increase in blood supply in the area of the growth plates. This inflammatory reaction causes accelerated ossification and growth of the epiphyses, leading to osseous problems and epiphyseal overgrowth in young patients.

Such bone hypertrophy can in turn lead to leg length discrepancies, angular deformity and alteration in structures, especially of the developing skeleton. The symptoms of chronic haemophilic arthropathy classically develop in patients with haemophilia during the second and third decades. In patients in whom growth has been completed and the epiphyses fused, the major effect is seen on the articular cartilage. As damage of articular cartilage progressively worsens, there is a progressive deterioration in the joint surface, with movement becoming increasingly limited and painful. Radiographs show a narrowing of the joint space and reduction and irregularity of the previously congruent joint surfaces [6].

Roosendaal reported that limited exposure of the healthy full-thickness articular cartilage to blood can induce long-lasting inhibition of the synthesis of proteoglycans, one of the main components of the extracellular matrix of cartilage. If this change in chondrocyte activity were to occur *in vivo*, it would inevitably result in irreversible damage to the articular cartilage and joint destruction in general [7]. Roosendaal has stated that blood has a direct harmful effect on cartilage irrespective of synovial changes, and that such cartilaginous changes coincide with the adverse changes in the synovium, but in the first instance precede these changes.

In other words, a single haemarthrosis has a long-lasting effect on the activity of chondrocytes and on the integrity of cartilage and synovium (see Plates 17.1–17.4, facing p. 118). These changes might be irreversible in a very early stage because of chondrocyte death. Moreover, Roosendaal has demonstrated that articular cartilage of younger individuals is more susceptible to blood-induced damage than is the cartilage of older individuals [7]. On the basis of Roosendaal's conclusions and my own personal clinical experiences, it is recommended that a joint should be aspirated (arthrocentesis) after a joint bleeding [8]. The main long-lasting effects of a single haemarthrosis on the activity of the chondrocytes and on the integrity of cartilage and synovium are summarized in Table 17.1.

Table 17.1 Main long-lasting effects of a single haemarthrosis on articular cartilage and synovium [7].

- As early as 4 days after the onset of a single haemarthrosis, the synovium often shows focal areas of villus formation
- Blood has a direct harmful effect on cartilage irrespective of synovial changes; such cartilaginous changes coincide with the adverse changes in the synovium, but in the first instance precede these changes
- Limited exposure of healthy articular cartilage to blood can induce long-lasting inhibition of the synthesis of proteoglycans
- These changes might be irreversible in a very early stage because of chondrocyte death
- Articular cartilage of younger individuals is more susceptible to blood-induced damage than is the cartilage of older individuals

Articular contractures

Joint flexion contractures in haemophilia are common and difficult problems to solve when they become fixed. The joints mainly involved are the elbow, knee and ankle (plantar flexion or equinus). The contracture is initially antalgic, and related to an acute bleeding episode. Therefore, during the first weeks the joint contracture can be reversed with suitable treatment. Later on, it can become a fixed flexion deformity with devastating functional implications. Prompt and efficient treatment of the haemarthrosis is essential, as previously described, and an intensive and long-term physiotherapy regimen is required for total recovery. Unfortunately, in many patients contractures cannot be completely avoided, and fixed-knee flexion deformity may take place. In many cases the fixed deformity at the knee is secondary to a plantar flexion deformity of the ankle. However, it is much more common that the fixed-knee flexion contracture occurs first and the ankle deformity develops later. Once the fixed deformity exists, it is essential to resolve it as soon as possible, and as efficiently as possible [8].

Angular deformities

During childhood, the hyperaemia produced by the hypervascularity of the synovium commonly results in an asymmetrical growth of the epiphyses. At the knee it frequently results in a valgus deformity. At the elbow a hypertrophic radial head is typical, as is a valgus joint deformity at the ankle. Therefore the haemophilic patient will present with a flexed and valgus knee, a valgus ankle with equinus position, and a flexed elbow. Theoretically, osteotomies around the knee and ankle could correct these angular deformities.

Leg length discrepancies

This complication is quite common but fortunately not very significant from the clinical point of view. It is caused by an asymmetrical vascularization of the epiphyses at the level of the knees. Leg length discrepancy of less than 2 cm can be treated simply by means of a 2-cm insole on the shorter leg. It is important to emphasize that leg length discrepancy used to be associated with a flexion deformity of the knee, and an equinus contracture of the ankle. In such a case, the insole should be placed at the level of the heel.

Haemophilic arthropathy

This term is used when a haemophilic joint presents with some radiological changes, indicating cartilaginous involvement. These findings are progressive throughout the patient's life, and eventually result in a severe destruction of the joint, mimicking degenerative osteoarthritis. The most important difference between haemophilic arthropathy and idiopathic osteoarthritis is the lack of correlation between pain and the radiographical signs. Many haemophilic patients will present with severe joint destruction over many years with little or no pain.

The most characteristic radiographical findings of haemophilic arthropathy are subchondral cysts, osteoporosis, enlarged epiphyses, irregular subchondral surface, narrowing of the joint space, erosions of joint margins, gross incongruence of the articulating bone ends, and joint deformity (angulation and/or displacement between articulating bones). In the final stages fibrous or osseous ankylosis will take place.

Muscular problems

More than 10% of bleeding complications in haemophilia occur within the muscles. Prompt treatment of an intramuscular bleed with clotting factors reduces the potential soft tissue complications and conservative orthopaedic management usually resolves the episode without any long-term sequelae. However, repeated and unresolved intramuscular haematomas may lead to the serious but fortunately rare complication of a pseudotumour.

Intramuscular bleeds are generally associated with direct trauma and the diagnosis is usually quite obvious because of the presence of pain, swelling, local warmth and frequently an overlying bruise. The vast majority of these muscle bleeds resolve spontaneously with no detrimental functional sequelae, but one must always carefully assess the patient to ensure that there is no danger of neurovascular compromise [9].

The extent and nature of the haematoma can be readily assessed by ultrasound, which is a reliable and inexpensive diagnostic aid. It is useful in demonstrating not only the size and distribution but can also determine whether the contents are in liquid or solid form. Magnetic resonance imaging and computed tomography scans, while obviously more precise, are expensive.

Treatment is generally conservative and relies upon restoration of normal clotting by administration of factor replacement. The limb is rested in a position of function and elevated if appropriate. Adequate analgesia is provided. Following the initial acute phase, physiotherapy should begin early in the rehabilitation in an attempt to restore full function, bearing in mind that secondary bleeds may subsequently occur.

The muscles of the forearm and of the lower leg in particular are enclosed in tight fascial compartments, and even a relatively small bleed can cause a disproportionate rise in intracompartmental pressure and risk the onset of compartment syndrome which, if left untreated, can progress to the classical Volkmann's ischaemic contracture [10]. It is important to differentiate between compartment syndrome and arterial occlusion, and if there is any doubt it is a relatively straightforward undertaking to measure the intracompartmental pressures. Under certain circumstances a trend of rising pressure is significant, rather than an absolute maximum. Confirmed compartment syndrome is a surgical emergency and requires prompt decompression. A single incision, if too small, may risk an incomplete decompression, and long incisions should be made whenever appropriate in order to achieve adequate decompression of all affected compartments [11]. It is safer dealing with late scarring as a result of a compartment decompression rather than the sequelae of an inadequate decompression.

The complications of compartment syndrome are well known and include reduced function of the nerves that traverse the compartment and late onset of ischaemic muscle contractures. Nerve decompression and later excision of fibrosed and contracted scar tissue may be effective in the management of late complications.

Bleeding into deeply situated muscles may be somewhat difficult to diagnose. Probably the most frequently affected muscle is the iliopsoas which classically presents with a painful flexion deformity of the hip, with any attempt to extend the hip joint resulting in an increase in the level of pain and a compensatory lumbar lordosis. It may also be associated with abdominal pain mimicking acute appendicitis, and signs of compression of the femoral nerve with reduction in sensation over the anterior aspect of the thigh.

A bleed into the iliopsoas has a similar clinical picture to an intra-articular bleed into the hip joint. It is important to distinguish between the two as the treatment differs considerably. Fortunately, ultrasound nicely differentiates the presence of a haemarthrosis in the hip joint, so distinguishing this from an iliopsoas bleed. An intra-articular bleed generally responds promptly to correction of any factor deficiency and possibly aspiration. An iliopsoas bleed does not usually respond promptly to infusion of factor concentrate and a flexion contracture of the hip may persist for several weeks. Secondary haemorrhages into the same area are common and prophylactic treatment is advisable under these circumstances. The leg should be rested and full weight bearing should be avoided, with care taken not to adversely stress the hip.

In the presence of HIV infection or AIDS, any common bacteria may result in sepsis, and abscess formation should be suspected, especially in patients who are immunocompromised and where the symptoms fail to respond to correction of normal clotting. Spontaneous infection of tissues has been reported in two HIV-positive patients with haemophilia [12].

Pseudotumours

In 1965, Fernandez de Valderrama and Matthews [13] first described a haemophilic pseudotumour as a progressive cystic swelling involving muscle. This they felt was the result of recurrent haemorrhage and was generally accompanied by radiographical evidence of bone involvement. They emphasized that the condition should be differentiated from simple bone cysts which occur in the fascial envelope of the muscle without evidence of radiographical change.

The presence of a slowly enlarging mass in the limb or especially the pelvis, in a patient with haemophilia, should raise the suspicion of the possibility of a pseudotumour, although there have been rare reports of malignant tumours simulating a pseudotumour. Koepke and Browner [14] described a chondrosarcoma of the scapula and recently, in the centre in Madrid, we have encountered a patient with liposarcoma presenting as a pseudotumour during an operation to remove a suspected haemophilic lesion.

The pattern of formation of a pseudotumour differs according to its anatomical site [15,16]. The majority of pseudotumours are seen in adults and occur in the long bones, where repeated and unresolved haematomas lead to subsequent encapsulation and calcification, with progressive enlargement of the mass and subsequent erosion of the adjacent bone. The muscles most frequently affected are iliacus, vastus lateralis and soleus. These have a large area of origin in common, they act across joints with a large range of motion, and insert into bones so as to enhance their lever effect. Indirect trauma to these muscles is the most likely cause of bleeding into them.

Gilbert [16] has described two different clinical features for proximal and distal pseudotumours. Proximal pseudotumours occur more frequently in the proximal axial skeleton, especially around the femur and the pelvis. They probably commence in the soft tissues and then secondarily erode the bones from the outside. They have a slow evolution. They occur more frequently in adults and do not respond to conservative treatment. Patients generally present with a painless firm expanding mass which may appear to be multilocular, non-tender and adherent to the deeper structures. Such pseudotumours frequently remain painless and asymptomatic, until the patient presents later with a pathological fracture.

The radiological features are typical with a large soft-tissue mass and areas of adjacent bone destruction. Calcification within the mass is a frequent finding. Pseudotumours of the ilium may cause significant bony erosion with little new periosteal bone formation.

Distal pseudotumours predominantly affect younger skeletally immature patients. They are seen more commonly in children and adolescents and are generally the result of direct trauma. In this group of patients it is not unusual to see such tumours distal to the wrist and the ankle. Unlike the proximal lesions, these distal pseudotumours develop rapidly and appear to be secondary to an intraosseous haemorrhage. They are seen especially in the small cancellous bones such as the calcaneus, talus and metatarsals of the feet, but seldom in the carpus. Occasionally, this appearance and resolution of these pseudotumours has been reported following radiotherapy [17].

A pseudotumour is basically an encapsulated haematoma. It has a thick fibrous capsule surrounding the haematoma which may be in varying states of organization (see Plates 17.5 and 17.6, facing p. 118). Calcification and later ossification may be seen within its wall. Smaller cysts may have a thinner but less adherent capsule containing more fluid blood.

The nature of the classic pseudotumour was accurately described by Duthie et al. [18] on analysing an amputation specimen. Sequential longitudinal sections showed that the haemorrhage was mainly extraosseous and loosely enclosed in fibrous tissue. There was evidence of extensive subperiosteal bleeding and reactive new bone formation which expands both externally and internally, leading to extensive destruction of the involved bone. Histological examination showed the presence of subperiosteal woven bone. There were accumulations in the inflammatory cells, many of which were histiocytes containing copious amounts of siderin.

Neurological problems

The most frequent cause of nerve palsy in haemophilia is compression by a large haematoma, a classical example being crural nerve palsy following an iliopsoas haematoma. Electromyography is a helpful tool for diagnostic and outcome purposes; however, periodic ultrasound scans will show the rate of reabsorption of the haematoma. Joint deformity in advanced cases of haemophilic arthropathy can be another cause of nerve compression. At the level of the elbow, ulnar nerve involvement has been found in some patients. Most nerve palsies in haemophilia are reversible (neuroapraxia) but some of them may require surgery (neurolysis), as in the case of ulnar nerve entrapment at the sulcus formed by the medial epicondyle and the olecranon at the elbow. Of vital importance in haemophilic nerve palsies is the early diagnosis of the original cause, and its resolution by medical measures or by surgery if necessary.

References

1 Safran MR, Johnston-Jones K, Kabo JM, Meals RA. The effect of experimental hemarthrosis on joint stiffness and synovial histology in a rabbit model. *Clin Orthop* 1994; **303**: 280–8.
2 Roosendaal G, Vianen ME, Van den Berg HM *et al*. Cartilage damage as a result of hemarthrosis in a human *in vitro* model. *J Rheumatol* 1997; **24**: 1350–4.
3 Arnold WD, Hilgartner MW. Hemophilic arthropathy. *J Bone Joint Surg Am* 1977; **59A**: 287–305.
4 Stein H, Duthie RB. The pathogenesis of chronic haemophilic arthropathy. *J Bone Joint Surg Br* 1981; **63B**: 601–9.

Plate 2.1 Crystal structure of the FVIII C2 domain bound to the Fab fragment of the human monoclonal inhibitor antibody BO2C11. (a) Ribbon structure of the complex between the C2 domain (red-brown), and BO2C11 heavy and light chains (green and yellow, respectively) [40]. (b) The C2 domain and the antigen binding site of BO2C11 are shown with a rotation of 90° with respect to (a). Residues of the C2 domain at the interface with BO2C11 are shown in ball and stick representation. The three groups of residues labelled in orange (M2199, F2200; L2251, L2252 and V2223) are hydrophobic residues which are predicted to insert into the phospholipid membrane. The blue residues are basic residues predicted to interact by electrostatic interaction with charged phospholipid head groups. Residues contributing to binding to BO2C11 but not directly involved in interactions with phospholipids are labelled in black. (Courtesy of K. Pratt, C. Spiegel and B. Stoddard.)

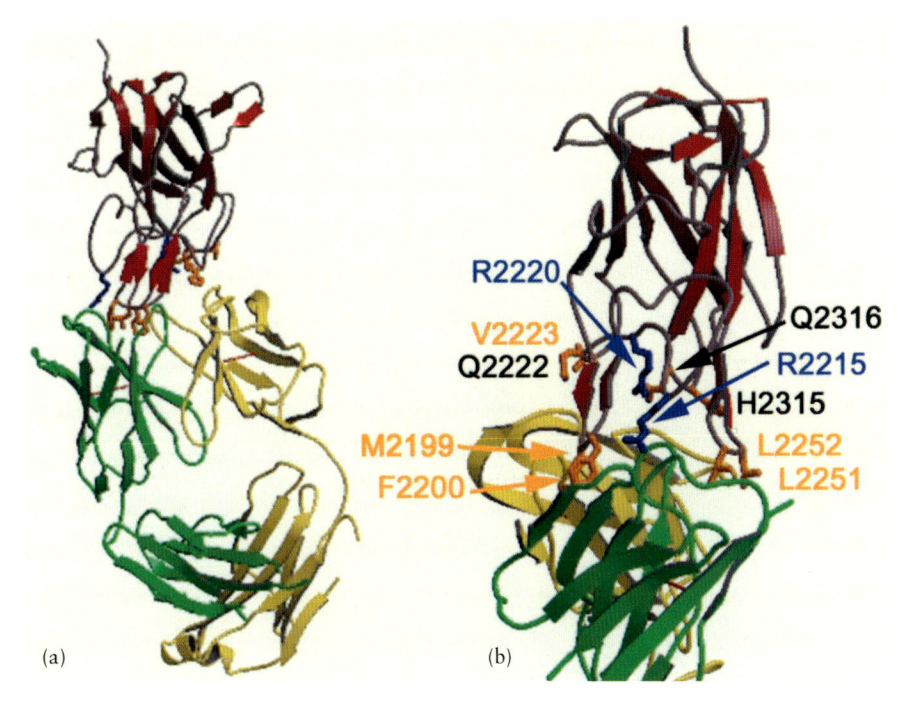

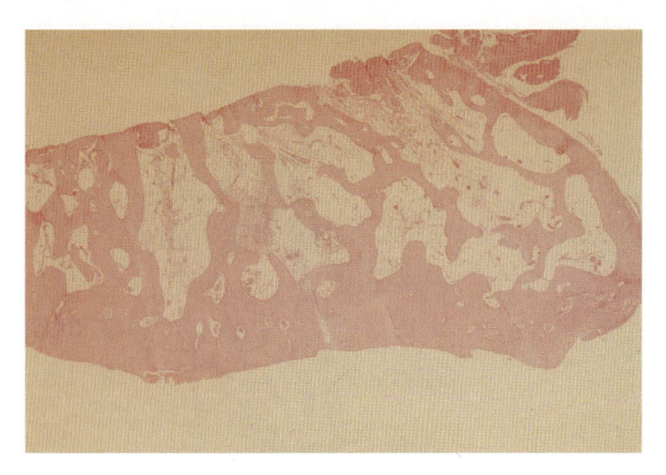

Plate 17.1 Microscopic view of a section of the articular end of a proximal tibia showing a severe degree of haemophilic arthropathy (stain: haematoxylin and eosin).

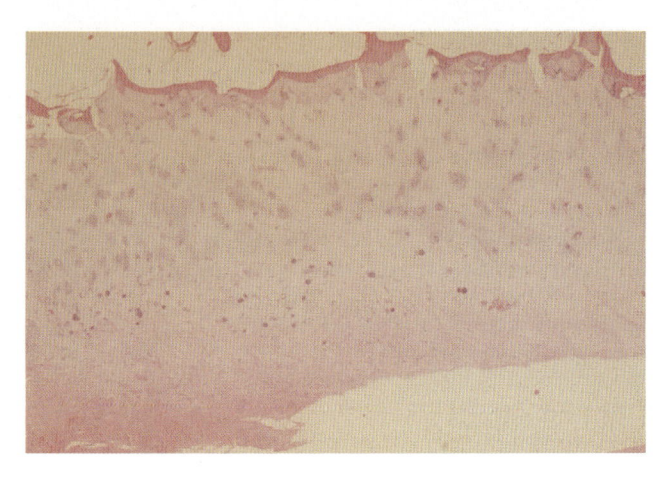

Plate 17.2 Microscopic image of the articular cartilage of a tibial plateau showing haemosiderin deposits within the cytoplasm of a small number of chondrocytes (stain: haematoxylin and eosin).

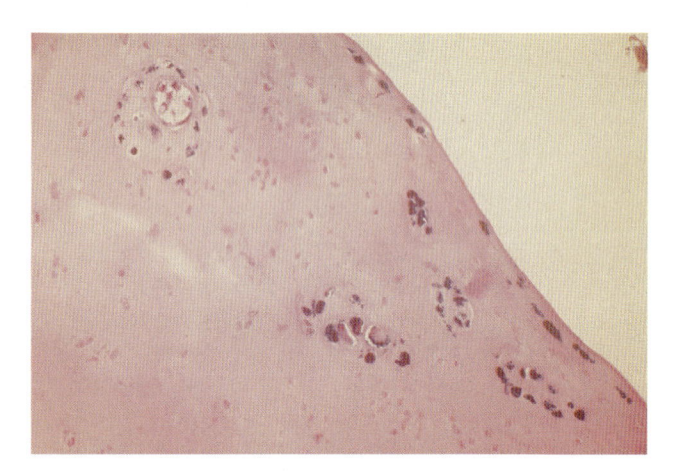

Plate 17.3 Microscopic view of a haemophilic articulating surface with signs of regeneration at the chondrocytes (cellular groups); intracytoplasmatic haemosiderin deposits can also be noted (stain: Perl).

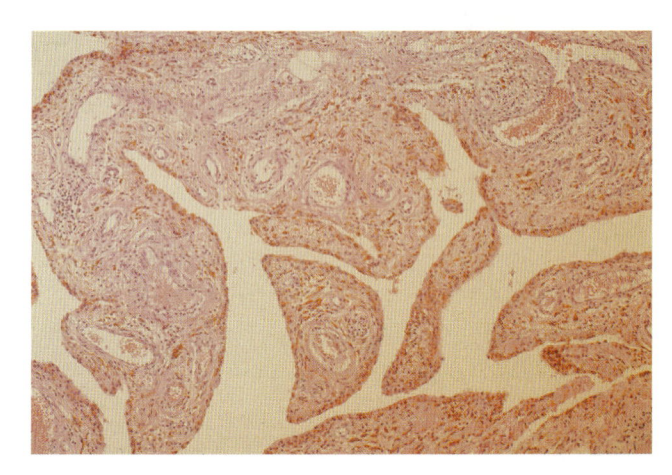

Plate 17.4 Microscopic image of the synovium showing its typical villous shape; intracellular haemosiderin deposits can be noted at the synovial epithelium, and also within the subepithelial macrophages (stain: haematoxylin and eosin).

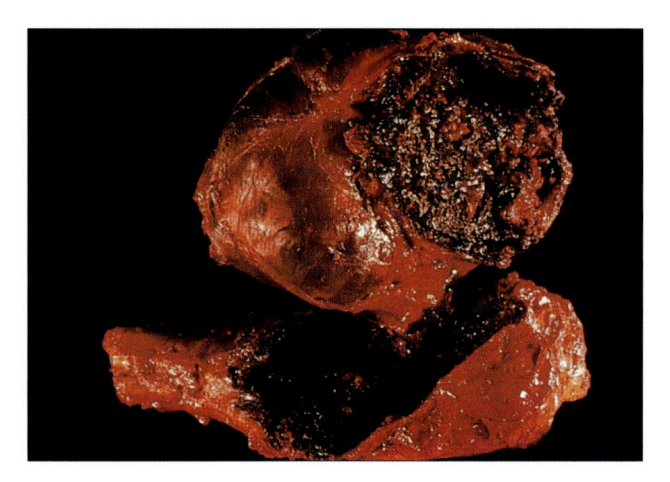

Plate 17.5 Macroscopic view of a surgically removed haemophilic pseudotumour. A large clotting haematic collection can be noted inside, surrounded by a thick fibrous capsule.

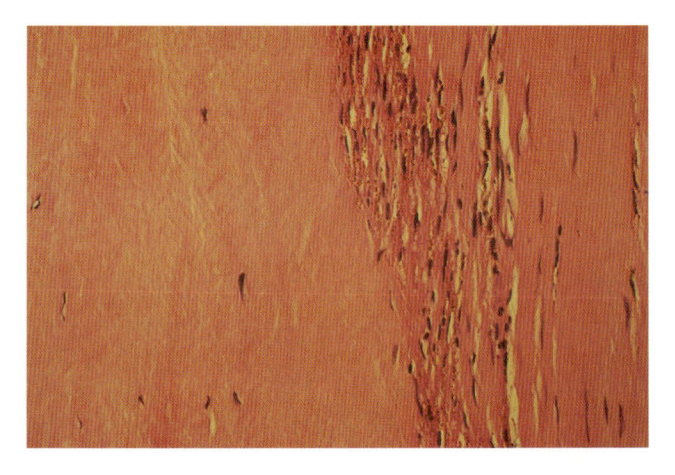

Plate 17.6 Microscopic image of the fibrous capsule of a haemophilic pseudotumour. Note the high amount of collagen within such a capsule as well as some groups of histiocytes containing intracytoplasmic haemosiderin (stain: haematoxylin and eosin).

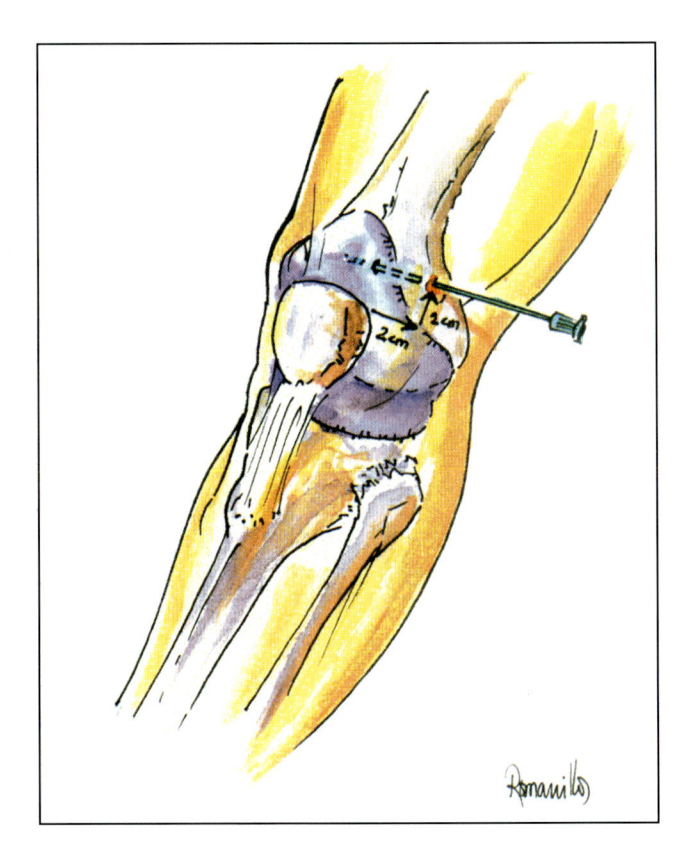

Plate 18.1 Knee arthrocentesis through the suprapatellar lateral route. The injection is made above the lateral corner of the patella and directly into the suprapatellar pouch.

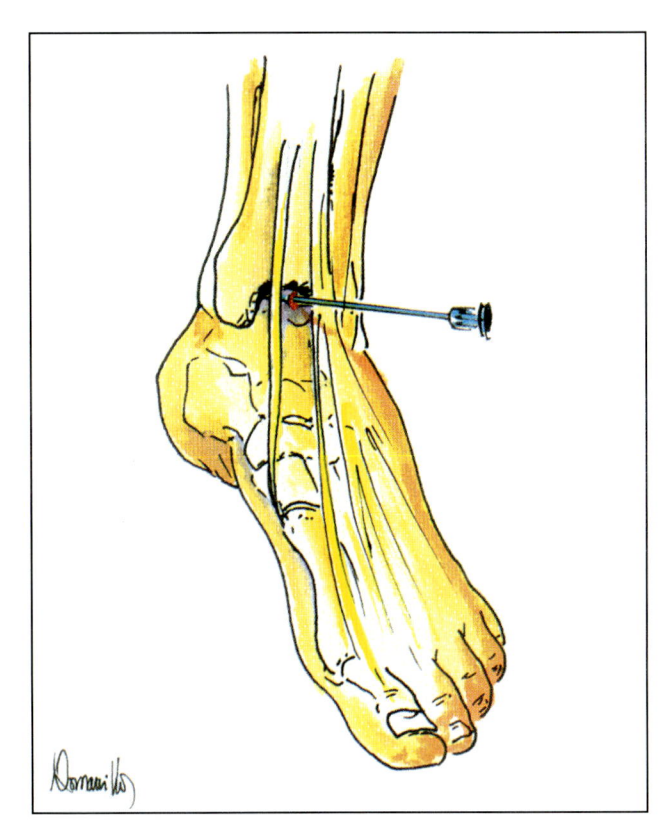

Plate 18.2 Ankle arthrocentesis through the anterior route. Note the site of injection between the tibialis anterior tendon and the extensor digitorum longus.

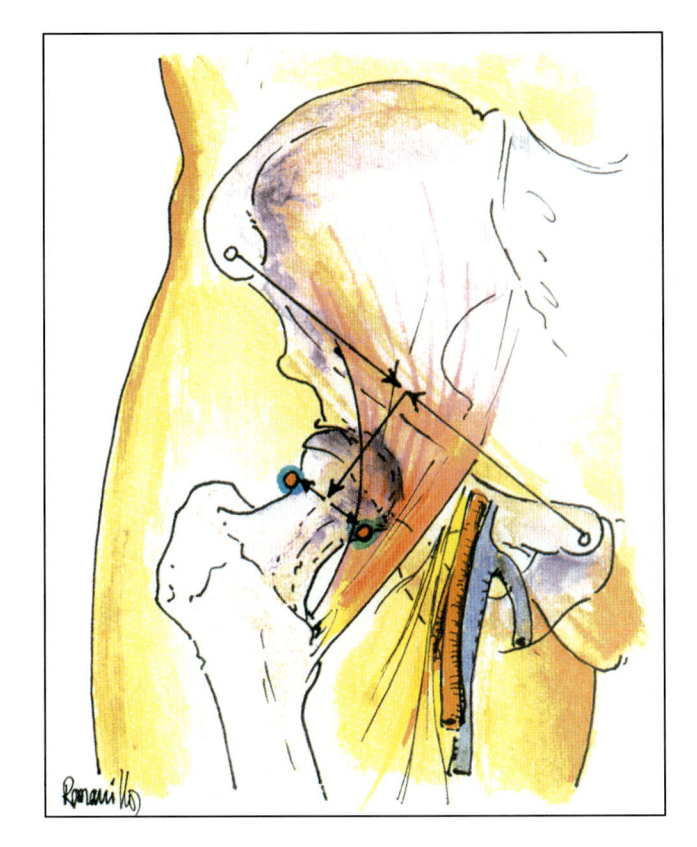

Plate 18.3 Hip arthrocentesis should be carried out with the help of two important cutaneous landmarks: the anterosuperior iliac spine and the pubic tubercle. In this way one can obtain the position of the centre of the femoral neck. Then one can go through the inferomedial route or through the superolateral route.

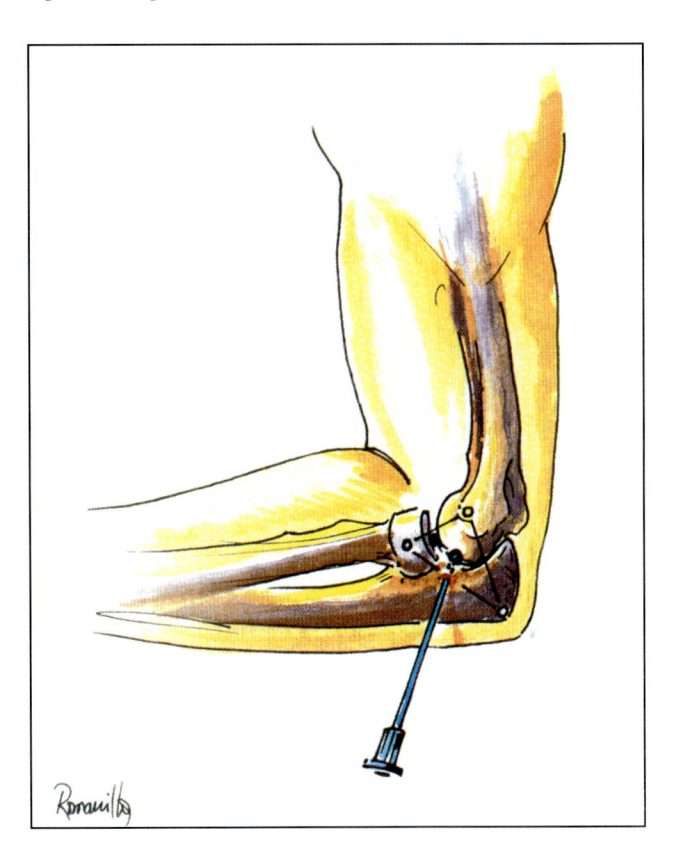

Plate 18.4 Elbow arthrocentesis; with the elbow in lateral view the needle should be inserted in the centre of the triangle formed by the olecranon, the radial head and the lateral epicondyle.

(a)

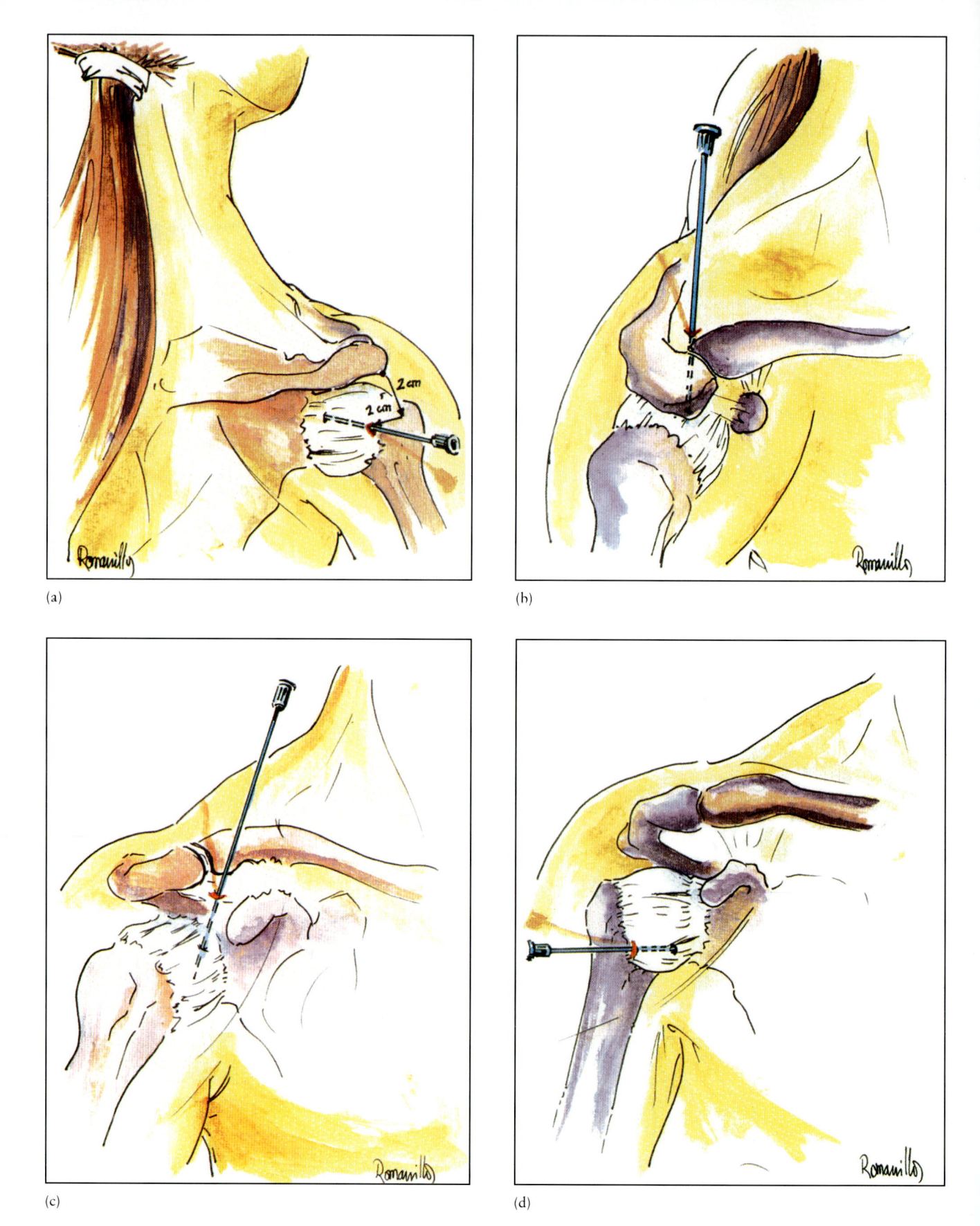

(b)

(c)

(d)

Plate 18.5 Shoulder arthrocentesis: (a) injection by the posterior route; (b) injection by the superior route; (c) injection by the anterosuperior route; (d) injection by the anteroinferior route.

5 Roosendaal G, Vianen ME, Wenting MJG et al. Iron deposits and catabolic properties of synovial tissue from patients with haemophilia. *J Bone Joint Surg Br* 1998; **80B**: 540–5.

6 Rodriguez-Merchan EC. Effects of hemophilia on articulations of children and adults. *Clin Orthop* 1996; **328**: 7–13.

7 Roosendaal G. *Blood-induced cartilage damage in hemophilia.* Doctoral thesis, Utrecht, The Netherlands, 1998.

8 Rodriguez-Merchan EC, Goddard NJ, Lee CA, eds. *Musculoskeletal Aspects of Haemophilia.* Oxford: Blackwell Science, 2000.

9 Heim M, Rodriguez-Merchan EC, Horoszowski H. Orthopedic complications and management of hemophilia. *Int J Pediatr Hematol Oncol* 1994; **1**: 545–51.

10 Heim M, Martinowitz U, Horoszowski H. The short foot syndrome: an unfortunate consequence of neglected raised intracompartmental pressure in a severely hemophilic child [Case report]. *Angiology* 1986; **37**: 128–31.

11 Cohen MS, Garfin SR, Hargens AR et al. Acute compartment syndrome: effect of dermotomy on fascial decompression in the leg. *J Bone Joint Surg Br* 1991; **73B**: 287–90.

12 Rodriguez-Merchan EC, Villar A, Magallon M, De Orbe A. Spontaneous infection of soft tissue haematomas in two HIV seropositive haemophilia patients. *Haemophilia* 1995; **1**: 137–9.

13 Fernandez de Valderrama JA, Matthews JM. The haemophilic pseudotumour or haemophilic subperiosteal haematoma. *J Bone Joint Surg Br* 1965; **47B**: 256–65.

14 Koepke JA, Browner TW. Chondrosarcoma mimicking pseudotumour of haemophilia. *Arch Pathol* 1965; **80**: 655–8.

15 Alberg AKM. On the natural history of hemophilic pseudotumor. *J Bone Joint Surg Am* 1975; **57A**: 1133–6.

16 Gilbert MS. The hemophilic pseudotumor. *Prog Clin Biol Res* 1990; **324**: 263–8.

17 Chen YF. Bilateral hemophilic pseudotumors of the calcaneus and cuboid treated by irradiation. *J Bone Joint Surg Am* 1965; **47A**: 517–21.

18 Duthie RB, Matthews JM, Rizza CR et al. Haemophilic cysts and pseudotumours. In: Duthie RB, ed. *The Management of Musculo-Skeletal Problems in the Haemophilias.* Oxford: Blackwell Scientific, 1972.

CHAPTER 18
Orthopaedic management of haemarthroses

E.C. Rodriguez-Merchan

The most typical manifestation of haemophilia is articular bleeding (haemarthrosis). When haemarthroses become frequent and/or intense, the synovium may not be able to reabsorb the blood. To compensate for such reabsorptive deficiency the synovium will hypertrophy, resulting in what is called chronic haemophilic synovitis. Thus, it is very important not only to avoid acute haemarthrosis, but also to manage it as efficiently as possible, with the aim of avoiding the development of cartilage damage and synovitis. Haematological prophylactic treatment from the age of 2 years to the end of skeletal maturity is the best way to avoid articular bleeds, or at least to diminish their intensity. However, one should remember that problems may be caused by the permanent intravenous infusion needed in such circumstances. Most haemophilia centres in industrialized countries have treatment on-demand, which consists of the administration of the deficient coagulation factor when haemarthrosis occurs.

The development of an inhibitor against factor VIII (FVIII) or factor IX (FIX) is the most common and most serious complication of substitutive therapy in haemophilia A or B patients, since the exclusive use of virus-inactivated plasma-derived concentrates or recombinant products. When present, the inhibitor inactivates the biological activity of infused FVIII or FIX, making the patient refractory to treatment. Moreover, even when the inhibitor titre is low, allowing some partial efficacy, the titre will increase within a few days of the use of FVIII- or FIX-containing products, making the treatment rapidly ineffective [1]. The management of bleeding episodes in haemophilia patients with inhibitors is particularly difficult [2].

Between 10 and 30% of patients with severe haemophilia A, and 2–5% of patients with severe haemophilia B or mild–moderate haemophilia A develop an inhibitor against FVIII or FIX after treatment with either plasma-derived or recombinant products. Inhibitor detection using the Bethesda assay, measured in Bethesda units (BU), is part of the regular follow-up for all haemophilia patients treated with such products. After the development of the inhibitor, the inhibitor titre decreases if no FVIII- or FIX-containing products are used for a long period so that the inhibitor may become undetectable. However, the inhibitor usually reappears after a new challenge with FVIII- or FIX-containing products (anamnestic response).

Two approaches for the management of patients with inhibitors have been proposed. Immune tolerance induction using high dose FVIII or FIX daily or twice daily for a period of a few months to several years may completely eliminate the inhibitor, allowing the patient to again be treated efficiently with FVIII or FIX [3,4]. However, immune tolerance induction fails in around 20% of cases and is not proposed for all patients because of the high probability of failure or adverse events. Furthermore, this procedure is very costly. The other possibility is to treat bleeding episodes with prothrombin complex concentrates (PCC), activated prothrombin complex concentrates (aPCC; Autoplex™, Feiba™) [5–7] or, more recently, with recombinant activated factor VIIa (rFVIIa; NovoSeven™) [8–11]. In case of failure of aPCC or rFVIIa in life- or limb-threatening bleeds or as first-line treatment for major bleeds, high dose human [12] or porcine FVIII [13] or human FIX may be efficacious if the inhibitor is low or is lowered using plasmapheresis [14] or protein A immune adsorption [15]. However, the anamnestic rise of the inhibitor will render treatment with FVIII or FIX ineffective within a few days, making the patient resistant to rescue with FVIII or FIX for months or even years.

Recombinant FVIIa has made joint aspiration (arthrocentesis) and even major elective orthopaedic surgery possible in patients with high titre inhibitors. Most of these procedures would not have been possible because it would have been difficult to overcome the inhibitor even with high doses of human or porcine FVIII or FIX. The reported experience with aPCC, for example, is minimal despite aPCCs being available for more than 20 years. There were no reported serious adverse events of rFVIIa in this data search, except for one episode of disseminated intravascular coagulation (DIC) in a patient who had an excision of an infected pseudotumour and was treated with rFVIIa by continuous infusion [16]. This large series of major elective orthopaedic surgical procedures [8,17–25] is the largest ever reported in haemophilia patients with inhibitors, despite the long-standing presence of other treatment modalities such as high dose human FVIII, porcine FVIII and aPCC. This means that rFVIIa is a novel and real alternative for major elective orthopaedic surgery in patients with inhibitors. The standard regimen is 90 µg/kg body weight

Table 18.1 The main approaches for the management of patients with haemophilia and inhibitors.

Immune tolerance induction (high doses of FVIII or FIX for months to years)
Prothrombin complex concentrates (PCC)
Activated prothrombin complex concentrates (aPCC)
Recombinant activated FVII (rFVIIa)
High dose human or porcine FVIII or human FIX (if the inhibitor is low or is lowered using plasmapheresis or protein A immunoadsorption)

---

every 2 h for the first 48 h with increasing interval between doses after this first postoperative period. Lower doses are much less efficient [21] and administration by continuous infusion is not yet approved for rFVIIa as a few bleeds and one episode of DIC were reported with continuous infusion of rFVIIa [16,20]. These main approaches for the management of haemophilia patients with inhibitors are summarized in Table 18.1.

Types of haemarthroses

It is important to differentiate between acute bleeding and subacute bleeding. Subacute haemarthrosis is generally associated with previous synovitis or arthropathy, while acute haemarthrosis commonly occurs in a previously healthy joint [26]. Acute bleeding is usually felt by the patient as a burning sensation in the joint. Haemarthrosis develops within a few hours; the joint becomes inflamed, tense and warm, and the skin becomes bright red. The affected joint is held in an antalgic flexion position, with painful and limited mobility.

After administration of the appropriate doses of factor concentrates, pain will rapidly diminish, although inflammation and limitation of articular mobility commonly disappear more slowly. The degree of inflammation and limitation of motion are always related to the amount of blood in the joint. Subacute haemarthroses commonly occur after two or three articular bleeding episodes and persist despite adequate haematological treatment. Pain can be tolerable and is commonly associated with hypertrophic synovium on palpation and a slight lack of joint mobility. When subacute haemarthroses recur for months and years, they will result in a state of haemophilic arthropathy. This usually occurs in young adults, who complain of persistent pain in the affected joint, not only with movement but also at rest. They may also have intermittent episodes of acute pain and inflammation related to synovitis or articular bleeding.

Treatment of haemarthroses

It is important to differentiate between acute and subacute haemarthroses.

Acute haemarthroses

Optimal treatment of acute haemarthroses involves a combination of haematological treatment, joint aspiration (arthrocentesis), rest (with or without splinting), ice, appropriate analgesia, and supervised rehabilitation once the acute phase has been controlled and the risk of bleeding reduced [27]. The objectives of treatment are to avoid muscular atrophy, maintain an adequate degree of articular mobility, control the recurrence of haemarthroses, and recover joint function if possible.

Joint aspiration (arthrocentesis)

Joint aspiration is not commonly performed, but in cases of severe bleeding it may relieve the patient's pain and speed up rehabilitation. There is a great deal of controversy on the role of arthrocentesis in haemophilia. The author's view is that minor bleeding episodes can be treated by other means. However, major bleeds may benefit from joint aspiration, providing it is performed within the first 12 h. Before deciding on joint aspiration, the presence of a circulating inhibitor must be investigated to determine the best treatment for it. In any case, 3–4 days of joint rest are recommended. When haemarthrosis does not respond to haematological treatment, septic arthritis must be suspected, especially if the patient is immunodepressed; joint aspiration and culture will allow a diagnosis to be reached. The five main joints—knee, ankle, hip, elbow and shoulder—should be approached for injection as follows.

Knee

Numerous injection sites can be used: medial infrapatellar, lateral infrapatellar, suprapatellar—either medially or laterally—and parapatellar. The most recommended are the lateral parapatellar route or the lateral suprapatellar injection. Both routes are approached with the patient lying supine with the knee as fully extended as possible. In the lateral parapatellar route the patella is pushed laterally and the injection is made in the depression created over the middle of its lateral margin, aiming for the centre of its posterior surface and pushing the needle slightly forward and downward until it just touches the patella. It is inserted about 3 cm. In the lateral suprapatellar route the knee is extended. The injection is made above the lateral corner of the patella and directly into the suprapatellar pouch (see Plate 18.1, facing p. 118). This route is the most commonly used by orthopaedic surgeons all over the world. After the procedure, a compressive bandage is used for 24 h.

Ankle

The site of injection is on the anteromedial aspect of the joint (see Plate 18.2, facing p. 118). Dorsiflexion of the foot to 90° relaxes the anterior cul-de-sac and brings into prominence the tendon of tibialis anterior, which is the principal landmark. Palpation reveals a depression between the tendon medially and the tendinous expansion of the toe extensors laterally. The injection of 1–2 mL of Scandicain™ without resistance confirms that the needle is in the joint. The alternative method is to carry out the procedure with radiographical control. One can mark the joint line with precision and the needle can be inserted slightly below the joint line. When the needle touches the bone, its position is adjusted with radiographical control. A compressive bandage is recommended for 24 h after the procedure.

Hip

Injection of the hip is dangerous because of the proximity of the vascular and lymphatic structures of the area, and the fact that the hip joint is particularly vulnerable to infection. Therefore strict asepsis is essential. The hip joint is deeply situated and its surface markings are not obvious. The site of injection must be chosen carefully under radiographical control. Local anaesthesia is adequate and a correctly performed injection inflicts minimal pain. General anaesthesia should be used in children. It is recommended to carry out the procedure in the operating theatre.

The patient should be supine with the leg internally rotated and the hip a little flexed; placing the patient on an orthopaedic table can be helpful. The cutaneous landmarks of the hip joint are as follows: one takes the middle of a line joining the anterosuperior iliac spine and the pubic tubercle. The femoral neck is situated three fingerbreadths below this point along a second line drawn perpendicular to the first. It is lateral to the great vessels, which can be palpated. The superior border of the neck is one fingerbreadth above this point and the inferior border a fingerbreadth below (see Plate 18.3, facing p. 118). Injection of 2 mL dye is recommended to begin to outline the joint cavity and judge immediately if the needle is in the joint. A period of 24 h bed rest is recommended followed by full range of motion and progressive weight bearing.

Elbow

The procedure can be performed in the office or at the patient's bedside. The patient is placed supine with the palm of the hand on the abdomen. The needle is inserted at the centre of a triangle formed by the olecranon, the lateral epicondyle, and the radial head (see Plate 18.4, facing p. 118). At this point, there is a small depression that can be marked with a skin pencil. The needle can be felt to penetrate the capsule. Any effusion is aspirated. If the needle is properly situated, the injection will be easy and without resistance. After withdrawal of the needle, a compressive bandage is recommended for 24 h, then allowing full range of motion of the elbow.

Shoulder

There are several injection routes for the shoulder: superior, posterior, anterosuperior and anteroinferior. In every case the injection should be carried out under radiographical control, and the needle can be felt to pass through the capsule. It is recommended that the procedure is carried out in the operating theatre. If there is an effusion it is aspirated. Otherwise, one injects 1–2 mL local anaesthetic to confirm that there is no resistance and that the needle is correctly situated within the joint. In case of doubt, injection of 2–4 mL of iodine-based contrast material can be helpful. After the procedure, 24 h of rest are recommended with the shoulder immobilized in a collar and cuff sling.

(a) Posterior route
The patient sits, or preferably lies face down, with the arm abducted to 20° (see Plate 18.5a, facing p. 118). A vertical line is drawn 2 cm medial to the lateral margin of the acromion and a horizontal line 2 cm below the inferolateral edge of the acromion. The injection is made at the point where these lines cross, the needle being introduced perpendicular to the skin until it touches the humeral head.

(b) Superior route
With the patient sitting, the arm is abducted to 45°. The injection is placed behind the acromioclavicular joint in the angle formed by the posterior edge of the clavicle and the medial border of the acromion. The needle is directed obliquely downwards and slightly outwards until it touches the humeral head (see Plate 18.5b, facing p. 118).

(c) Anterosuperior route
The patient lies on his or her back with the shoulder rotated outwards. The injection is made 1 cm below the acromioclavicular joint with the needle inclined slightly downwards until it reaches the bone (see Plate 18.5c, facing p. 118).

(d) Anteroinferior route
Position the patient on his or her back, tilted 30° toward the shoulder to be injected and supported by a sandbag (see Plate 18.5d, facing p. 118). The shoulder is laterally rotated and slightly abducted to about 20°.

In clinical practice, the intra-articular injection of elbows, knees and ankles can be easily performed in the office or at the patient's bedside by the general practitioner or the haematologist. On the other hand, the injection of shoulder and hip should be performed in the operating theatre under radiographical control with the help of contrast material, if possible by an orthopaedic surgeon. Although all these procedures can be carried out under local anaesthesia, in children general anaesthesia can sometimes be necessary.

If haemarthrosis does not respond to haematological treatment, one must suspect haemophilic synovitis, which can be detected by clinical examination. Ultrasonography and magnetic resonance imaging will help confirm the occurrence of synovitis. In such cases only aggressive treatment of synovitis will allow control of articular bleeding, which is secondary to synovial hypertrophy. Synovitis can be controlled with early prophylactic treatment or by synovectomy (radionuclide synovectomy or surgical synovectomy). Diagnostic imaging is paramount to assess the response to any type of treatment.

Heim et al. [28] reported an interesting case of a patient with haemophilia who had a fixed flexed hip and intractable pain. This clinical picture was suggestive of haemorrhage in that area. Ultrasonography confirmed the diagnosis of acute hip haemarthrosis. Narcotic drugs failed to alleviate the severe pain. Joint aspiration produced dramatic pain relief and early joint rehabilitation. However, Heim et al. did not suggest that every coxhaemarthrosis should be aspirated. It should be remembered that raised intra-articular pressure may contribute to femoral head necrosis in adults, or to Perthes' disease in children. It is important to emphasize that while arthrocentesis of the elbow, knee and ankle are quite simple procedures that can be performed at the outpatient clinic, both shoulder and hip joint aspirations require sedation and radiographical control by an image intensifier, that is to say, they are surgical procedures to be carried out in an operating room, with an anaesthetic and by an orthopaedic surgeon.

Rest and splinting

Rest for lower limb bleeding episodes should include bed rest (1 day) followed by avoidance of weight bearing and the use of crutches when ambulating, and elevation when sitting (3–4 days). For the knee a compressive bandage is adequate, although in very painful cases the bandage should be supplemented with a long-leg posterior plaster splint. For the ankle, a short-leg posterior plaster splint is recommended. For the upper limb, usually a sling (for the shoulder) or a long-arm posterior plaster splint (for the elbow) will provide sufficient rest, support and protection. Lifting and carrying heavy items should be avoided until the bleeding has resolved (4–5 days).

Ice

Ice therapy relieves pain and reduces the extent of bleeding by promoting vasoconstriction. Ice therapy can be applied to the affected joint in various ways: cold packs, wet towels, crushed ice and Cryocuff™. Applications of ice over 24–48 h can help control volume of blood and pain. Ice should not be applied directly to the skin, but wrapped in a thick towel, because it burns and prolonged application can cause skin damage. The effectiveness of ice as a treatment for acute bleeds lies not only in its physical effects but also in its ease and simplicity of application.

Analgesia

Depending on the degree of pain, paracetamol or a combination of paracetamol and dextropropoxyphene should be administered; they usually provide adequate relief. Aspirin-containing products and non-steroidal anti-inflammatory drugs must be avoided.

Subacute haemarthroses

It is advisable to treat subacute haemarthroses with haematological substitutive therapy, with 2–3 weeks of immobilization by means of a semiflexible splint. Some studies recommend 6–8 weeks of prophylaxis with physiotherapy. It is recommended to administer enough deficient factor, three times a week, to obtain 20–30% of the normal level. After each transfusion the patient should complete an exercise programme focusing on active joint mobility, under the surveillance of an expert physiotherapist. If such mobility exercises are painful, only isometric exercises should be performed.

When a flexion contracture does appear, it should be treated early and aggressively by conservative means to avoid it becoming irreversible. Conservative measures include Oxford's inverted dynamic splints, extension–desubluxation hinged casts, dynamic splints, and traction followed by a polypropylene orthosis. Oxford's technique was specially designed for the knee joint and requires admitting the patient to hospital. The lower limb is put in a balanced traction on a semicircular Thomas' splint which has a knee flexion Pearson's device. Soft traction is put on the calf with the heel free; a posterior force is applied on the thigh by means of a cushioned spring located on the distal part of the thigh, which is connected by means of a string to a 3-kg weight. Such a posterior force counteracts the anterior force produced by the springs located on the posterior part of the calf. Both the longitudinal traction and the thigh weight are progressively increased. When the knee becomes fully extended, or if the technique does not work after 1 week of treatment, the patient is mobilized with a Böhler's cast which is open in its anterior part.

The hinged extension–desubluxation cast can be made of plaster of Paris or of a thermoplastic material; it should be open in its anterior part. The hinge is adjusted once or twice a day to correct the deformity. When the contracture is less than 20°, the cast can be removed and replaced with a plaster splint. Haematological substitutive therapy is necessary during the procedure. The dynamic splint is adjustable and allows a low intensity but long duration force through the knee joint. A gain of 5–10° of knee extension can be expected in 6–9 months with this procedure. However, many patients may have haemarthrosis during the follow-up. Traction followed by orthosis is another alternative.

Flexion contracture has a different treatment and prognosis, depending on its chronicity and other associated deformities [29]. A flexion contracture of few days' duration can be corrected by means of a traction followed by rehabilitation and orthosis. A flexion contracture with a duration ranging from weeks to months may require surgery: hamstring release and/or supracondylar extension osteotomy. A flexion contracture associated with osseous or fibrous ankylosis may also require a patello-femoral osteotomy.

Table 18.2 Optimal management of acute haemarthroses in haemophilia.

Adequate haematological treatment
Joint aspiration (arthrocentesis)
Rest and splinting of the affected joint
Ice therapy
Analgesia

Conclusions

Optimal treatment of acute haemarthroses should allow rapid resolution of the bleeding episode with minimal risk of long-term problems (Table 18.2). A review of the literature on inhibitors has shown that, with the availability of rFVIIa, haemophilic patients with high inhibitor titres requiring arthrocentesis or even elective orthopaedic surgery can undergo such surgery with a high expectation of success. Recombinant FVIIa appears to be an efficient haemostatic product for surgery in patients with haemophilia A and B and inhibitors. Such a product makes arthrocentesis and elective orthopaedic surgery viable options, leading to an improved quality of life for individuals with haemophilia. The affected joint should remain at rest for a short period of time (4–5 days): bed rest for the hip, a sling for the shoulder, and a compressive bandage and plaster splint for the elbow, knee and ankle. Ice therapy helps to relieve pain and reduce the extent of bleeding. Analgesics (paracetamol) may also be required, depending on the degree of pain.

An early and progressive physical therapy is then required to recover the full range of movement and the strength of periarticular muscles. While some authors recommend joint aspiration to remove the blood as an important therapeutic measure, others do not routinely perform this procedure in patients with haemophilia. Currently, arthrocentesis is one of the most controversial issues regarding treatment of haemarthrosis in haemophilia. My view is that arthrocentesis should always be performed in major haemarthrosis (voluminous, very tense and painful joint). Minor haemarthosis commonly responds to haematological treatment and rest. Arthrocentesis of the hip and shoulder should be carried out under radiographical control in an operating room by an orthopaedic surgeon. Aspiration of the elbow, knee and ankle are quite simple procedures that can be performed in the outpatient clinic, not necessarily by an orthopaedic surgeon. Joint aspiration should always be carried out under factor coverage and in aseptic conditions, in order to avoid recurrence or septic arthritis.

References

1. Rodriguez-Merchan EC, de la Corte H. Orthopaedic surgery in haemophilic patients with inhibitors: a review of the literature. In: Rodriguez-Merchan EC, Goddard NJ, Lee CA, eds. *Musculoskeletal Aspects of Haemophilia*. Oxford: Blackwell Science, 2000:136–42.
2. The United States Pharmacopeial Convention. Hemophilia management. *Transfus Med Rev* 1998; **12**: 128–40.
3. Brackmann HH, Gormsen J. Massive factor VIII infusion in a haemophilia with factor VIII inhibitor: high response. *Lancet* 1977; **ii**: 933.
4. Brackmann HH, Oldenburg J, Schwaab R. Immune tolerance for the treatment of factor VIII inhibitors: twenty years 'Bonn Protocol'. *Vox Sang* 1996; **70** (Suppl. 1): 30–5.
5. Abildgaard CF, Penner JA, Watson-Williams EJ. Anti-inhibitor coagulant complex (Autoplex) for treatment of factor inhibitors in hemophilia. *Blood* 1980; **56**: 978–84.
6. Hilgartner MW, Knatterud GL, and the Feiba Study Group. The use of factor eight inhibitor by-passing activity (Feiba Immuno) product for treatment of bleeding episodes in hemophiliacs with inhibitors. *Blood* 1983; **61**: 36–40.
7. Hilgartner M, Aledort L, Andes A, Gill J and the Members of the Feiba Study Group. Efficacy and safety of vapor-heated anti-inhibitor coagulant complex in hemophilia patients. *Transfusion* 1990; **30**: 626–30.
8. Hedner U, Glazer S, Pinkel K et al. Successful use of recombinant factor VIIa in a patient with severe haemophilia A during synovectomy. *Lancet* 1988; **ii**: 1193.
9. Hedner U, Glazer S, Falch J. Recombinant activated factor VII in the treatment of bleeding episodes in patients with inherited and acquired bleeding disorders. *Transfus Med Rev* 1993; **7**: 78–83.
10. Hedner U, Ingerslev J. Clinical use of recombinant FVIIa (rFVIIa). *Transfus Sci* 1998; **19**: 163–76.
11. Roberts HR. Clinical experience with activated factor VII: focus on safety. *Blood Coag Fibrinolysis* 1998; **9** (Suppl. 1): S115–18.
12. White GC, Taylor RE, Blatt PM, Roberts HR. Treatment with a high titer anti-factor VIII antibody by continuous factor VIII administration: report of a case. *Blood* 1983; **62**: 141–5.
13. A multicenter US experience: the use of porcine factor VIII concentrate (Hyate:C) in the treatment of patients with inhibitor antibodies to factor VIII. *Arch Intern Med* 1989; **149**: 1381–5.
14. Bona RD, Pasquale DN, Kalish RI, Witter BA. Porcine factor VIII and plasmapheresis in the management of hemophiliac patients with inhibitors. *Am J Hematol* 1986; **21**: 201–7.
15. Nilsson IM, Berntorp E, Freiburghaus C. Treatment of patients with factor VIII and IX inhibitors. *Thromb Haemost* 1993; **70**: 56–9.
16. Schulman S, d'Oiron R, Martinowitz U et al. Experiences with continuous infusion of recombinant activated factor VII. *Blood Coag Fibrinolysis* 1998; **9** (Suppl. 1): S97–101.
17. Schulman S, Lindstedt M, Alberts KA, Agren PH. Recombinant factor VIIa in a multiple surgery [Letter]. *Thromb Haemost* 1994; **41**: 154.
18. O'Marcaigh AS, Schmalz BJ, Shaughnessy WJ, Gilchrist GS. Successful hemostasis during a major orthopaedic operation by using recombinant activated Factor VII in a patient with severe hemophilia A and a potent inhibitor. *Mayo Clin Proc* 1994; **69**: 641–4.
19. Ingerslev J, Freidman D, Gastineau D et al. Major surgery in haemophilic patients with inhibitors using recombinant factor VIIa. *Haemostasis* 1996; **26** (Suppl. 1): 118–23.
20. Schulman S, Bech Jensen M, Varon D et al. Feasibility of using recombinant factor VIIa in continuous infusion. *Thromb Haemost* 1996; **75**: 432–6.
21. Shapiro AD, Gilchrist GS, Hoots WK, Cooper HA, Gastineau DA. Prospective, randomised trial of two doses of rFVIIa (NovoSeven) in haemophilia patients with inhibitors undergoing surgery. *Thromb Haemost* 1998; **80**: 773–8.
22. Vermylen J, Peerlinck K. Optimal care of inhibitor patients during surgery. *Eur J Haematol* 1998; **61** (Suppl. 63): 15–17.

23 Mauser-Bunschoten EP, de Goede-Bolder A, Wielenga JJ et al. Continuous infusion of recombinant factor VIIa in patients with haemophilia and inhibitors: experience in The Netherlands and Belgium. *Neth J Med* 1998; **53**: 249–55.

24 Goudemand J. Treatment of patients with inhibitors: cost issues. *Haemophilia* 1999; **5**: 397–401.

25 Scharrer I on the behalf of the German NovoSeven Study Group. Recombinant factor VIIa for patients with inhibitors to factor VIII or IX or factor VII deficiency. *Haemophilia* 1999; **5**: 253–9.

26 Heim M, Rodriguez-Merchan EC, Horoszowski H. Current trends in hemophilia and other coagulation disorders: orthopaedic complications and management. *Int J Pediatr Hematol Oncol* 1994; **1**: 545–51.

27 Ribbans WJ, Giangrande P, Beeton K. Conservative treatment of hemarthrosis for prevention of hemophilic synovitis. *Clin Orthop* 1997; **343**: 12–18.

28 Heim M, Varon D, Strauss S, Martinowitz U. The management of a person with haemophilia who has a fixed flexed hip and intractable pain. *Haemophilia* 1998; **4**: 842–4.

29 Rodriguez-Merchan EC. Management of orthopaedic complications of haemophilia. *J Bone Joint Surg Br* 1998; **80-B**: 191–6.

CHAPTER 19

Chemical synoviorthesis

H.A. Caviglia, F. Fernandez-Palazzi, C. Pascual-Garrido, N. Moretti and R. Perez-Bianco

Haemophilic synovitis is still an unsolved problem which is even more complex when patients experience an associated high titre inhibitor. This chapter explains the authors' approach to treatment of this condition. It is important to find out about the degree of arthropathy of the patient and the duration of the synovitis as the older and bigger the synovitis, the worse the prognosis for the patient's recovery. The 'golden moment' for treatment is when no associated arthropathy is present, because synovitis and muscular hypotrophy are reversible and the patient can be cured of the lesion; arthropathy cannot be reversed [1].

Physical examination will evaluate mobility, the degree of hypotrophy of the muscle and joint perimeter. It is important to instruct the patient's parent or carer on the evaluation of joint perimeter as they will have to measure it twice a day and draw the corresponding curve. The clinical classification of synovitis used is the one by Fernandez-Palazzi and Caviglia, divided into four grades (Table 19.1) [2].

The statistical evaluation is essential and is performed by means of radiography and echography. A radiograph shows the degree of arthropathy (Table 19.2). An echogram tells us about the characteristics of the synovitis: its thickness and about the existence of nodules and bridles. It also allows for the evaluation of joint content and its dimensions. Rodriguez-Merchan *et al.* [4] used ultrasonography in the diagnosis of the early stages of haemophilic arthropathy of the knee. An echogram should be performed every 2 weeks to permit an evaluation of synovial tissue response to treatment. During phase I of the treatment, oral rofecoxib and rehabilitation are indicated. Rofecoxib 0.6 mg/kg/day is prescribed for 6 weeks. This dosage has been tried in previous regimens [5].

Rehabilitation consists of isometric exercises, active exercises with assistance and physiotherapy. Electrostimulation is started during the third week. At the end of the second week of treatment the joint perimeters taken by the patient's parent or carer are evaluated. If an increase >15% is found, the patient is considered hyper-active and a flexion-controlling splint is used. The splint is set at 15% flexion. Two weeks later the patient is

Table 19.1 Clinical classification of synovitis of Fernandez-Palazzi and Caviglia [2].

Grade I	Transitory synovitis post bleeding without sequelae Indicated when >3 episodes occur in 3 months
Grade II	Permanent synovitis with enlarging of the joint, synovial thickening and limited range of movement
Grade III	Chronic arthropathy with axial deformities and muscular arthropathy
Grade IV	Ankylosis. Can be fibrous or osseous

Table 19.2 Radiographical classification of Pettersson and Gilbert [3].

Type of change	Finding	Score
Osteoporosis	Absent	0
	Present	1
Enlarged epiphysis	Absent	0
	Present	1
Irregular subchondral bone	Absent	0
	Surface partly involved	1
	Surface totally involved	2
Narrowing of joint space	Absent	0
	Joint space >1 mm	1
	Joint space ≤1 mm	2
Subchondral cyst formation	Absent	0
	1 cyst	1
	>1 cyst	2
Erosions at joint margins	Absent	0
	Present	1
Incongruences of joint surfaces	Absent	0
	Slight	1
	Pronounced	2
Joint deformity (angulation and/or displacement between the articulating bony ends)	Absent	0
	Slight	1
	Pronounced	2

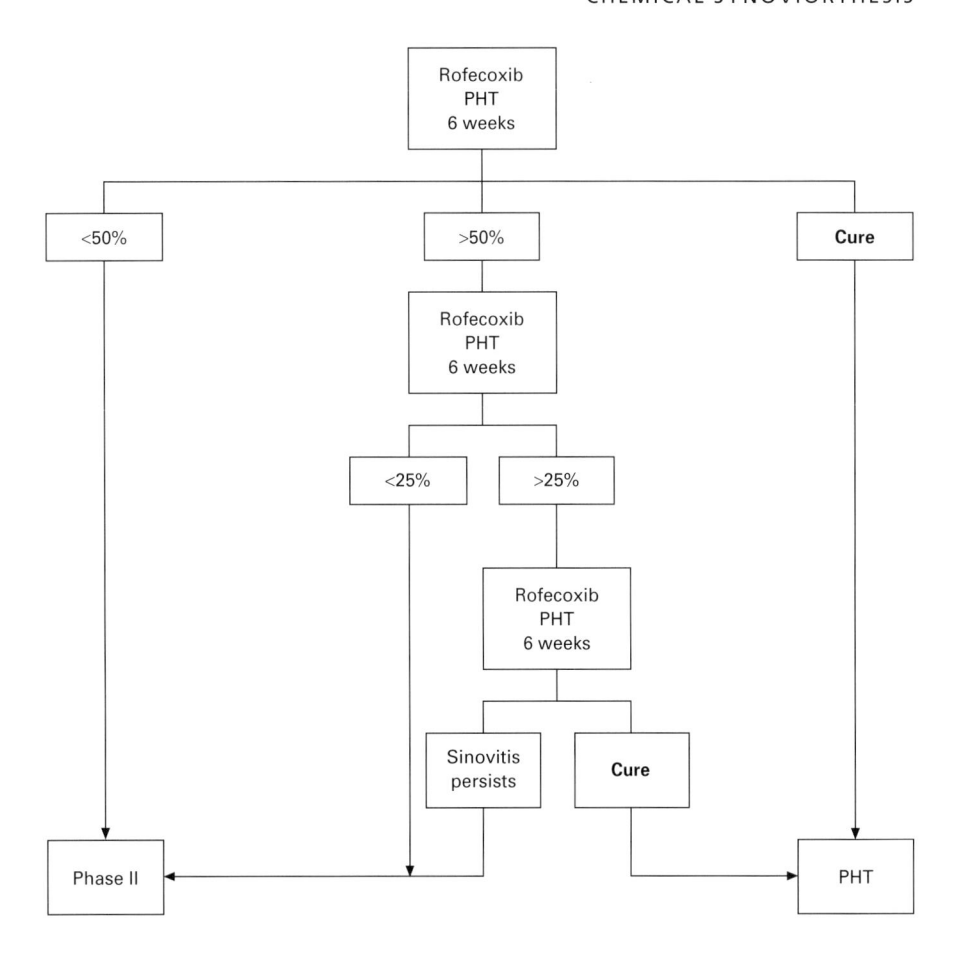

Fig. 19.1 Algorithm of chemical synoviorthesis with rofecoxib. PHT, physiotherapy.

evaluated with an echogram and the joint perimeters. If the synovitis has decreased in size, 20° of movement may be allowed, from 0 to 20°. The same procedure is repeated during the sixth week.

A clinical and echographical evaluation is performed at the end of the sixth week. Three possibilities exist:
1 the synovitis may have been cured, in which case only the rehabilitation programme is necessary;
2 the synovitis may have improved <50%, measured echographically, and treatment therefore progresses to phase II; or
3 the synovitis may have improved >50%, in which case the same treatment continues.

The treatment is repeated for 6 more weeks and then another echographical evaluation is carried out. If the synovitis has not been reduced by 25% after 12 weeks, the patient goes on to phase II. If the 25% reduction level has been reached, the same treatment is repeated for another 6 weeks. If at the end of the repeat period the synovitis persists, the patient goes on to phase II (Fig. 19.1). Consequently, when the patient moves on to phase II, it is a result of failure of conservative treatment. At this point, the availability of factor VIIa is important. If it is available and if the patient's synovitis is corrected by its infusion, the procedure will vary according to joint size. In smaller joints, such as the elbow or ankle, a chemical synoviorthesis with rifampicin is performed because the average cure with this procedure is 1.5 injections for the ankle and three for the elbow [6]. The use of radioactive synoviorthesis in smaller joints can lead to actinic necrosis. The factor VIIa dosage to perform chemical or radioactive synoviorthesisis is 90 μg/kg every 2–3 h (two doses).

Chemical synoviorthesis with rifampicin is a highly satisfactory procedure with appropriate patient selection; grade I–II synovitis is considered ideal. This procedure was performed jointly in Brazil and Venezuela. The procedure consists of aspirating joint content and then injecting 250 mg rifampicin in 1 mL of local anaesthesia intra-articularly. This is followed by good joint compression and immediate mobilization. The session may be repeated weekly, depending on the clinical and echographical results observed. Its disadvantage is that it may require more than one session, which would cause more antihaemophilic factor consumption. Also, the first injection may be painful. It is contraindicated in acute and chronic cases with 'full joints'. The results may be fully superposable with those obtained with radioactive synoviorthesis.

In 1975, Tezanos-Pinto et al. [7] introduced the use of radioactive gold in the treatment of patients with haemophilia in Argentina. Radiosynoviorthesis has also been used in Israel; Spain; Caracas, Venezuela; Los Angeles, USA; and South Africa. For sinovitis in knees, we prefer to use radioactive synoviorthesis with ^{189}Re injection because it requires a single

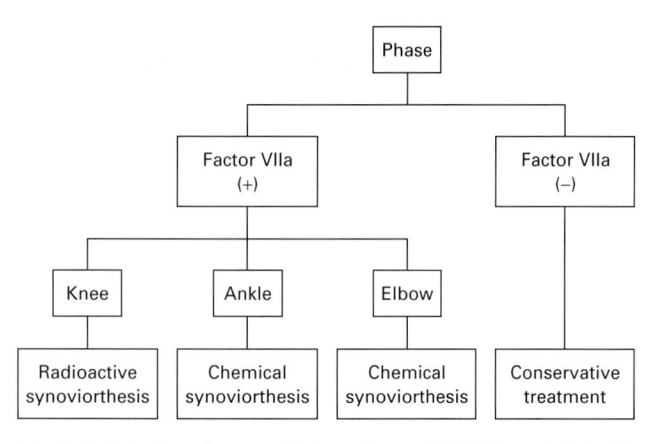

Fig. 19.2 Algorithm for synovitis in patients with inhibitors. FVIIa (+), factor VIIa available; FVIIa (–), factor VIIa not available.

injection and there is no danger of actinic necrosis. This procedure does not limit movement. Results are similar to those published for surgical synovectomy. As for synoviorthesis, there is an 80% chance of a successful outcome [7–14].

To obtain an objective image of the swelling, joint control with ^{99m}Tc is performed before the session, factor VIIa is then infused, and finally the injection of radioactive material after arthrocentesis with local anaesthetic. After injection and before removal, the needle is washed with anaesthetic to prevent transfer of material and a possible burn. When a colloid with gamma radiation has been used (^{198}Au, ^{189}Re) a gammagram and scintigraphy are performed to verify the injection was intra-articular. Regular clinical controls are performed to evaluate possible recurrence [15].

Chromosomal tests are carried out before and after the injection to evaluate whether chromosomal aberrations are present. The possibility of chromosomal damage to patients subject to this procedure has been suggested, but previous tests have proved that these chromosomal changes are transitory and are not therefore considered to be premalignant [16].

The materials used for treatment are gold (^{198}Au), rhenium (^{189}Re) and yttrium (^{90}Y). The dosage is 5 millicuries in 10–15 cc physiological saline solution. When there is no factor VIIa available, it is preferable to continue with conservative treatment, that is to say, a flexion-controlling splint and physiotherapeutic treatment. The flexion-controlling splint is used for 1 month and set at 15° flexion. Its clinical and echographical response is evaluated every 30 days, with a 20° release of flexion.

This treatment (Fig. 19.2) was presented in Athens, in 1992, at the Congress of the World Federation of Haemophilia, where good results were noted except for those patients who presented a complex psychological profile. We consider this to be the best form of treatment for patients with synovitis and who have high titre inhibitors.

References

1. Caviglia HA. Hemophilic synovitis: is rifampicin an alternative to synovectomy? XXIII International Congress of the World Federation of Hemophilia, The Hague, the Netherlands, 17–22 May, 1988.
2. Fernandez-Palazzi F, Caviglia HA. On the safety of synoviorthesis in haemophilia. In: Rodriguez-Merchan EC, Goddard NJ, Lee CA, eds. *Musculoskeletal Aspects of Haemophilia*. Oxford: Blackwell Science, 2000: 50–6.
3. Pettersson H, Gilbert MS. *Diagnostic Imaging in Hemophilia*. Springer-Verlag, Berlin, 1985.
4. Rodriguez-Merchan EC, Gago J, Orbe A. Ultrasound in the diagnosis of the early stages of haemophilic arthropathy of the knee. *Acta Orthop Belg* 1992; 58: 122–6.
5. Pascual-Garrido C, Galatro G, Daffunccio C, Pérez Ballester G, Caviglia HA. Oral rofecoxib in hemophilic chronic synovitis. Seventh Musculoskeletal Congress of World Federation of Haemophilia, Lahore, Pakistan, March 2001.
6. Caviglia H, Galatro G, Moretti N, Perez Bianco R. Intra-articular rifampicin therapy of chronic hemophilic synovitis. Fifth Musculoskeletal Congress of the World Federation of Hemophilia. Sydney, Australia, 14–16 April, 1999.
7. Tezanos-Pinto M, Perez-Bianco R, Bengo RM. La artropatía hemofílica: tratamiento con oro radioactivo. *Bol Acad Nacional Med Buenos Aires*, 1975; 53: 324–5.
8. Fernandez-Palazzi F. Treatment of acute and chronic synovitis by non-surgical means. *Haemophilia* 1998; 4: 518.
9. Fernandez-Palazzi F, de Bosch N, de Vargas A. Radioactive synovectomy in hemophilic hemarthrosis: follow up of 50 cases. *Scand J Haematol* 1984; 33: 291–300.
10. Fernandez-Palazzi F, Rivas S, Ciberia JL, Dib O, Viso R. Radioactive synoviorthesis in haemophilic haemarthrosis: materials, techniques and dangers. *Clin Orthop* 1995; 328: 14–18.
11. Rivard G, Girard M, Belanger R *et al.* Sinoviortesis en artropatía hemofílica. Di Gráfica Gomez. Caracas: 61–75, 1986.
12. Rodriguez-Merchan EC, Goddard NJ. Chronic haemophilic synovitis. In: Rodriguez-Merchan EC, Goddard NJ, Lee CA, eds. *Muskuloskeletal Aspects of Haemophilia*. Oxford: Blackwell Science, 2000: 43–9.
13. Viso R, Fernandez-Palazzi F, Rivas S *et al.* Sinovectomy with rifampicin in haemophilic haemartrosis. II. Encontro de Florianopolis, Santa Catarina. Brasil, November 1995.
14. Rodriguez-Merchan EC, Caviglia HA, Magallon M, Perez-Bianco R. Chemical synovectomy vs. radioactive synovectomy for the treatment of chronic haemophilic synovitis: a prospective short-term study. *Haemophilia* 1997; 3: 118–22.
15. Fernandez-Palazzi F, Caviglia HA, Bernal R. Problemas ortopédicos en el niño hemofílico. *Rev Ortop Traumatol* 2001; 45: 144–50.
16. Fernandez-Palazzi F, Caviglia HA. On the safety of synoviorthesis. In: Rodriguez-Merchan EC, Goddard NJ, Lee CA, eds. *Musculoskeletal Aspects of Haemophilia*. Oxford: Blackwell Science, 2000: 50–3.

CHAPTER 20
Radiosynoviorthesis in patients with haemophilia and inhibitors

C.J. Petersson and E. Berntorp

There exists a relationship between recurrent joint bleedings, synovitis and the development of destructive arthritis in patients with haemophilia. Various methods to stop bleeding and synovitis have been suggested, including both medical and surgical techniques. Prophylactic replacement of the missing factor and maintaining a factor level above 1% has been shown to be an efficient way of avoiding joint bleeding episodes and arthropathy [1].

A serious complication in haemophilia treatment is the development of antibodies (inhibitors) to factor VIII (FVIII) and factor IX (FIX). If the antibody titre rises to high levels after challenge with factor concentrates, the patient can become resistant to conventional replacement therapy. In such cases, with high-responding inhibitors, haemostasis can be achieved to some extent by using so-called bypassing agents such as activated recombinant factor VII (rFVIIa) or activated prothrombin complex concentrates (aPCC) [2,3]. However, it is preferable to induce immune tolerance with the goal of permanently eradicating the inhibitor and making it possible for the patient to be treated with ordinary replacement therapy with FVIII or FIX. Several immune tolerance treatment protocols have been described especially for FVIII inhibitors, using FVIII in low, intermediate or high doses. A fourth protocol is that developed in Malmö (the Malmö protocol) which combines intravenous immunoglobulin G (IgG) and cyclophosphamide with FVIII or FIX, depending on type of inhibitor [4]. A prerequisite for the Malmö protocol is that the inhibitor titre is low at the start of treatment, so that free plasma levels of FVIII:C/FIX:C are reached during the first week of treatment. Therefore extracorporeal adsorption of plasma to protein A is sometimes used immediately prior to the start of factor infusion. This also means that procedures requiring good haemostasis can be performed concomitantly with the start of the Malmö protocol. In patients with low-responding inhibitors, below 10 Bethesda units (BU), haemostasis can usually be achieved by increasing the dose of FVIII or FIX.

In 1968, at the Malmö Haemophilia Centre, Ahlberg started to treat recurrent bleeding episodes and persistent joint effusions in haemophilia patients with radioactive ^{198}Au, inspired by the encouraging results of Ansell *et al.* [5] using radioactive gold in treating persistent joint effusion in rheumatoid knee joints. Ahlberg's first report in 1971 [6] on gold synoviorthesis treatment of an 11-year-old boy met with much criticism. However, Ahlberg was able to demonstrate that the radioactive gold colloid was anchored to the synovial layer of the injected joint and scintigrams showed that there was no escape route for the isotope from the joint to the regional lymph nodes, so the risk of side-effects was small. In 1979, Ahlberg and Pettersson [7] reported the results of radioactive gold treatment of 27 patients with haemophilia followed for 3–9 years. They observed no negative effect of the radioactive gold on the joint or on the growing zone. In 1986, Löfqvist and Petersson [8] reported on 42 patients with severe haemophilia, mean age 15 years, treated with intra-articular gold and followed on average for 95 months without any observed side-effects of the isotope. However, in 1990 the Malmö Haemophilia Centre changed the isotope used in radiosynoviorthesis from ^{198}Au to ^{90}Y, being a pure β emitter.

The introduction of modern treatment options in patients with haemophilia and inhibitory antibodies has also made radiosynoviorthesis available for this cohort of patients. This is a retrospective report of our experiences at the Malmö Haemophilia Centre of radioactive gold and yttrium isotope treatment of haemophilia patients with inhibitory antibodies.

Materials and methods

Eighteen joints in eight patients with haemophilia A and inhibitors, and eight joints in five patients with haemophilia B and inhibitors were treated with synoviorthesis at the doses given in Table 20.1. The age of the patients ranged from 3 to 40 years. The indication for synoviorthesis was recurrent joint bleeding episodes or chronic synovitis.

The type of haemostatic treatment used has evolved over the years and has been individualized depending on the type of haemophilia and response of inhibitor (Table 20.1). Those patients treated prior to the early 1980s usually received treatment with FVIII or FIX with or without cyclophosphamide, depending on inhibitor titre and anticipated anamnestic response. As the haemophilia B patients with inhibitors received prothrombin complex concentrate, antithrombin concentrate

Table 20.1 Radiosynoviorthesis in patients with haemophilia and inhibitors.

Pat.	Type	High (H)/low (L) responding inhibitor*	Age	Joint (L, left; R, right)	Radioactive dose (Mci)†	Coagulation treatment‡	Bleed-free interval (months)	Long-term results	Joint score	Surgical procedures
I	B	L	39	L knee	5 Au	1	<6	Symptoms	10	Arthroplasty
II	B	H	38	R shoulder	5 Au	2	<6	–		
			40	R shoulder	5 Au	2	<6	Symptoms	5	
III	A	L	12	L knee	5 Au	2	>12	Symptoms	9	
IV	A		20	R elbow	3,5 Au	2	MI	Symptoms	7	
V	A	H	18	L elbow	2 Au	1	MI	MI		
			19	L knee	5 Au	3	MI	MI		
VI	A	L	8	L ankle	3 Au	2	>12	No symptoms	4	
			13	R elbow	3 Au	1	>12	Symptoms	5	
			15	L knee	4 Au	4	>12	No symptoms	4	
VII	A	L	3	L ankle	2 Au	1	>12	Symptoms	6	
			5	L ankle	2 Au	2	>6	No symptoms	6	
			7	R elbow	2,5 Au	2	>12	Symptoms	8	
			11	L elbow	3 Au	2	MI	Symptoms	8	
VIII	A	L	9	R ankle	2 Au	2	>6	No symptoms	4	
			9	L ankle	2 Au	2	>12	No symptoms	4	
			10	R knee	4 Au	4	>12	No symptoms	5	
IX	B	H	11	R knee	3 Au	2	<6	Symptoms	7	Synovectomy
			12	R elbow	2 Au	2	>12	No symptoms	4	
			13	L elbow	2 Au	4	MI	No symptoms	5	
X	A	H	11	L knee	4 Y	5	>12	Symptoms	6	
			14	R knee	4 Y	5	>6	Symptoms	6	
			14	R elbow	3 Y	5	>6	Symptoms	6	
XI	B	H	10	R knee	4 Y	5	<6	Symptoms	10	
XII	A	H	5	L knee	3 Y	1, 5	<6	Symptoms	8	Extension cast
XIII	B	H	5	R knee	3 Y	5	>12	No symptoms	4	

*High responding inhibitor ≥10 BU.
†Au, gold; Y, yttrium.
‡1 High dose factor VIII or factor IX.
 2 High dose factor VIII or IX + cyclophosphamide.
 3 Immunotolerance induction according to the Malmö model with protein A adsorption.
 4 Immunotolerance induction according to the Malmö model without protein A adsorption.
 5 Recombinant factor VIIa (NovoSeven™).
Mci, millicuries; MI, missing information.

was often given concomitantly to prevent the risk of thromboembolic complications caused by replacement with the full complement of prothrombin complex factors. Later, some treatment episodes were combined with immune tolerance induction according to the Malmö treatment model. This treatment was sometimes combined with extracorporeal protein A adsorption because of the high inhibitor titre at the start of treatment. From the mid 1990s, rFVIIa (NovoSeven™, Novo-Nordisk, Denmark) has been successfully used.

Results

Of the 26 joints treated, bleed-free intervals of more than 6 months were obtained in 15, 10 of which remained bleed-free for more than 12 months. In six joints, bleeding tendency and synovitis were not influenced by the synoviorthesis and for five joints the registration of haemarthroses after treatment were missing or insufficient (Table 20.1) The treated joints were followed for 5 years on average: nine joints were then symptom-free and the joint score ranged between 4 and 6 [9]; 14 joints were symptomatic and had a score of between 5 and 10.

One symptomatic knee in a patient with haemophilia B was later surgically synovectomized and another patient with haemophilia B and progressive arthropathy after radiosynoviorthesis received a total knee replacement. A 5-year-old boy with haemophilia A had recurrent joint bleeding episodes and flexion contracture of the knee very soon after radiosynoviorthesis and was treated with an extension cast. No adverse side-effects of the radioisotope treatment were observed in any patient.

The coagulation treatment gave sufficient haemostatic coverage for all procedures, even if the inhibitor reappeared within a week in several of the high-responding cases, resulting in unmeasurable FVIII or FIX plasma activity.

Discussion

In 1979, Ahlberg and Pettersson [7] published their results using a new sensitive radiological classification system of joint changes in patients with haemophilia. They described a subgroup of 27 haemophilia patients treated with radioactive gold synoviorthesis and followed for 3–9 years. They found decreasing bleeding frequency in the treated joints and also inhibition of the progress of arthropathy if the isotope treatment was carried out at an early stage when the arthropathy was still reversible. If the treatment was instituted at a later stage, the arthropathy seemed to progress independently of the effect on the bleeding tendency [7]. In 1986, Löfqvist and Petersson [8] reported on a group of 42 patients with severe haemophilia (37 A, 5 B), mean age 15 years, treated with radiosynoviorthesis: 23 knees, 22 elbows, 20 ankles and 7 shoulders. The early effects of the treatment were assessed by estimation of substitution-demanding bleeding episodes in the treated joints 1 year before and 1 year after synoviorthesis. There was a significant reduction in bleeding episodes in the treated ankles, elbows and shoulders, but not in the knees and also five synoviorthesized knees had to be surgically synovectomized. At the follow-up evaluation, after an average of 95 months, 32% of knees, 33% of shoulders, 40% of elbows and 65% of ankles were symptom-free. This is in accordance with the results of the present group of radiosynoviorthesis-treated patients with haemophilia and inhibitors. About one-third of knees and elbows and three out of four ankles were symptom-free at follow-up, indicating the same natural course of arthropathy after radiosynoviorthesis in patients with inhibitors as in those without. A consequence of effective prophylactic factor replacement has been that among ordinary patients with haemophilia today, the need for such complementary orthopaedic surgical measures as synoviorthesis is minimal.

The situation is different where patients with haemophilia and inhibitors are concerned. Immune tolerance induction is not always successful and is, moreover, an expensive procedure and not a generally applicable treatment method. Thus, in several patients with inhibitors, especially high responders, treatment options are limited in the event of recurrent bleeding episodes and synovitis.

Conclusions

The results of our study, although of a small series, seem to suggest that a decrease in the frequency of bleeding episodes can be obtained for 6–12 months in more than half of cases. Long-term improvement can be obtained in about one-third of treated knees and elbows and even better figures can be expected in synoviorthesized ankles. Against this background, radiosynoviorthesis seems to be a viable treatment alternative in patients with haemophilia and inhibitors. Additional advantages with radiosynoviorthesis are its ease of operation, low cost and reduced need of peroperative hospital care.

References

1 Nilsson IM, Berntorp E, Löfqvist T, Pettersson H. Twenty-five years' experience of prophylactic treatment in severe haemophilia A and B. *J Intern Med* 1991; **232**: 25–32.
2 Key NS, Aledort LM, Beardsley D *et al*. Home treatment of mild and moderate episodes using recombinant factor VIIa (Novoseven) in haemophiliacs with inhibitors. *Thromb Haemost* 1998; **80**: 912–18.
3 Négrier C, Goudemand J, Sultan Y *et al*. Multicenter retrospective study on the utilization of FEIBA in France in patients with factor VIII and factor IX inhibitors. *Thromb Haemost* 1997; **77**: 1113–19.
4 Freiburghaus C, Berntorp E, Ekman M *et al*. Tolerance induction using the Malmö treatment model 1982–95. *Haemophilia* 1998; **5**: 32–9.
5 Ansell BM, Crook A, Mallard JR, Bywaters EGL. Evaluation of intra-articular colloidal Gold Au 198 in the treatment of persistent knee effusions. *Ann Rheum Dis* 1963; **22**: 435–9.
6 Ahlberg Å. Radioactive gold in the treatment of chronic synovial effusion in haemophilia. International Congress Series No. 252 *Haemophilia*. Amsterdam: Excerpta Medica, 1971: 212–15.
7 Ahlberg Å, Pettersson H. Synoviorthesis with radioactive gold in haemophiliacs. *Acta Orthop Scand* 1979; **52**: 513–17.
8 Löfqvist T, Petersson CJ. Experience with colloid 198 Au synoviorthesis in haemophiliacs. XVII International Congress of the World Federation of Hemophilia, Milan, 8–13 June 1986 [Abstract].
9 Petersson C. Orthopedic joint evaluation in hemophilia: prophylactic treatment of hemophilia A and B current and future perspectives. *Sci Med* 1994: 35–9.

CHAPTER 21

Musculoskeletal magnetic resonance imaging

R. Nuss, R.F. Kilcoyne and J.D. Wiedel

It is widely accepted that recurrent haemorrhage into joints induces haemophilic arthropathy. The underlying pathological process causing haemophilic arthropathy is not well understood. However, synovial findings in haemophilic arthropathy have been likened to those occurring in rheumatoid arthritic joints [1] and bone and cartilage changes are similar to those found in osteoarthritic joints [1]. Traditionally, it was thought that synovitis preceded bone and cartilage damage in haemophilic arthropathy but, more recently, it has been suggested that cartilage damage may ensue prior to synovial and bone damage [2]. In either case, the radiological findings of haemophilic arthropathy include haemarthrosis, effusion, synovial hyperplasia, osteophyte formation, osteoporosis, subchondral cysts, erosions, pseudotumour formation, cartilage loss and ankylosis [3]. Figure 21.1 is a magnetic resonance image showing some findings of haemophilic arthropathy.

The conventional radiological method for evaluation of haemophilic arthropathy is by plain X-ray. Advantages of plain X-ray evaluation of a joint include its speed and relative cost. Also, access to equipment for plain X-ray is readily available and results are easily interpreted. However, it has long been recognized that plain X-rays underestimate joint pathology found at surgical inspection in people with haemophilic arthropathy. Synovial hyperplasia plays a critical part in the development of haemophilic arthropathy, but with plain X-ray synovial hyperplasia cannot be distinguished from an effusion. With plain X-ray, cartilage loss is assessed by loss of the normal joint space rather than by direct visualization. In general, the downfall of plain X-ray for evaluation of haemophilic arthropathy is its inability to effectively assess soft tissue findings such as synovium and cartilage.

Although magnetic resonance imaging (MRI) of a joint is more expensive, less accessible, and more difficult to interpret than plain X-ray evaluation, it is an excellent means of evaluating haemophilic arthropathy. MRI is performed by placing the joint in a homogeneous magnetic field which is orientated along an axis from the person's head to foot [4]. The protons within the joint (especially those in lipid and water) generate their own small magnetic field which aligns with the machine's magnetic field (longitudinal magnetization). Radiofrequency waves or echoes are superimposed to disrupt the alignment of protons.

T1 relaxation time (spin–lattice relaxation) reflects the time it takes for the protons to realign with the static magnetic field after excitation. T2 relaxation time (spin–spin relaxation) reflects how quickly the transverse magnetization between protons decays over time. Each tissue within the joint has its own characteristic proton density, T1 and T2 relaxation times. MRI converts the voltage changes received by the coil from the joint into digital data which are then converted into images. Time to echo (TE), the time from the start of the study to the peak signal intensity, and time of significant repetition of echoes (TR) determine the appearance of the final images.

The final images are multiplanar and are in an area (the field-of-view) which is divided into many small picture elements (voxels), which are composed of two-dimensional picture elements (pixels), in a certain thickness of tissue. Images with small fields-of-view, small pixels and thin slices have higher spatial resolution which allows small structures that are close together to be distinguished from each other.

Although T1-weighted images (fat-sensitive) are important for showing anatomic detail, most pathological findings have an increase in tissue T2 values so T2-weighted images (water-sensitive) are better for musculoskeletal imaging. The increase in T2 is brought about by the very bright signal which is made by fluid which is especially relevant in imaging haemophilic joints because fluid is often present. Conventional T2-weighted spin–echo images require a long acquisition time. An alternative technique is the use of gradient echo pulse sequences. Gradient echo pulse sequences use a radiofrequency pulse with a flip angle of less than 90° and do not refocus a 180° pulse as is used in the spin–echo technique. Gradient echo imaging, indicated as T2*, supports shorter repetition times so imaging time is reduced. A repetition time of 20–60 ms makes it possible to obtain three-dimensional gradient echo images with thin contiguous sections which have high spatial resolution so small structures and lesions can be appreciated.

As MRI has the ability to visualize synovial hyperplasia, it was postulated that MRI might be of value in the evaluation of haemophilic joints. The visualization of synovial hyperplasia is of great significance in haemophilic joints because it precedes bone destruction. Synovial hyperplasia is often the earliest visible pathological abnormality in haemophilic joints.

MUSCULOSKELETAL MRI

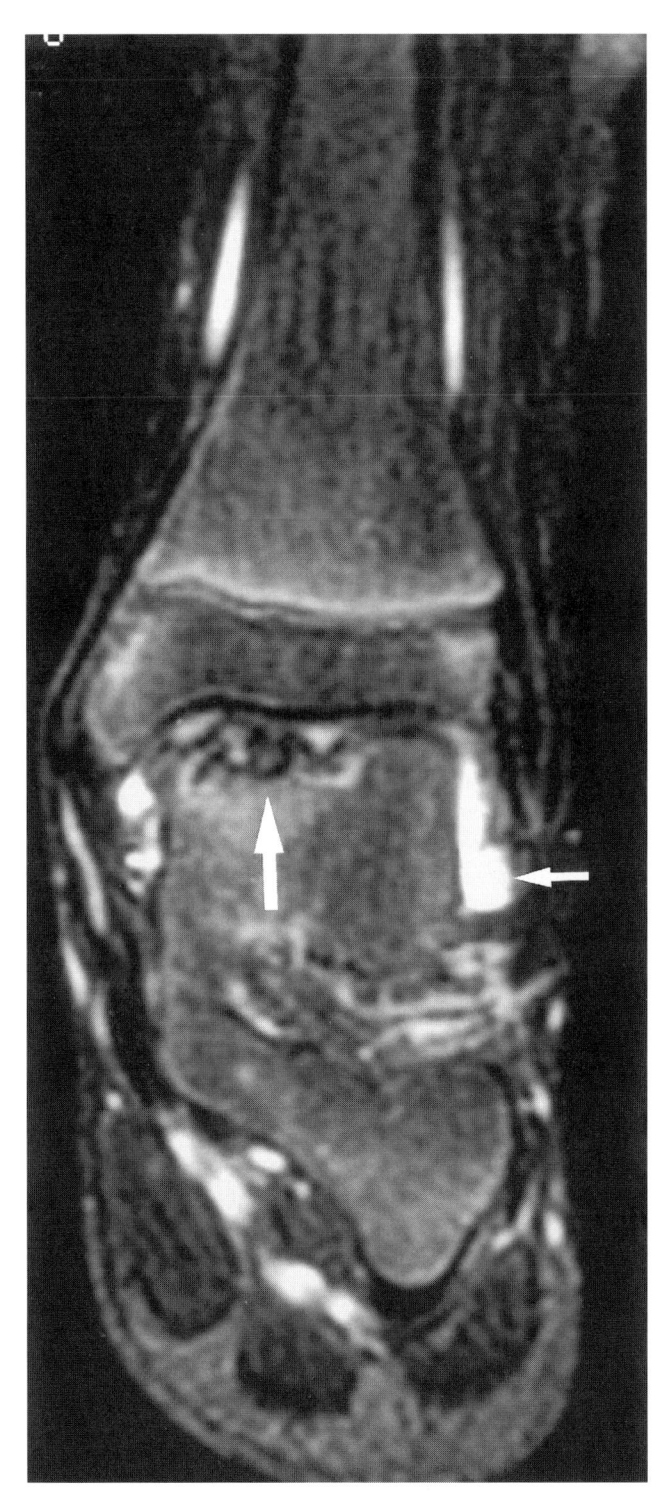

Fig. 21.1 Gradient echo magnetic resonance image (MRI) coronal view of the ankle of a 9-year-old boy with severe factor VIII deficiency and high titre inhibitor demonstrates an effusion (small arrow) and cysts and erosions (large arrow).

Visualization of hypertrophied synovium is significant because its presence signals the need to consider intervention in an attempt to prevent bone and cartilage damage. It is thought, but not proven, that intervention at the time of synovial hyperplasia but prior to bone and cartilage damage may save a joint from developing those late changes. Synovitis is postulated, but not proven, to be reversible, whereas bone and cartilage damage is considered irreversible.

Because MRI is capable of directly visualizing cartilage, it was postulated to be of value in the evaluation of haemophilic joints. In haemophilic arthropathy, cartilage may be directly destroyed or lost secondary to the inflammatory changes associated with synovitis. Whereas articular cartilage thickness is assessed only indirectly with plain X-ray by evaluating loss of joint space width, with MRI cartilage is directly visualized as an intermediate to high intensity line next to the low intensity subchondral bone [5]. MRI supports direct visualization of cartilage in all compartments of a joint, whereas plain X-rays only show a two-dimensional view of a joint, primarily the bone structure. In individuals with osteoarthritis, MRI has been shown to be better than plain X-ray at delineating joint pathology [6].

There have been at least seven series and one case report demonstrating the general applicability of MRI to the evaluation of haemophilic arthropathy. In 1986, Steudel *et al.* [7] and Kulkarni *et al.* [8] were the first investigators to report that MRI was a valuable tool for the evaluation of haemophilic arthropathy. Steudel *et al.* studied 40 knees in patients with haemophilia and reported that MRI was particularly good at demonstrating intra-articular findings including synovium and cartilage [7]. Kulkarni *et al.* studied 11 knees and reported that MRI aptly demonstrated periarticular and subchondral abnormalities including subchondral haemorrhage, cysts and fibrosis [8]. Because MRI allows differentiation of blood from other effusions, it was deemed valuable in the assessment of haemophilic joints.

In 1987 Pettersson *et al.* evaluated 16 knees in patients with haemophilia using MRI [9]. In the same year Yulish *et al.* [10] studied 35 haemophilic joints (14 knees, 15 ankles and six elbows) with MRI. Both investigative teams reported MRI to be effective for assessing cartilage, synovial hyperplasia, bone and cysts. In addition, Pettersson *et al.* also reported that MRI could aptly assess menisci and ligaments in joints with haemophilic arthropathy.

In a series of inflammatory synovial processes evaluated by MRI, Sanchez and Quinn [11] included one haemophilic joint. They concluded that the soft tissue contrast resolution of MRI and its multiplanar capability made it 'suitable' for evaluating mass-like or fluid inflammatory processes, determining the extent of disease and, for some diseases, determining the composition of the inflammatory process.

Baunin *et al.* [12] in 1991 reported the first series confined to haemophilic joints in children. Twenty-eight joints (17 knees, 10 ankles and one elbow) were studied in this series where MRI findings were compared with plain X-ray findings. MRI revealed cartilage changes, effusions, synovial hyperplasia and early bone lesions better than plain X-ray. The authors concluded that MRI findings should be useful for developing treatment plans and monitoring response to treatment.

In 1991 Armstrong and Watt [13] reported a 23-year-old man with haemophilia evaluated for recurrent intermittent swelling above his painful left knee. MRI demonstrated a suprapatellar pouch filled with blood which was separated from the knee joint by an intact suprapatellar plica.

The largest series of haemophilic joints evaluated by MRI was reported by Briukhanov et al. [14] who evaluated 320 joints. Findings on MRI were compared with those on plain X-ray and MRI was again reported as superior in its ability to distinguish haemarthroses and haemophilic arthropathy which were not appreciated by plain X-ray examination.

The most recent series evaluating MRI and haemophilic joints was published by Mathew et al. [15] in 2000. They studied 17 joints (2 knees, 10 ankles and 5 elbows) in 11 children with haemophilia. Four joints in three children were normal by plain X-ray evaluation but had substantial bone and cartilage findings on MRI evaluation. They concluded that MRI was effective for the evaluation of haemophilic arthropathy but its applicability would be limited until a universally accepted MRI scale was developed.

To date, one series has focused specifically on imaging bone cysts in haemophilic joints by MRI. Subchondral cysts are often missed on plain X-ray because of their position adjacent to osteopenic areas. Osteopenic areas produce juxta-articular radiolucent areas that obscure the cyst margins on plain X-ray. In 1992 Idy-Peretti et al. [16] studied 128 knees in 64 people with haemophilia and based on T1 and T2 findings could classify cysts into four stages by their signal intensity. The joints of 25 people were followed up 10–30 months later which demonstrated that the four stages were successive. They concluded, with MRI T2*–weighted gradient echo images, that cysts have higher spatial intensity and are better differentiated from the adjacent low intensity marrow and subchondral bone.

In an attempt to optimize the capabilities of MRI for evaluating haemophilic arthropathy, investigators have studied various different MRI techniques and the administration of contrast agents. Erlemann et al. [17] reported the use of two elaborate MRI techniques, flash (FLA) and fast imaging with steady state precession (FISP-3D), in evaluation of 52 knees and ankles. These techniques are based on gradient echo imaging. They concluded that these techniques were better than plain X-ray for evaluation of all joint findings except bone damage.

Rand et al. [18] studied 21 joints, including 10 symptomatic knees and ankles, in 16 children with haemophilia. They used gradient echo imaging with three-dimensional fat saturation and another technique called Flash 2D GE or Hemo Flash. They concluded that MRI with three-dimensional gradient echo fat-suppression sequences was advantageous in allowing thin high resolution contiguous slices with an increased signal : noise ratio to be obtained. It also supported high quality image reformation. It was especially good for cartilage imaging because it increased the contrast : noise ratio between cartilage and joint fluid and between cartilage and subchondral bone. The net result was cartilage as a high signal intensity band.

Flash 2D GE supported evaluation of blood products in the joint; six joints had blood evident on MRI despite no clinical signs of bleeding.

There are four single case reports and two series demonstrating the value of MRI in visualizing haemophilic pseudotumours. Wilson and Prince [19] published the first case report of a person with haemophilia found to have multiple pseudotumours in both thighs. Ehara and Kattapuram [20] reported the diagnosis of a pseudotumour of the sacrum in an adult with haemophilia by MRI. Hermann et al. [21] published a pictorial essay of haemophilic arthropathy including two pseudotumours shown by MRI. Case reports by the authors in 1994 [22] and Gaary et al. [23] in 1996 showed MRI to be of value in detecting pseudotumours in people with haemophilia. The largest series to date is by Jaovisidha et al. [24] who reported 12 pseudotumours in five people with haemophilia detected by MRI. They reported MRI that was very accurate in the localization of pseudotumour in fat, muscle, fascia, superiosteum or bone. It was useful for treatment planning and for following the course of pseudotumours.

Whether or not the administration of a contrast agent, such as gadolinium, improves MRI evaluation of a haemophilic joint is controversial. Gadolinium is a paramagnetic agent that acts to decrease the T1 but increase T2 relaxation of tissue. Gadolinium-enhanced tissues appear brighter on T1-weighted images but it is less useful for enhancing T2-weighted images.

Two groups have published studies evaluating intravenous administration of gadolinium for haemophilic joint imaging. Nagele et al. [25] studied 17 knees in patients with haemophilia following intravenous gadolinium and found that synovial uptake of gadolinium was good in haemophilic joints with only minimal uptake by muscle, fat, tendons, marrow or effusion. They concluded that gadolinium helped to delineate and quantify synovium. Rand et al. [18] gave gadolinium intravenously to 16 children with haemophilic arthropathy. They concluded that gadolinium helped to demarcate synovial hyperplasia from cartilage and fluid.

We and others (Bjorn Lundin, personal communication) have not found gadolinium to be of value in imaging haemophilic arthropathy if gradient echo images are obtained. Haemophilic arthropathy differs from other forms of arthritis because of the presence of haemosiderin deposits in much of the inflamed synovium. The presence of this low signal substance enhances the detection of synovitis on MRI. Reasons to avoid gadolinium include the possibility of allergic reactions and the need for venous access which is often problematic, especially in young children.

Although it has often been suggested that MRI could be useful in treatment planning for people with haemophilia, only a few series have studied MRI in that capacity. In 1993 we postulated that MRI findings might globally affect management plans based on conventional plain radiographical data. We studied 14 target joints in 13 children with haemophilia aged 7–16 years [26,27]. For this study a target joint was defined as having sustained two or more haemorrhages per month in the preceding 6 months. After reviewing the history, examining the target joint, and obtaining a plain X-ray of the target joint, a treatment plan was developed. Only then was an MRI of the target joint reviewed, which resulted in a change of treatment plan in 40% of joint assessments.

The Rand series [18] compared the number of bleeding episodes, joint physical examination and pain scores, and plain X-ray findings with MRI assessments of 21 joints in 16 children. Those authors found that the joint examination and pain scores underestimated joint changes identified with MRI. Based on MRI findings, the treatment plans were intensified for 40% of the children, a figure comparable to our study.

We next evaluated the utility of MRI in planning treatment for patients with haemophilic joints by studying whether synovial findings by MRI were predictive of response to radionuclide synovectomy and whether MRI joint findings changed following a radioactive synovectomy [28]. In 2000, we studied 21 joints (10 ankles, 6 knees and 5 elbows) in 17 people with haemophilia. Although synovial hyperplasia varied, we found the degree of synovial hyperplasia did not predict response to the procedure as 17 of the 21 joints had a decrease in bleeding frequency of 50% or greater at the 6-month MRI evaluation. There was progressive joint damage in 28% of joints studied with MRI at 6 months' follow-up. We found that joint damage was underestimated by plain X-ray when compared to MRI prior to and 6 months after the injections. Seven joints were reimaged at 1 and 2 years postprocedure. With MRI, progressive cartilage and bone damage could be appreciated in 6 joints. One joint had only synovial hyperplasia preprocedure, which diminished at follow-up.

In 2000, we reported the anatomical outcome in haemophilic joints in patients who had undergone radiosynoviorthesis at a median of 16 years earlier [29]. We were interested in determining whether the widely accepted postulate that synovial hyperplasia regresses in severely damaged haemophilic joints was true. We studied 13 joints with extensive bone and cartilage loss, and found that 11 joints had synovial hyperplasia on MRI. Both clinical examination and plain radiographs underestimated the presence of synovial hyperplasia and effusion compared with MRI.

Wallny et al. [30] reported on their use of MRI to assess 21 knees in 20 patients with haemophilia prior to intra-articular administration of hyaluronic acid. An MRI of the joints was obtained prior to weekly injection of 20 mg hyaluronic acid for 5 weeks and a repeat MRI of the joints was obtained 3 months after the first hyaluronic acid injection. There were no significant differences between the pre- and post-treatment MRI findings.

By 2000 there was enough evidence demonstrating MRI to be useful that we designed and published a provisional MRI scale for the evaluation of haemophilic joints [28]. The haemophilic MRI arthropathy scale was designed to be clinically useful but also effective for research trials. We modelled the scale after the Arnold–Hilgartner plain X-ray scale in that we made it progressive rather than additive. Information in each of the categories (effusion/haemarthrosis, synovial hyperplasia/haemosiderin, cysts/erosions, and cartilage loss) is collected but the final score is the highest number or greatest degree of joint damage present. We chose complete loss of cartilage to represent the maximum damage because functionally this is the most significant pathology.

Publications on the use of MRI in evaluating haemophilic arthropathy have generally been published in the radiological literature, and often do not indicate the type of haemophilia, severity, or inhibitor status in the people whose joint findings are reported. In our limited experience, joint findings by MRI in patients with inhibitors do not appear to differ from findings in individuals without an inhibitor.

Medical and surgical management has significantly improved for patients with inhibitors, so MRI is as useful for evaluating joints in people with inhibitors as for those without inhibitors. As is true for plain X-ray findings, MRI anatomical findings must be interpreted in conjunction with a detailed history of bleeding, joint physical examination and pain assessment [31]; ideally, an assessment of quality of life would also be available. Treatment planning should not be based solely on MRI findings.

In the medical management of a patient with inhibitors, MRI findings can be helpful in localizing the precise site of haemorrhage. This may have an impact on the intensity and duration of treatment as well as the choice of treatment with standard factor concentrates, aPCC or rVIIa. For example, demonstration of haemorrhage into a joint such as the hip would indicate that more intense treatment with factor concentrate is warranted than haemorrhage into soft tissue or muscle. The finding of osteonecrosis of the hip on MRI rather than haemorrhage into the hip in an inhibitor patient would lead to management with anti-inflammatory drugs rather than prescription of expensive factor concentrates. Similarly, for people with an inhibitor in whom haemorrhagic joint pain and pain secondary to osteoarthritis cannot be differentiated, MRI without synovium spares the expense and exposure to factor concentrate.

Activated prothrombin complex concentrate (aPCC) is sometimes prescribed as a prophylactic programme to prevent joint haemorrhage in people with inhibitors [32]. MRI could be used in a treatment trial of aPCC to obtain baseline joint imaging and then to study the efficacy of concentrate administration at a predetermined date. The ideal use of MRI is in conjunction with plain X-ray in medical treatment or surgical trials.

Historically, surgical intervention was not an option for people with haemophilia who had an inhibitor because of concerns with refractory haemorrhage. Since rVIIa concentrate has become available as a treatment option for patients with inhibitors, elective surgery is now a potential option. Successful major orthopaedic procedures have been performed in patients with inhibitors with rVIIa coverage [33]. To date, reported

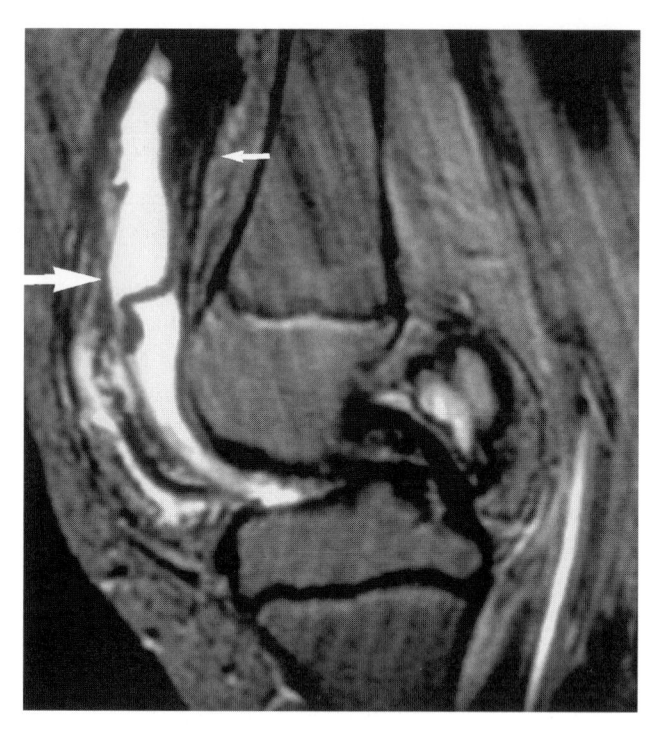

Fig. 21.2 T2-weighted MRI sagittal view shows synovial hyperplasia (small arrow) and an effusion (large arrow) in the ankle of an 11-year-old boy with severe factor VIII deficiency and high titre inhibitor. No bone or cartilage damage is present.

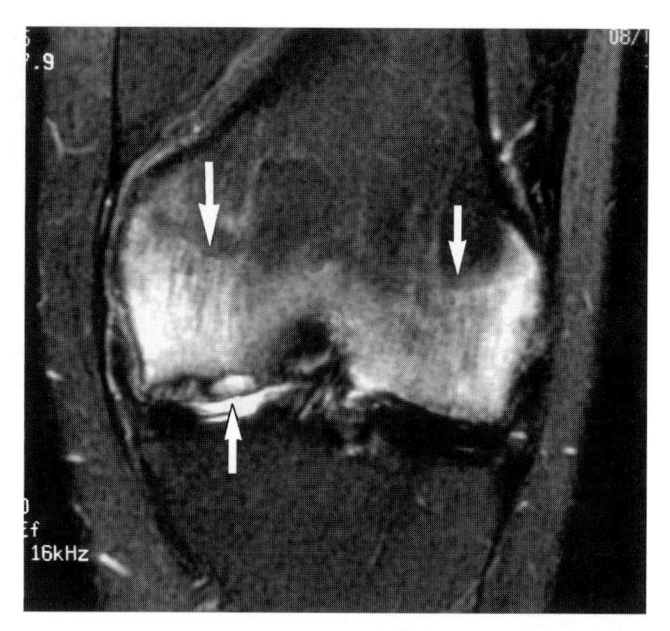

Fig. 21.3 T2-weighted sagittal MRI view of the knee of a 32-year-old man with severe factor VIII deficiency and high titre inhibitor demonstrates bone oedema of the femur (large arrows) and a cyst (mixed arrow). No cartilage can be visualized.

procedures include knee, hip and ankle arthroplasties, knee, ankle and elbow arthroscopic and open synovectomies, radionuclide synovectomies, pseudotumour resections and osteotomies.

MRI can have a role in the evaluation of the anatomical joint status of the patient who has an inhibitor contemplating surgical intervention. For the patient with an inhibitor and a painful joint, who is unable to differentiate between arthritic and haemorrhagic pain, as for the non-inhibitor patient MRI can be useful in determining whether or not synovium is present. The extent of haemophilic arthropathy in the joint can also be determined. This information can be used to make an informed decision about treatment options. If synovial hyperplasia is present but limited, intervention could be warranted in the absence of bone or cartilage damage in an attempt to preserve joint integrity. Treatment by arthroscopic, open, or radionuclide synovectomy may be indicated (Fig. 21.2 shows an MRI of a joint that may benefit from intervention). If bone or cartilage damage is present, radionuclide synovectomy may be preferred because it is less invasive and is associated with less postprocedural morbidity. If synovial hyperplasia is extensive, the decision becomes more difficult because treatment is less likely to be effective and may need to be repeated. If synovial hyperplasia is absent, no surgical intervention may be indicated; Fig. 21.3 shows an MRI of an endstage joint which would not be alleviated by any type of synovectomy.

MRI can reveal early bone cysts in joints of people with haemophilia, including those with an inhibitor. Early detection may support local intervention which may be performed by puncturing the cyst with a trocar, aspirating the lesion, then sealing it with fibrin sealant or bone graft rather than waiting until the cyst is large and more extensive surgery is indicated [33].

For the patient with an inhibitor suspected of having a pseudotumour, MRI will be effective in delineating the number, size, extent, neurovascular involvement and impact on surrounding structures. This information could be used to determine whether or not the benefits outweigh the risks of surgical intervention. MRI could be useful in the inhibitor patient considering joint debridement because of pain or locking of a joint. Figure 21.4 shows the MRI of a joint that was evaluated for possible surgical intervention. Visualization of the joint with MRI could aid the surgeon in determining how extensive surgical debridement would need to be to improve joint pain and function.

Finally, with the recent advances in treatment, people with inhibitors are more active than they have been in the past. They may suffer from traumatic events related to increased activity and further damage their joints. Even if they do not increase their activity they, like anyone else, may be involved in trauma as a consequence of daily life. Like others without haemophilia, they may benefit from the ability of MRI to visualize joint anatomy, especially ligament and meniscal pathology [34–38].

In summary, studies of MRI show it to be superior to plain X-ray, especially in visualizing synovium and cartilage in haemophilic joints. With recent advances, people with inhibitors now have more and better treatment options and, most importantly, surgery is now a possibility. With MRI, anatomical joint pathology can be accurately assessed so that the best medical or surgical intervention for a person with an

MUSCULOSKELETAL MRI

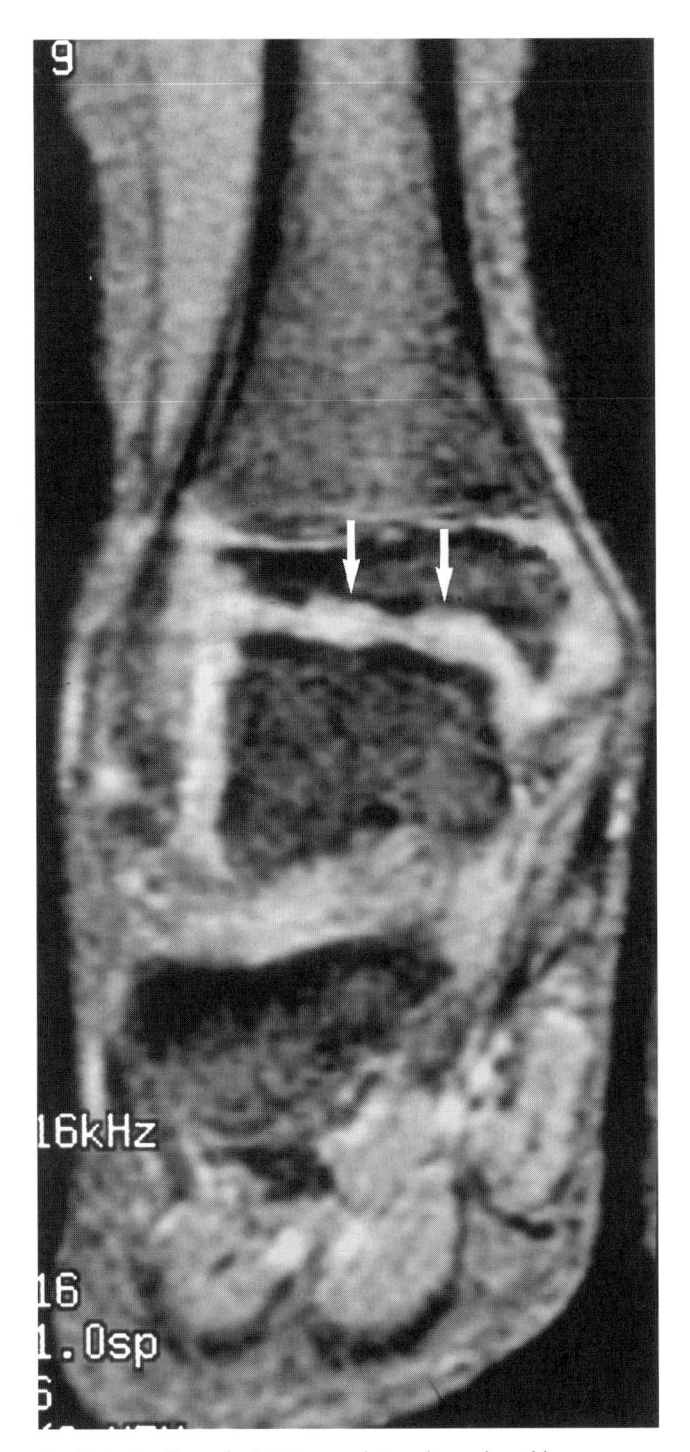

Fig. 21.4 Gradient echo MRI coronal view shows the ankle of an 11-year-old boy with severe factor VIII deficiency who was diagnosed with a high titre inhibitor at 3 years of age. Immune tolerance was achieved by the time he was 5 years of age. There are erosions (arrows) and very little cartilage. No synovium can be seen.

inhibitor can be determined. MRI findings are best used to determine treatment plans in conjunction with joint physical examination findings, pain assessment, plain X-ray and quality of life determinations.

References

1 Roosendaal G, van Rinsum AC, Vianen ME et al. Haemophilic arthropathy resembles degenerative rather than inflammatory joint disease. *Histopathology* 1999; **34**: 144–53.
2 Roosendaal G, TeKoppele JM, Vianen ME et al. Articular cartilage is more susceptible to blood induced damage at young than at old age. *J Rheumatol* 2000; **27**: 1740–4.
3 Pettersson H. Radiographic scores and implications. *Semin Hematol* 1993; **30**: 7–9.
4 Mitchell DG, ed. *MRI Principles*. W.B. Saunders, London: 1999.
5 Disler DG, Recht MP, McCauley TR. MR imaging of articular cartilage. *Skeletal Radiol* 2000; **29**: 367–77.
6 Buckland-Wright C. Current status of imaging procedures in the diagnosis, prognosis and monitoring of osteoarthritis. *Baillière's Clin Rheumatol* 1997; **11**: 727–48.
7 Steudel A, Clauss G, Traber F, Nicolas V, Lackner K. MR tomography of hemophilic arthropathy of the knee joint. *Rofo Fortschr Geb Rontgenstr Neuen Bildgeb Verfahr* 1986; **145**: 571–7.
8 Kulkarni MV, Drolshagen LF, Kaye JJ et al. MR imaging of hemophiliac arthropathy. *J Comput Assist Tomogr* 1986; **10**: 445–9.
9 Pettersson H, Gillespy T, Kitchens C et al. Magnetic resonance imaging in hemophilic arthropathy of the knee. *Acta Radiol* 1987; **28**: 621–5.
10 Yulish BS, Lieberman JM, Strandjord SE et al. Hemophilic arthropathy: assessment with MR imaging. *Radiology* 1987; **164**: 759–62.
11 Sanchez RB, Quinn SF. MRI of inflammatory synovial processes. *Magn Reson Imaging* 2001; **7**: 529–40.
12 Baunin C, Railhac JJ, Younes I et al. MR imaging in hemophilic arthropathy. *Eur J Pediatr Surg* 1991; **1**: 358–63.
13 Armstrong SJ, Watt I. Case report 661. *Skeletal Radiol* 1991; **20**: 369–71.
14 Briukhanov AV, Fedorov VV, Mikhal'kov DF. Magnitno-rezonansnaia tomografiia v diagnostike gemofilicheskikh porazhenii sustavov [Magnetic resonance imaging in the diagnosis of hemophilic lesions of joints]. *Vestn Rentgenol Radiol* 1998; **5**: 25–9.
15 Mathew P, Talbut DC, Frogameni A et al. Isotopic synovectomy with P-32 in paediatric patients with haemophilia. *Haemophilia* 2000; **6**: 547–55.
16 Idy-Peretti I, Le Balc'h T, Yvart J, Bittoun J. MR imaging of hemophilic arthropathy of the knee: classification and evolution of the subchondral cysts. *Magn Reson Imaging* 1992; **10**: 67–75.
17 Erlemann R, Pollmann H, Vestring T, Peters PE. MR-Tomographie der hamophilen Osteoarthropathie unter besonderer Berucksichtigung der synovialen und chondrogenen Alterationen [The MR tomography of hemophilic osteoarthropathy with special reference to the synovial and chondrogenic changes]. *Rofo Fortschr Geb Rontgenstr Neuen Bildgeb Verfahr* 1992; **156**: 270–6.
18 Rand T, Trattnig S, Male C et al. Magnetic resonance imaging in hemophilic children: value of gradient echo and contrast-enhanced imaging. *Magn Reson Imaging* 1999; **17**: 199–205.
19 Wilson DA, Prince JR. MR imaging of hemophilic pseudotumors. *Am J Roentgenol* 1988; **150**: 349–50.
20 Ehara S, Kattapuram SV. Hemophiliac pseudotumor of the sacrum. *Radiat Med* 1989; **7**: 214–16.
21 Hermann G, Gilbert MS, Abdelwahab IF. Hemophilia: evaluation of musculoskeletal involvement with CT, sonography and MR imaging. *Am J Roentgenol* 1992; **158**: 119–23.

22 Nuss R, Kilcoyne R, Geraghty S et al. Magnetic resonance imaging visualization of hemorrhage into a suprapatellar pouch in a child with hemophilia. *Am J Pediatr Hematol Oncol* 1994; **16**: 183–5.

23 Gaary E, Gorlin JB, Jaramillo D. Pseudotumor and arthropathy in the knees of a hemophiliac. *Skeletal Radiol* 1996; **25**: 85–7.

24 Jaovisidha S, Ryu KN, Hodler J et al. Hemophilic pseudotumor: spectrum of MR findings. *Skeletal Radiol* 1997; **26**: 468–74.

25 Nagele M, Kunze V, Hannan M et al. Hemophiliac arthropathy of the knee joint: Gd-DTPA-enhancing MRI—clinical and roentgenological correlation. *Rofo Fortschr Geb Rontgenstr Neuen Bildgeb Verfahr* 1994; **160**: 154–8.

26 Nuss R, Kilcoyne RF, Geraghty S, Wiedel J, Manco-Johnson M. Utility of magnetic resonance imaging for management of hemophilic arthropathy in children. *J Pediatr* 1993; **123**: 388–92.

27 Manco-Johnson MJ, Nuss R, Geraghty S, Funk S, Kilcoyne R. Results of secondary prophylaxis in children with severe hemophilia. *Am J Hematol* 1994; **47**: 113–17.

28 Nuss R, Kilcoyne RF, Geraghty S et al. MRI findings in haemophilic joints treated with radiosynoviorthesis with development of an MRI scale of joint damage. *Haemophilia* 2000; **6**: 162–9.

29 Nuss R, Kilcoyne RF, Rivard GE, Murphy J. Late clinical, plain X-ray and magnetic resonance imaging findings in haemophilic joints treated with radiosynoviorthesis. *Haemophilia* 2000; **6**: 658–63.

30 Wallny T, Brackmann HH, Semper H et al. Intra-articular hyaluronic acid in the treatment of haemophilic arthropathy of the knee: clinical, radiological and sonographical assessment. *Haemophilia* 2000; **6**: 566–70.

31 Gilbert MS. Prophylaxis: musculoskeletal evaluation. *Semin Hematol* 1993; **30** (Suppl. 2): 3–6.

32 Leissinger CA. Use of prothrombin complex concentrates and activated prothrombin complex concentrates as prophylactic therapy in haemophilia patients with inhibitors. *Haemophilia* 1999; **5** (Suppl. 3): 25–32.

33 Rodriguez-Merchan EC, de la Corte H. Orthopaedic surgery in haemophilic patients with inhibitors: a review of the literature. In: Rodriguez-Merchan EC, Goddard NJ, Lee CA, eds. *Musculoskeletal Aspects of Haemophilia*. Oxford: Blackwell Science, 2000: 136–59.

34 Chen HC, Hsu CY, Shih TT, Huang KM, Li YW. MR imaging of displaced meniscal tears of the knee: importance of 'disproportional posterior horn sign'. *Acta Radiol* 2001; **42**: 417–21.

35 Kathira K, Yamashita Y, Takahashi M et al. MR imaging of the anterior cruciate ligament: value of thin sliced direct oblique coronal technique. *Radiat Med* 2001; **19**: 1–7.

36 Vande Berg BC, Poilvache P, Duchateau F et al. Lesions of the menisci of the knee: value of MR imaging criteria for recognition of unstable lesions. *Am J Roentgenol* 2001; **176**: 771–6.

37 Rosenberg ZS, Beltran J, Bencardino JT. MR Imaging of the ankle and foot. *Radiographics* 2000; **20**: S153–S179.

38 Anderson SE, Otsuka NY, Steinbach LS. MR Imaging of pediatric elbow trauma. *Semin Musculoskelet Radiol* 1998; **2**: 185–98.

CHAPTER 22

Treatment of iliopsoas haematomas and compartment syndromes in patients with haemophilia who have a circulating antibody

M. Heim, M. Warshavski, Y. Amit and U. Martinowitz

The treatment of compartment syndromes is an area where the term 'orthopaedic emergency' is particularly apt. There are two factors which are critical: the extent of the intracompartmental pressure and the time. Muscles tolerate 4 h of ischaemia well, but by 6 h the results are uncertain and by 8 h the effects are irreversible [1]. Nerves are far more sensitive to ischaemia than muscles. One out of 10 patients presenting with a compartment syndrome has a bleeding disorder or is on anticoagulants [2]; the literature abounds with reports of compartment syndromes that have occurred as a result of anticoagulation medications [3–5].

Once the problem of coagulation has been brought under control it becomes necessary to determine whether the raised pressure in the compartment has reached such proportions that surgery is required. Compartment syndromes have been reported in patients where the cause was intraoperative hypotension [6], prolonged pressure from the intraoperative lithotomy position [6] and prolonged lack of movements in comatosed patients [2]. Compartment syndromes occur not only in anticoagulated, hypotensive or surgical patients but they may also occur in young muscular men as a result of strenuous exercise. This has been referred to as 'exercise induced' [7] or exertional [3] compartment syndrome.

The compression of the muscles within the compartments results in muscle necrosis with a concomitant elevation of serum creatine kinase and lactate dehydrogenase [7]. These two blood tests can provide information appertaining to the state of the intracompartmental musculature. The key to effective management is early diagnosis. Intracompartmental pressure needs to be measured and in areas where the recorded pressure reaches 20 mmHg below the diastolic blood pressure, decompression is urgently required [1,4,7,8]. Forearm and shin closed fascial compartments provide easy access for measuring intracompartmental pressure. The iliopsoas complex is deep and access is difficult, with decompression even more difficult. Symptoms of iliopsoas bleeding occur late and only do so if the haemorrhage is extensive. Furthermore, the limbs being visible make observation and palpation possible and hence the swollen, tense and painful limb can be easily assessed by clinical examination.

Unfortunately, the iliopsoas complex lies deep within the body and a diagnosis can only be suspected by indirect symptoms and inferences. For these deep pathologies investigation by the use of ultrasound becomes mandatory [9,10]. Not only can the lesions be visualized but the examination adds a quantitative factor, which can be used to monitor progress or regression. It has been recognized that neural elements are more sensitive to ischaemia than muscles and hence monitoring of neural conduction may contribute another modality for monitoring the effects of raised intracompartmental pressure. Electromyography is routinely used in the diagnosis of carpal tunnel syndrome, entrapped ulnar nerves in the area of the elbow joint and the diagnosis of tarsal tunnel syndrome.

Management of a person with haemophilia and inhibitors who has developed a suspected compartment syndrome

The initial mandatory action is a prompt ascertainment of the patient's coagulation status, and this depends upon the availability of products. The drug of choice is recombinant factor VIIa, the next is activated prothrombin complex or porcine factor VIII (FVIII) if there is no cross-reactivity between the inhibitor and the porcine FVIII in the particular patient and the titre of the antibody is not too high. Further down the list appear prothrombin complex preparations. The most acceptable method of assessment of the status within a compartment is by direct measurement of intracompartmental pressure. Many companies have prepared kits (Stryker Corp., Kalamazoo, MI, USA) where a needle is inserted into the tense compartment and connected to a measuring device, which records the intracompartmental pressure. This can be used not only to assess the pressure at a given time but also as a monitoring device to record improvement or deterioration [7].

Diagnosis of deep compartment syndromes is more reliant upon indirect assessment. With iliopsoas haematomas the degree of hip joint flexion and the extent of femoral nerve sensory changes in the anterior aspect of the thigh help in the ongoing assessment. It was previously thought that the femoral nerve was compressed against the inguinal ligament as the size of the iliopsoas increased but this is unlikely and it is more feasible

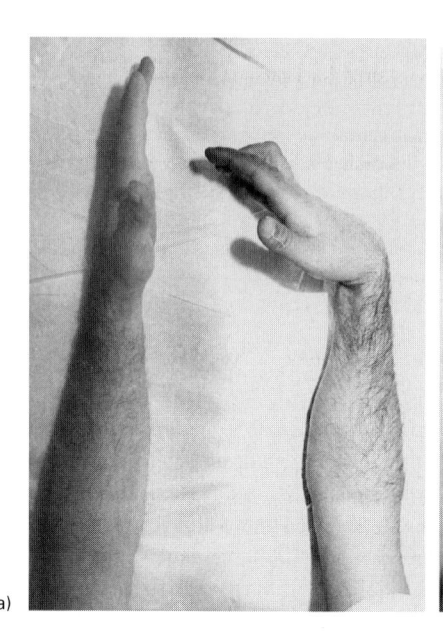

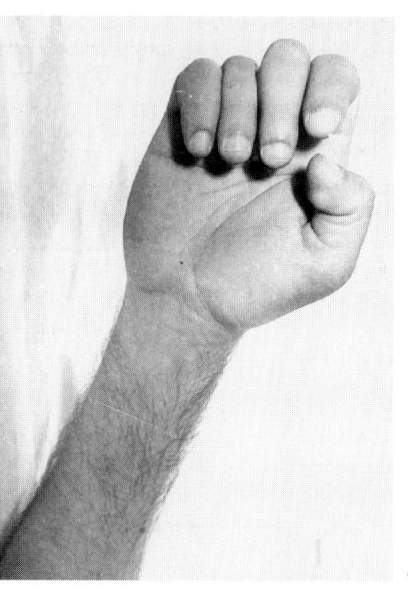

Fig. 22.1 (a) Note the different shapes of the forearms. The right hand demonstrates extension of most of the fingers while the wrist is flexed to about 70°. (b) When the wrist is extended to the neutral position, the fingers bend and cannot be extended. This is a result of intracompartmental bleeding in the forearm.

that the femoral nerve is compressed within the pelvis itself. The femoral nerve lies within a groove between the iliacus and the psoas muscles. As the haematomas within the psoas muscle seep down the sheath—pulled by gravity—the psoas muscle expands and compresses the nerve against the iliacus. To our knowledge, there has been no report of surgery being carried out because of an iliopsoas bleed. These cases have been treated conservatively [11] by factor replacement, best in the Fowler position, gentle exercises comprising passive range of motion (ROM) followed by active ROM, non-weight-bearing ambulation through partial weight-bearing to full weight-bearing. The rehabilitation programme is extended and over-zealous treatment may result in a secondary haemorrhage [12].

Classically, the literature with regard to compartment syndromes emphasizes the mnemonic of the 'five Ps': pain, pressure, paraesthesis, paresis, pulses. Treatment is aimed at the prevention of the late sequelae of raised intracompartmental pressure. Volkmann contractures of the hand (Fig. 22.1) have resulted from late fibrosis of the musculature within the affected compartment and a similar pathology may occur in the foot [13].

Decompression

Once it has been established that the intracompartmental pressure is of a magnitude that endangers the integrity of the compartment contents, then decompression is mandatory. Established haemostasis is a prerequisite and anaesthesia is required. Timing is a critical factor and if problems arise with regard to general anaesthesia then the procedure may be undertaken under local anaesthesia [14]. What is required is the incision of the fascial envelope that surrounds the muscles. A small incision is generally insufficient, and as the fascia is incised the muscles under pressure herniate through the incision, expanding to such an extent that the closure of the overlying skin becomes impossible. The muscles are rinsed repeatedly in sterilized water, dispersing all visible blood clots, and thereafter the limb is draped in sterile dressings. The spraying of the exposed tissues with fibrin glue is recommended to reduce the risk of bleeding. Antibiotics and analgesics are provided over the ensuing days. The muscle mass subsides but never, in our experience, to a size where the fascial compartment can be reconstituted and the overlying skin sutured closed. Wound closure is usually delayed and achieved by the application of a split-thickness skin graft. The aesthetic appearance of the post-surgical limb is grossly affected but, with timely treatment, the limb regains its functional capacity. When a compartment syndrome has been missed and the correct diagnosis is made at a later stage, surgery should not be carried out. The chances of neural recovery are minimal and the risk of serious infection of the necrotic muscles high [1]. Time is no longer a critical issue and surgery can be delayed.

The decision regarding surgical treatment in patients with haemophilia depends upon the availability of coagulation products. The patient should be covered by replacement therapy for relatively long periods, until closure of the wounds, which may take place weeks after the initial surgery. The local application of fibrin glue spray (every 2–3 days) allows early cessation of systemic therapy and hence markedly reduces costs. Factor replacement is necessary should bleeding re-occur.

Conclusions

The elevation of pressure within a closed fascial compartment may cause serious irreversible damage. Intracompartmental pressures need to be assessed and, should the pressure reach 20 mmHg below the diastolic blood pressure, surgery is indicated. Patients with haemophilia who have an antibody to their missing coagulation factor require the prompt attention of a haematologist who has ample supplies of clotting products as the patient may require coagulation maintenance for many weeks. The intra- and postsurgical use of fibrin glue may be beneficial. In summary, acute atraumatic compartment syndrome is a surgical emergency [15]. Primary care and emergency physicians must have a high index of suspicion to generate prompt recognition and appropriate treatment of this problem.

References

1. Whitesides TE, Heckman MM. Acute compartment syndrome: update on the diagnosis and treatment. *J Am Acad Orthop Surg* 1996; **4**: 209–18.
2. McQueen MM, Gaston P, Court-Brown CM. Acute compartment syndrome. *J Bone Joint Surg Br* 2000; **82B**: 200–3.
3. Kapoor V, Featherstone J, Jeffery ITA. Acute anterior compartment syndrome of the forearm following lifting a heavy weight. *Injury* 2000; **31**: 212–15.
4. Russell GV, Pearsall AW, Caylor MT, Nimityongskul P. Acute compartment syndrome after rupture of the medial head of the gastrocnemius muscle. *South Med J* 2000; **93**: 247–9.
5. Vijayakumar R, Nesathurai S, Abbott KM, Eustace S. Ulnar neuropathy resulting from diffuse intramuscular hemorrhage: a case report. *Arch Phys Med Rehabil* 2000; **81**: 1127–30.
6. Verdolin MH, Toth AS, Schroeder R. Bilateral lower extremity compartment syndrome following prolonged surgery in the low lithotomy position with serial compression stockings. *Anesthesiology* 2000; **92**: 1189–91.
7. Nau T, Menth-Chiari WA, Seitz H, Vecsei V. Acute compartment syndrome of the thigh associated with exercise. *Am J Sports Med* 2000; **28**: 120–2.
8. Roach R, Perkins R. Acute compartment syndrome. *J Bone Joint Surg Br* 2000; **82B**: 932–3.
9. Kotak BP, Bendall SP. Recurrent acute compartment syndrome. *Injury* 2000; **31**: 66–7.
10. Graif M, Martinowitz U, Strauss S, Heim M, Itzchak Y. Sonographic localization of hematomas in hemophilic patients with positive iliopsoas sign. *Am J Roentgenol* 1987; **148**: 121–3.
11. Rodriguez-Merchan EC, Goddard NJ. Muscular bleeding, soft-tissue haematomas and pseudotumours. In: Rodriguez-Merchan EC, Goddard NJ, Lee CA, eds. *Musculoskeletal Aspects of Haemophilia*. Oxford: Blackwell Science, 2000: 85–91.
12. Heim H, Horoszowski H, Seligsohn U, Martinowitz U, Strauss S. Iliopsoas hematoma: its detection, and treatment with special reference to hemophilia. *Arch Orthop Trauma Surg* 1982; **99**: 195–7.
13. Heim M, Martinowitz U, Horoszowski H. The short foot syndrome: an unfortunate consequence of neglected raised intracompartmental pressure in a severe hemophilic child. *Angiology* 1986; **37**: 128–31.
14. Yabuki S, Kikuchi S. Dorsal compartment syndrome of the upper arm. *Clin Orthop* 1999; **366**: 107–9.
15. Franc-Law JM, Rossignol M, Vernec A, Somogyi D, Shrier I. Poisoning in acute atraumatic compartment syndrome. *Am J Emerg Med* 2000; **18**: 616–21.

CHAPTER 23

A rational approach to the treatment of haemophilic blood cyst (pseudotumour) in patients with inhibitors

M.S. Gilbert and A. Forster

A chronic expanding blood cyst with the ability to displace and destroy adjacent tissues was described in 1856 by Erichsen [1]. Sixty-two years later, Starker [2], in a classic article, described a similar lesion in a patient with haemophilia and termed it a haemophilic subperiosteal haematoma. Before the availability of adequate replacement therapy, the prognosis of these lesions was poor and their nature remained poorly understood. For obvious reasons, the designation 'pseudotumour' was applied to these lesions. Gilbert summarized his experience with these lesions in 1975 [3] and again in 1997 [4]. For a description of the natural history and pathology of these lesions, the reader is referred to these chapters and to important publications by Fernandez de Valderrama and Matthews [5], Duthie [6] and Steel et al. [7]. It has now become clear that these lesions are not tumorous in nature and this chapter uses the more proper designation of 'haemophilic blood cyst' [8,9].

Experience has shown that the established haemophilic blood cyst in an adult patient will continue to expand and destroy adjacent tissue, albeit at a variable and irregular rate [3]. Therefore, most orthopaedic surgeons who care for a significant number of patients with haemophilia have concluded that early intervention to either ablate or excise the lesion is indicated. However, in the presence of an inhibitor, the treatment options must be modified because of the less predictable control of coagulation and the increased surgical risk. In addition, the therapeutic modalities used in patients with inhibitors are costly and economic factors make their use impossible in many areas of the world.

The authors would be remiss if they did not first address the issue of prevention of these lesions. In 1966, Gunning [10] estimated the prevalence of these lesions at 1% of all haemophilia patients in the UK. With greater availability of treatment options for patients both with and without inhibitors, the authors feel that the prevalence of these lesions has decreased. However, the authors are not familiar with any specific data that estimate the prevalence of these lesions in patients with inhibitors.

The local pathophysiology that results in the encapsulation of a haematoma remains unclear, but most authors agree that the lesion starts with bleeding into the soft tissue with failure of complete resorption. Most of these lesions occur in large proximal muscle masses adjacent to bone, most commonly in the lower extremity. It has been suggested that encapsulation occurs following subperiosteal bleeding but such factors as the site and size of the initial haematoma, pressures within an enclosed compartment and adequacy of treatment have not been fully evaluated.

Review of the literature

The serious nature and need for early treatment of intramuscular bleeding episodes is frequently not appreciated either by the patient or physician. Intra-articular bleeding has been well discussed in the scientific and lay literature. The need for early treatment to prevent and delay the development of haemophilic arthropathy has been well documented. Prophylaxis has been shown to prevent the development of arthropathy [11]. However, soft tissue bleeding is less well understood and has received less attention. Because of this and because the initial bleeding may not cause a great deal of pain, treatment is frequently delayed or even not initiated. This results in progression and persistence of the haematoma. As previously stated, for reasons still unknown, certain haematomas encapsulate and continue to expand as a haemorrhagic cyst. At this point, the lesions do not respond to control of the coagulation mechanism alone and will require intervention.

It is obvious why attempts at controlling the coagulation mechanism must be carried out before haematoma progresses to an encapsulated blood cyst. It is a common mistake to withhold costly factor replacement for relatively asymptomatic bleeding episodes, and it has been repeatedly shown that such thinking is only being 'penny wise and pound foolish'.

Frequently, the haematoma is deep, as for example in the iliopsoas, and the size cannot be clinically appreciated. In these circumstances ultrasonography should be used to follow the lesion until it disappears. If the haematoma persists or continues to expand despite adequate treatment, the lesion should be evaluated by computed axial tomography or magnetic resonance imaging. If the haematoma is encapsulated, it must be treated as if it was an established blood cyst and the options for treatment are discussed later in the chapter.

In an early analysis of haemophilic pseudotumours, we noted a different clinical pattern in cysts that occur in adults from those that occur in children [3]. Lesions in the immature skeleton frequently occur distal to the elbow and knee, are fast-growing and frequently respond to conservative modalities such as factor replacement, immobilization and even radiation therapy. Lesions in the mature skeleton are usually found in the proximal skeleton, more often in the lower extremity, are slow-growing and do not respond to non-surgical treatment: it is these lesions that we discuss in this chapter.

In order not to repeat the mistakes of the past, we first undertake a quick historical review. Before the advent of adequate factor replacement, most of these lesions were treated expectantly with disastrous consequences: progression with resulting neuropathy; pathological fracture; erosion into major vessels, or even through the skin with subsequent infection, septicaemia and death. In the past, attempts to treat these lesions with radiation therapy resulted in shrinkage of these lesions but subsequent follow-up by the authors of several of the patients reported by Brant and Jordan [12] showed continued expansion following a period of initial shrinkage. Occasional reports using this therapeutic strategy appear in the literature but, in general, it is not the treatment of choice if the lesion is resectable. There have been attempts at angiography and embolization of the cyst [13] but a single feeding vessel is rarely found and the technique, if used, is probably most useful to decrease bleeding at the time of surgery.

Authors' experience

Because of this, the authors have suggested that an encapsulated blood cyst be excised as completely as possible, leaving only those parts of the wall that involve vital structures. This should be carried out as soon as possible after diagnosis. Gilbert has resected 19 large haemophilic blood cysts and the data have been collected for presentation in a separate publication; however, some preliminary data are presented here. All surgery was carried out in patients who did not have an inhibitor. Despite the fact that the coagulation mechanism could be fully controlled, there was a complication rate of 36% (7 of 19 surgeries) including three infections, two vascular injuries requiring intraoperative vascular repair, one nerve injury, one intraoperative fracture and one postoperative iliac artery and vein thrombosis. There were no major haemorrhagic complications but all patients required multiple transfusions. One patient required 16 units of packed red blood cells during the perioperative period. There was a recurrence rate of 16% (three of 19 surgeries).

Discussion

Recently, alternatives to complete excision have been presented. Fernandez-Palazzi and Rivas [14] advocated the aspiration of smaller bone cysts and injected them with fibrin glue to prevent further bleeding and to obliterate the cavity. Caviglia [15] questioned the need to resect the fibrous wall in larger, more established soft tissue and bone cysts and introduced a laparascope into the cyst cavity and after evacuation of the contents filled the cavity with fibrin glue and bone graft. Good results with smaller bone cysts encouraged use of this technique in larger lesions and results have been very encouraging. The authors of this chapter have utilized this technique in two patients. The first was a patient who had severe arthropathy of his knee and a large destructive cyst in the proximal tibia and soft tissues of the calf. Despite the severe arthropathy, the patient, a talented songwriter and singer, was able to stand and work full-time with his band. Because of progression of the cyst, it was treated in a manner described by Caviglia [15]. Two small anterior portals were made, and arthroscope and curettes were inserted through. Bone grafts were introduced through the portals and radiography 1 year following surgery showed excellent obliteration of the lesions. Clinically, there was complete resorption of the soft tissue component. The patient's radiographical studies are shown in Figs 23.1–23.3. The authors used a similar approach in a much larger multi-eloculated cyst which replaced most of the musculature of the proximal thigh and had caused erosion of the distal femur. The lesion was considered unresectable and an amputation had been offered to the patient at another institution. In an attempt to salvage the limb, the larger cysts were evacuated through a two-portal approach. Despite adequate haemorrhagic control following

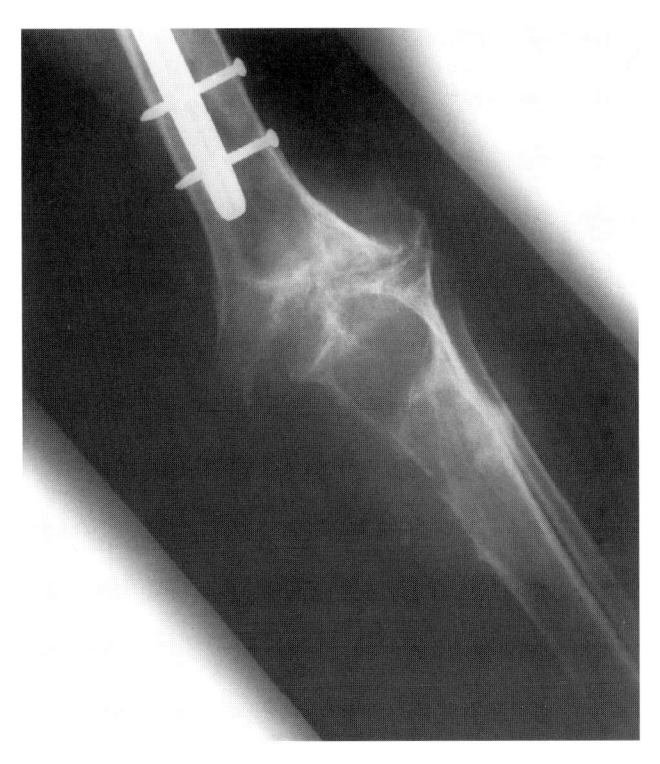

Fig. 23.1 Radiograph of an established blood cyst involving the bone and soft tissues of the proximal tibia.

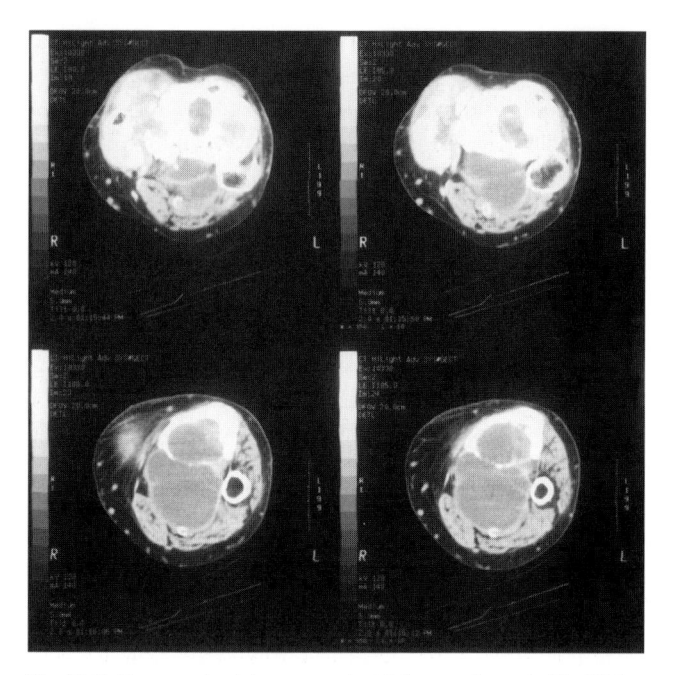

Fig. 23.2 Computed axial tomography of the cyst shown in Fig. 23.1.

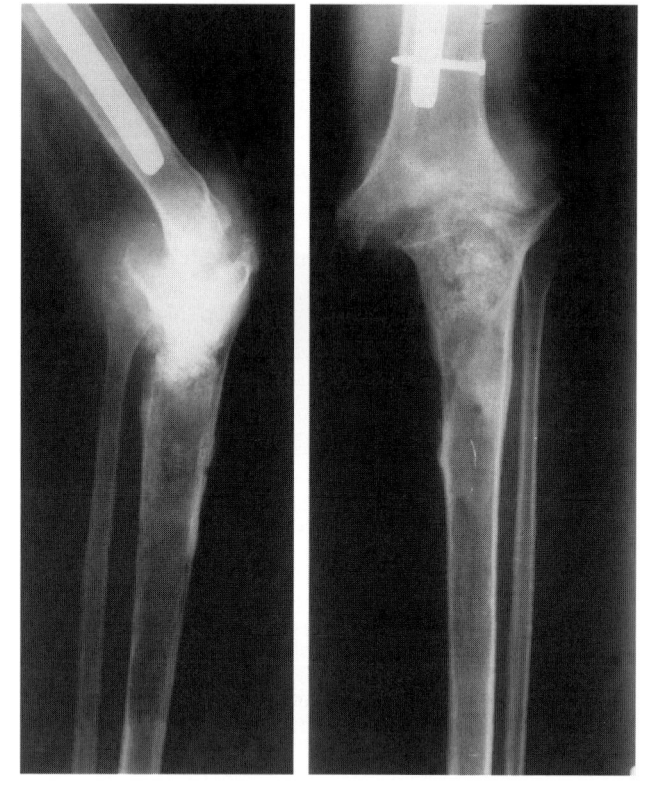

Fig. 23.3 Radiograph of the blood cyst 1 year following curettage and bone grafting of the lesion.

surgery, the cysts refilled with blood and subsequent infection required drainage and long-term antibiotic therapy.

With these experiences as a background, the authors summarize the treatment of the haemophilic blood cyst in the patient with an inhibitor. First and foremost it must be emphasized that the treatment of these lesions is a cooperative effort involving decisions by the haematologist, the orthopaedic surgeon and, most importantly, by the patient. The haematologist must first decide if the inhibitor can be controlled sufficiently to undertake an extensive surgical procedure. Hoyer [16] has written an excellent summary of the strategies that can be utilized. These are beyond the scope of this chapter, but it should be noted that there have recently been several reports documenting the successful use of recombinant activated factor VII in surgery [17–19].

Each haemorrhagic blood cyst must be individually evaluated. As previously noted, there is a spectrum ranging from an isolated soft tissue cyst to massive lesions involving muscle, bone and vital soft tissue structures. If the coagulopathy can be controlled, the surgeon must determine if the lesion can be obliterated by a less invasive surgery or whether surgical excision is required. In these patients, amputation may be a viable alternative. As these undertakings are potentially serious and not without risk, even in patients without an inhibitor, the patient must be fully informed of the nature of the procedure, the risks of the procedure and the risks of leaving the lesion untreated. If the patient elects not to undergo surgery, the lesion should be re-evaluated at regular intervals even if it remains asymptomatic. Radiation therapy can be considered, although the most successful reports of this are in lesions found in the immature skeleton. An alternative approach is arterial embolization, although the efficacy of this remains unproven.

At some point, the lesion may progress to the 'at risk' cyst. 'At risk' lesions include:
1 bone involvement with risk of pathological fracture;
2 nerve involvement, most commonly seen with iliopsoas lesions which compress the femoral nerve;
3 lesions that are about to rupture through the skin; and
4 lesions that have regressed to a size such that resection and obliteration of the dead space would be surgically difficult and risky.

In these patients partial excision may have to be considered. Long bones at risk of fracture will require stabilization, most commonly with intramedullary rods, and bone grafting. The patient must be informed that amputation is a real possibility if the procedure is compromised by uncontrolled bleeding and/or by infection.

Conclusions

It must be emphasized that the treatment of these lesions is a major undertaking that must only be performed at a centre with experience in the comprehensive care of patients with bleeding and coagulation disorders. At the present time, progress in the prevention and treatment of these lesions must come from the haematologist. Until such progress has been made, excruciating decisions must still be made by the haematologist, the orthopaedic surgeon and the patient.

References

1 Erichsen. Extravasation of blood into the calf of the leg. *Lancet* 1856; May 10: 511–13.
2 Starker L. Knochenusur durch ein hamophiles subperiostales hamatom. *Mitt Grenzegebieten Med Chirurgie* 1918; **31**: 381–415.
3 Gilbert MS. Hemophilic pseudotumor. In: Brinkhoust KM, Hemker HC, eds. *Handbook of Hemophilia*. Amsterdam: Excerpta Medica, 1975: 435–46.
4 Gilbert MS. *Pseudotumors: Hemophilia Blood Cysts in Hemophilia*. In: Forbes CD, Aledort L, Madmok R, eds. London: Chapman and Hall, 1997: 123–31.
5 Fernandez de Valderrama JA, Matthews JM. The haemophilic pseudotumour or haemophilic subperiosteal haematoma. *J Bone Joint Surg Br* 1965; **47B**: 256.
6 Duthie RB. Haemophilic cyst formation. In: Duthie RB, Rizza CR, Giangrande PLF, Dodd CAF, eds. *The Management of Musculoskeletal Problems in the Haemophilias*. Oxford: Oxford University Press, 1994: 139–58.
7 Steel WM, Duthie RB, O'Connor BT. Haemophilic cysts. *J Bone Joint Surg Br* 1969; **51B**: 614.
8 Ahlberg AK. On the natural history of hemophilic pseudotumor. *J Bone Joint Surg Am* 1975; **57A**: 1133–6.
9 Magallon M, Monteagudo J, Altisent C *et al*. Hemophilic pseudotumor: multicenter experience over a 25-year period. *Am J Hematol* 1994; **45**: 103–8.
10 Gunning AJ. The surgery of haemophilic cysts. In: Biggs R, McFarlane RG, eds. *Treatment of Haemophilia and Other Coagulation Disorders*. Oxford: Blackwell, 1966.
11 Nilsson IM, Berntorp E, Lofqvist T, Petterson H. Twenty-five years' experience of prophylactic treatment in severe hemophilia A and B. *J Intern Med* 1992; **232**: 25–32.
12 Brant EE, Jordan MH. Radiologic aspect of hemophilic pseudotumors in bone. *Am J Roentgenol* 1972; **114**: 525.
13 Pisco JM, Garcia VL, Martins JM, Mascarenhas AM. Hemophilia pseudotumor treated with transcatheter arterial embolization: case report. *Angiology* 1990; **41**: 1070–4.
14 Fernandez-Palazzi F, Rivas S. The use of 'fibrin-seal' in surgery of coagulation diseases with special reference to cysts and pseudotumors. In: Dohring S, Schulitz KP, eds. *Orthopedic Problems in Hemophilia*. Munich: W. Zuckschwerdt, 1985: 170–9.
15 Caviglia M. Hemophilic pseudotumor: is it necessary to resect the pseudocapsule? [Abstract] XXI Congress of the World Federation of Hemophilia, Mexico City, 1994.
16 Hoyer LW. *Inhibitors in Hemophilia in Hemophilia*. In: Forbes CD, Aledort L, Madmok R, eds. London: Chapman and Hall, 1997: 213–27.
17 Schulman S, Lindstedt M, Alberts KA, Gren PH. Recombinant factor VIIa in multiple surgery. *Thromb Haemost* 1994; **71**: 154.
18 O'Marcaigh AS, Schmalz BJ, Shaughnessy WJ, Gilchrist GS. Successful hemostasis during a major orthopedic operation by using recombinant activated factor VII in a patient with severe hemophilia A and a potent inhibitor. *Mayo Clin Proc* 1994; **69**: 641–4.
19 Tagariello G, De Biasi E, Gajo GB *et al*. Recombinant FVIIa (NovoSeven) continuous infusion and total hip replacement in patients with haemophilia and high titre of inhibitors to FVIII: experience of two cases. *Haemophilia* 2000; **6**: 581–3.

CHAPTER 24

Haemophilia patients with inhibitors: a rheumatologist's point of view

J. York

The development of inhibitors to blood coagulation proteins may occur in the inherited haemophilias or be acquired later in life, often in association with diffuse diseases of connective tissue, the most common of which is rheumatoid arthritis [1].

The pathophysiology of inhibitor formation in the haemophilias has been described earlier in this volume and it is postulated that environmental factors may be more important than genetic factors in the antibody response to factor VIII [2]. The incidence of inhibitor formation in haemophilia B is significantly less than in haemophilia A but the management problems are identical, although anaphylactic reactions at the time of inhibitor formation appear to be a unique complication restricted to haemophilia B [3].

Despite progressive improvements in the products available for the management of haemophilia, inhibitor development is now the main complication of replacement therapy, occurring in up to 30% of patients with haemophilia A [3], although the inhibitor level varies considerably. In high responders, the anamnestic immunological response to re-exposure to infused factors VIII or IX has severely limited the use of replacement therapy and any surgical intervention in such patients.

Treatment options include the use of heterologous products, such as porcine factor VIII, prothrombin complex materials and activated prothrombin complex materials. The induction of immune tolerance by large doses of factor VIII associated with corticosteroid and immunosuppressive agents has been demonstrated to be effective but is not always successful in the long term and is very costly [4]. Recombinant factor VIIa (rFVIIa) is also available in industrialized countries for use in patients with haemophilia A and B and inhibitors, to treat bleeding episodes or to cover surgery but is very expensive [5].

Against this background, the geographical situation may limit access to highly specialized care and the lack of availability or the risks of appropriate factor replacement strategies, and the relative expense involved often preclude optimum management. The haemophilia patient with a coagulation inhibitor in these circumstances is placed at great risk from bleeding episodes and injudicious therapeutic interventions, particularly surgery.

The rheumatologist must be guided by haematological and orthopaedic surgical colleagues with regard to overall management, the treatment of musculoskeletal bleeding episodes, and the advisability and risks of any surgical procedures.

Principles of treatment

As in the management of musculoskeletal bleeding in haemophilia generally, the basic principles are:
1 control of bleeding;
2 physiotherapy; and
3 joint protection.

These principles apply at all three stages of haemophilic joint disease: acute, subacute and chronic arthropathy.

The development of a target joint(s), often after one or several haemarthroses is an important cue to the onset of the subacute stage of synovial hypertrophy and progessive cartilage damage and, if inadequately treated, may go on to damage the joint severely.

Control of bleeding

In the context of musculoskeletal bleeding, the decision as to whether appropriate factor replacement or alternative products are likely to be effective or not must be made in consultation with the haematologist. The severity and situation of the bleeding, the inhibitor titre and the history of previous bleeding episodes are important determinants, as well as any evidence of subacute joint disease. For example, where haemorrhage has occurred into a muscle group constrained by dense fascia producing a compartment syndrome, such as in the forearm or lower leg, ischaemic muscle atrophy is very likely. Femoral nerve damage may follow bleeding into the psoas muscle or hip joint and the decisions about management are more critical than with an uncomplicated haemarthrosis.

Other contributing therapies to control bleeding include splinting and immobilization, the use of cold packs or ice, and rest.

Since the days of Rasputin there has been interest and controversy about the role of psychological factors in bleeding control but adequate pain relief, confident sympathetic management and reassurance are essential.

While desmopressin (DDAVP) may be used subcutaneously, intravenously or intranasally in this situation, its effects are unpredictable [6].

Physiotherapy

The physiotherapist provides management and advice to the patient and trains carers and family members to maintain strong muscles around the affected joint, to preserve function, and to prevent deformity and this role is of critical importance in patients with inhibitors. The use of resting splints may minimize bleeding at night but long-term splinting is counterproductive and results in muscle wasting and should only be used to control unstable joints. Hydrotherapy is an excellent form of supervised exercise and can be continued later outside the clinic setting together with recreational swimming if facilities are available.

Joint protection

Common sense restrictions on certain activities should be part of an agreement between the patient and the treatment team, and the consequences of continued overstressing of specific joints clearly explained.

Control of synovitis

The early recognition of the development of subacute arthropathy is important, although therapeutic options are limited. It is characterized by the presence of 'boggy' synovial thickening and a persistent joint effusion with associated muscle wasting; pain is often absent except in bleeding episodes. Regular prophylactic factor replacement is not possible in most circumstances and non-steroidal anti-inflammatory drugs (NSAIDs) have not proved effective in our experience at this stage of the disease. Oral corticosteroid agents in short courses of 1–2 mg/kg or up to 20–40 mg prednisone/day for 5–7 days given on two or three occasions several weeks apart may suppress inflammation but repeated use is hazardous.

Intra-articular injections of corticosteroids or more effective agents such as ^{90}Y and other radiocolloids; and sclerosants such as rifampicin or tetracycline have been shown to reduce the frequency of bleeding episodes and thus suppress synovitis. As intrasynovial iron deposition is important in the inflammatory process [7], these procedures are advantageous but the presence of a high titre inhibitor is a contraindication and the ability to prevent further cartilage deterioration is greatly limited.

When chronic arthropathy has supervened, joint bleeding is usually not frequent but the differentiation of pain because of chronic arthritis from that caused by bleeding is often difficult. A therapeutic trial of appropriate factor replacement is not an option. In this situation NSAIDs are of more value than in the subacute stage. The newer COX-2 selective NSAIDs are preferable and there is little to choose between the two most commonly used agents, celecoxib and rofecoxib, both of which have no effect on platelet function. While gastrointestinal bleeding is significantly less common with these agents compared with conventional non-selective drugs, they still have the potential to cause bleeding which in the presence of inhibitors can be very dangerous. Their ability to cause idiosyncratic liver damage is relevant in a group of patients with a high prevalence of chronic liver disease and liver function tests should be regularly monitored. The effects of the newer agents on blood pressure control in hypertension, and renal function—particularly in older patients—is not clearly documented at this stage and both these parameters must be carefully documented. While celecoxib has a reported incidence of skin reactions because of sulphonamide sensitivity, both agents have been associated with allergic responses including bronchospasm and urticaria in common with non-selective NSAIDS.

It is particularly important at this stage of the disease in patients with inhibitors, as in all patients with haemophilia, that firm but sympathetic and compassionate staff are involved in their chronic care. Pain relief is a major problem and oral analgesics and, where necessary, more potent agents are essential in conjunction with pain management services. Liaison psychiatry and skilled counselling provide the opportunity for the patient to vent their frustration and anxieties about the future; antidepressants and muscle relaxants are of limited value. The trial use of non-pharmacological methods of pain control utilizing relaxation and pain control techniques, hypnotherapy, and forms of electrical pain relief, such as transcutaneous nerve stimulation, are to be encouraged.

Other management problems

HIV and AIDS

Although patients with inhibitors may have been exposed to fewer human blood products, a significant number are still affected by HIV and other viruses such as hepatitis B and C. In these patients their general condition and extensive list of medications mean that potential drug interaction and toxicity must be anticipated.

Sepsis

Unrecognized and untreated sepsis in a joint is a disaster in the patient with haemophilia and an inhibitor. If there is any doubt about possible infection in a joint then aspiration and culture of the synovial fluid to identify and determine the sensitivity of the organism responsible is necessary, followed by appropriate antibiotic therapy. In industrialized countries, haematological control is possible in most major centres, allowing diagnosis and treatment, but generally with very expensive agents. However, in developing nations this remains an unattainable goal with severe consequences for life and limb, and antibiotic treatment must be empirical. The incidence of sepsis is increased in patients with HIV and AIDS [8].

Orthopaedic surgery

Surgery may be possible in industrialized countries where any of the abovementioned techniques to produce haemostasis may allow surgery including joint replacement. The risks are greater and the costs of haematological control sometimes high enough to compromise hospital budgets. Understandably, hospital administrators and government providers are often very resistant to the procedures being undertaken.

Osteoporosis

The recently reported increased incidence of generalized osteoporosis in patients with haemophilia (M.T. Sohail and J. York, unpublished data, 6th Musculoskeletal Congress of the World Federation of Haemophilia, Lahore, 2001) emphasizes the fact that patients with inhibitors tend to have more advanced joint disease and therefore be at greater risk. The availability of effective medications to prevent or minimize such bone loss is a further challenge to the comprehensive care of patients with haemophilia and inhibitors.

Conclusions

The rheumatological problems presented by patients with haemophilia who have developed significant inhibitor levels are challenging in centres with access to the most advanced medical care. It is sobering to reflect that most patients with haemophilia in the world do not have access to even basic care. Little is known about the prevalence of inhibitors in this patient population but the presumably high death rate in infancy and early life and the absence of any exposure to human blood products would suggest that the number of people with inhibitors is small. The contrast between their predicament and the highly sophisticated treatments available in the industrialized world is stark and highlights the importance of the World Federation of Haemophilia Decade Plan which has identified encouraging the growth of haemophilia care in developing countries as its primary objective [9,10].

References

1 Green D, Schuette PT, Wallace WH. Factor VIII antibodies in rheumatoid arthritis: effect of corticosteroids and cyclophosphamide. *Arch Intern Med* 1980; **140**: 1232–5.
2 Tuddenham EGO, McVey JH. The genetic basis of inhibitor development in haemophilia A. *Haemophilia* 1998; **4**: 543–5.
3 Warrier I. Management of haemophilia B patients with inhibitors and anaphylaxis. *Haemophilia* 1998; **4**: 574–6.
4 Di Michele DM. Immune tolerance: a synopsis of the international experience. *Haemophilia* 1998; **4**: 568–73.
5 Schulman S. Safety, efficacy and lessons from continuous infusion with rFVIIa. *Haemophilia* 1998; **4**: 564–7.
6 White B, Lee CA. The diagnosis and management of inherited bleeding disorders. In: Rodriguez-Merchan EC, Goddard NJ, Lee CA, eds. *Musculoskeletal Aspects of Haemophilia*. Oxford: Blackwell Science, 2000: 3–10.
7 Rosendaal G. *Blood Induced Cartilage Damage in Haemophilia: Monograph*. Utrecht: Elinwijk, 1998.
8 Ragni MV, Handley EN. Septic arthritis in haemophilic patients and infection with human immunodeficiency virus. *Ann Intern Med* 1989; **110**: 169–70.
9 Lee CA. Towards achieving global haemophilia care: World Federation of Haemophilia Programs. *Haemophilia* 1998; **4**: 463–73.
10 Jones P. Haemophilia: a global challenge. *Haemophilia* 1995; **1**: 11–13.

CHAPTER 25

Rehabilitation of patients with haemophilia and inhibitors

F. Querol, J.A. Aznar, S. Haya and A.R. Cid

The development of inhibitors against factor VIII (FVIII) or factor IX (FIX) is considered to be the most serious complication in the treatment of haemophilia as they make prevention and management of bleeding episodes more difficult [1]. As a result, greater care must be taken in order to prevent damage to the musculoskeletal system and with the physiotherapeutic management of patients with inhibitors. Between 24 and 33% of patients with haemophilia A, and between 2 and 4% of patients with haemophilia B develop inhibitors. We are monitoring 269 patients who are receiving treatment in our unit: 250 with haemophilia A, and 19 with haemophilia B. Data on our patients are shown in Table 25.1.

Incidence of bleeds

The presence of an inhibitor does not entail a higher incidence of bleeds. However, inhibitors aggravate bleeding episodes because they render treatment with FVIII and FIX concentrates inefficacious, making integral management of bleeds more difficult. In addition, bleeds affecting a physical activity, such as deambulation, increase the risk of triggering bleeding episodes in other muscles or joints because of the added strain to which they are subjected to compensate for the deficiency of the initially affected body segment. We reviewed the bleeding episodes occurring between July 1998 and July 2001

Table 25.1 Haemophilic population monitored at the Congenital Bleeding Disorder Unit of the University Hospital La Fe, Valencia.

	Haemophilia A ($n = 250$)	Haemophilia B ($n = 19$)
Severe (levels <1%)	79	3
Moderate (levels 1–5%)	38	12
Mild (levels >5%)	133	4

(Table 25.2); eight patients with inhibitors who received treatment during this period were found to have 35 joint lesions and 41 muscular lesions (Table 25.3).

Aims of basic treatment

The primary aim of treatment of patients with inhibitors is the elimination of the inhibitor by means of immune tolerance treatment (ITT). In the event of an unsuccessful outcome, alternative treatment is available: activated prothrombin complex concentrates (aPCC), porcine FVIII concentrate, activated recombinant factor VII (rFVIIa), or immune absorption [1,2]. These aspects are dealt with at greater length in other chapters, but here we would like to point out that none of these

Table 25.2 Bleeding episodes affecting the musculoskeletal system in inhibitors patients. Data for July 1998 to July 2001.

Case	Haemophilia A	Age	Inhibitor titre range (BU)	Treatment	Haemarthroses	Haematomas
1	Severe	5	1–6	FVIII	8	5
2	Severe	5	8–73.6	rFVIIa	1	11
3	Severe	24	1–8	FVIII/rFVIIa	1	1
4	Severe	3	0.8–11	FVIII	5	4
5	Moderate	16	7.6–16.9	rFVIIa	3	1
6	Moderate	70	0.7–19	FVIII/rFVIIa	1	6
7	Severe	24	0–27.5	FVIII/rFVIIa	13	3
8	Severe	31	0.4–19.2	FVIII/rFVIIa	3	10

BU, Bethesda units; FVIII, factor VIII; rFIIa, recombinant factor VII activated.

Table 25.3 Areas of bleeding episodes affecting the musculoskeletal system.

Haemarthroses	
Elbow	15
Knee	7
Ankle	5
Shoulder	4
Hip	1
Hand	2
Foot	1
Haematomas	
Leg	12
Hip	12
Thigh	6
Arm	5
Trunk	3
Hand	2
Foot	1

alternatives is as efficacious as FVIII or FIX cover in patients without inhibitors. Further, haemostasis induced by alternative treatments does not last as long, impedes prophylactic treatment, and makes rehabilitation more difficult. This chapter therefore aims to describe the specific precautions to be adopted when rehabilitating patients with inhibitors.

Aims of rehabilitation in patients with inhibitors

In patients with inhibitors who also have lesions to the musculoskeletal system, rehabilitation pursues two objectives:
1 the restoration of normal activity; and
2 the prevention and early detection of disabilities.

In addition to diagnosing and treating lesions affecting the musculoskeletal system, the rehabilitation specialist gives advice and guidance concerning the type of activities of daily living that the patient may safely perform. The specialist is part of a multidisciplinary team of professionals whose cooperation is essential for the restoration of the quality of life of the patient with haemophilia [3].

With this end in view, we first give a brief evaluation of the most frequent musculoskeletal problems so as to obtain a better understanding of their evolution. Next, we describe the procedures common to all these problems adopted in physiotherapy, followed by a description of the techniques specific to the rehabilitation of each joint.

Physical examination of the musculoskeletal system

Any evaluation of the joint in the patient with haemophilia must consider range of motion (ROM) (Figs 25.1 and 25.2) and

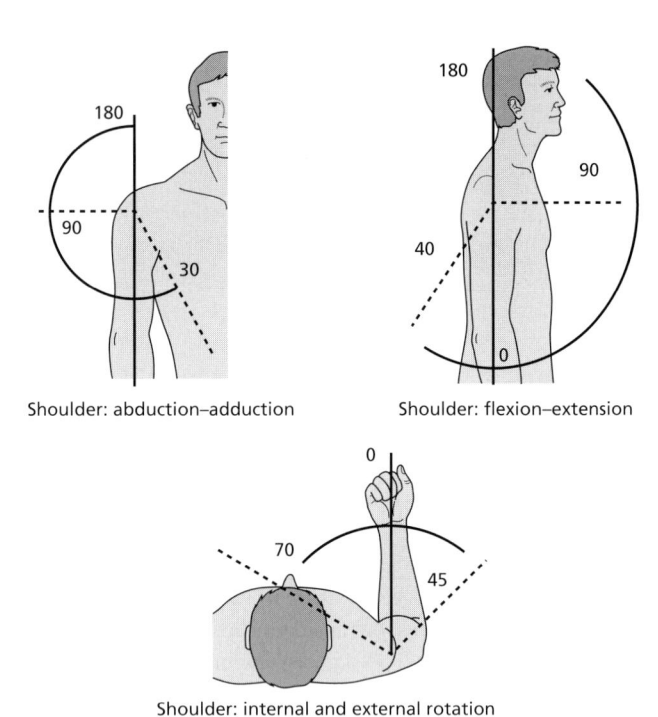

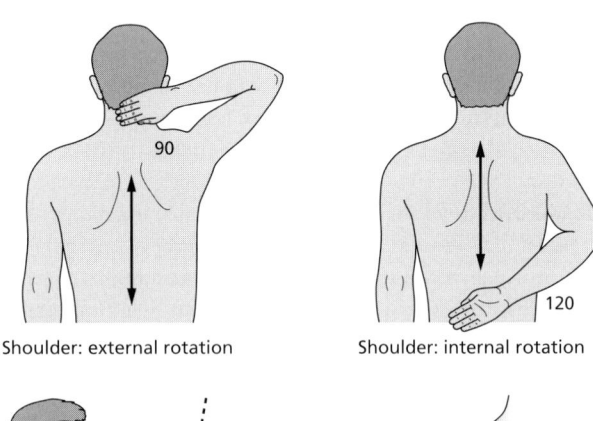

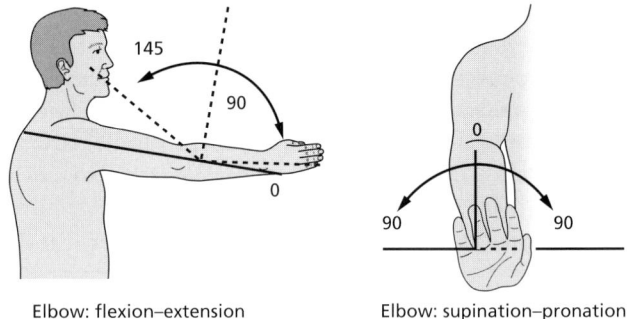

Shoulder: abduction–adduction

Shoulder: flexion–extension

Shoulder: internal and external rotation

Shoulder: external rotation

Shoulder: internal rotation

Elbow: flexion–extension

Elbow: supination–pronation

Fig. 25.1 Range of motion (ROM) of the upper limbs.

muscle balance (Table 25.4). These two factors are important because they enable the specialist to determine functional limitations affecting the joint. The guidelines for a physical evaluation include the following factors:
1 pain scores;
2 absence or presence of swelling;
3 ROM expressed in percentage of movement;
4 absence or presence of flexion contracture;

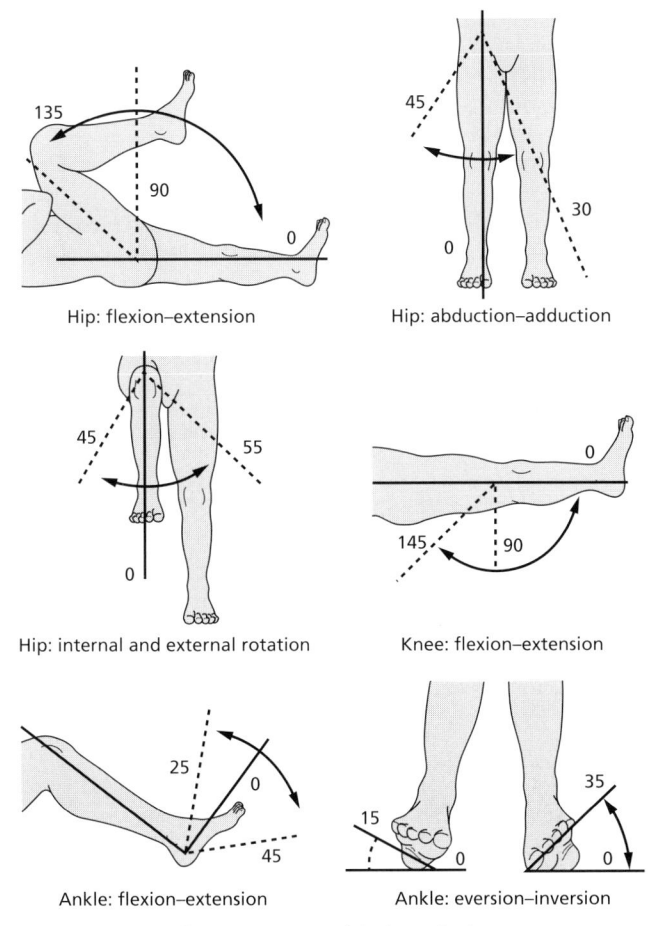

Fig. 25.2 Range of motion (ROM) of the lower limbs.

5 absence or presence of axial abnormality;
6 absence or presence of crepitus on motion;
7 joint stability measured in terms of functional limitation and recourse to orthotics;
8 muscle atrophy—none, minimal or significant; and
9 muscle balance to assess musculoskeletal function.
 Muscle balance is measured on a scale of 0–5:
0 total absence of muscle contraction;
1 possibility of muscle contraction but inability to move body segment;
2 contraction that only allows complete joint motion in suspension or in water;
3 complete joint motion is possible only in the absence of any external resistance other than the force of gravity;
4 a sufficiently strong contraction to enable the muscle to make an acceptable physical effort, such as accomplishing full joint motion while overcoming moderate resistance (an 'acceptable physical effort' is a concept that varies according to the patient's age and constitution); and
5 normal function.

The Musculoskeletal Committee of the World Federation of Hemophilia developed a model of clinical evaluation described by Gilbert [4], which was subsequently adopted. Manco-Johnson et al. [5] proposed a number of changes to the model, which include a scoring system to distinguish pathological from normal gait.

Techniques used in rehabilitation and physiotherapy

Although the techniques used in physiotherapy are the subject of Chapter 26, they will also be dealt with here, albeit briefly, because of the role that physiotherapy has in rehabilitation. Haemarthroses and muscle haematomas cause the patient with haemophilia pain and functional limitations, with negative repercussions on his or her activities of daily living and quality of life, and in the worst case they can lead to total disability. Many techniques have been developed in physiotherapy with the aim of speeding up the beneficial effects of rehabilitation [3,6–8], some of these techniques being highly specialized so as to remedy specific problems (Table 25.5). We point out the relevance of these techniques and detail some of the most frequently used ones whose efficacy has been established.

All lesions causing functional limitations require physiotherapy because there is no medication capable of increasing strength or improving motion. In patients with inhibitors even greater care must be taken with their rehabilitation, given the different treatment they receive to obtain effective haemostasis. Routine physiotherapy sessions therefore entail greater complications in such patients; for instance, physiotherapy requiring the use of crutches brings greater strain to bear on the upper

Table 25.4 Evaluation of muscle balance and appropriate basic kinesiotherapy.

Muscle balance score	Description	Basic kinesiotherapy
0	Absence of muscle contraction	Passive motion to maintain ROM
1	Muscle contraction without motion	Isometric/assisted motion
2	Complete ROM without gravity	Assisted motion, in suspension or kinesiohydrotherapy
3	Complete ROM without resistance (weights)	Active exercise or kinesiohydrotherapy
4	Complete ROM against moderate resistance	Endurance exercises
5	Normal motion. Increased muscle endurance possible	Muscle strengthening

ROM, range of movement.

Table 25.5 Physiotherapy techniques: general effects and indications. Musculoskeletal problems of patients with inhibitors need special rehabilitation care. The programme of therapy should be individually tailored by a specialist in physical therapy.

Therapy	Effects, actions and indications	Indications in haemarthroses and haematomas		
		Acute phase	Subacute phase	Chronic phase
Kinesiotherapy	Correction and prevention of joint deformities. Improve contractures. Analgesia. Trophism and strength	Light isometric exercises and ice	Correction of muscle imbalances. Proprioceptive training. Isotonic and isodynamic exercises	Appropriate movement according to muscular balances. Strengthening. Isokinetic training
Laser	Bioenergetic. Biostimulant. Analgesia. Improve degenerative changes. Reabsorptive oedema	Dose: 10–13 J/cm^2	Dose: 15 J/cm^2	Dose: 20 J/cm^2
Infrared	Contusions. Causalgias. Sedation. Relief of pain in arthritis and arthrosis	Non-indicating	Feasible a few minutes before kinesiotherapy	Feasible chronic lesions
Iontophoresis	Analgesia. Reabsorption bleeds. Haematomas. Arthritis. Arthrosis	Non-indicating	Feasible with caution	Indicating with fibrinolytic and analgesic or anti-inflammatory topical drugs
Ultrasound	Vasomotor. Fibrinolytic. Analgesic. Anti-inflammatory	Non-indicating	Feasible after 48 h. Dose 0.5–1 W/cm^2. Impulses	Indicating. Dose 1–2 W/cm^2. Impulses
Magnetic waves Low frequency	>50 G: Analgesic. Anti-inflammatory Anti-oedema. Bone consolidation <50 G: Relaxation muscle. Vasodilator	Non-indicating	Feasible after 24–48 h Therapeutic ranges 10–100 Hz	Indicating
Electrotherapy: Interferential currents (TENS) Functional electric stimulation	Analgesia. Restoring trophism. Strengthen muscles. Improve joint stability	Non-indicating	Feasible with caution	Indicating

limbs, with the concomitant risk of joint and muscle bleeds in patients with inhibitors.

Specialized techniques for the rehabilitation of each joint are described below. However, kinesiotherapy requires a prior study of physiological movements that takes into account muscle balance and the risk of bleeds entailed by normal activities, such as walking or lifting heavy objects.

In normal circumstances, lesions to the musculoskeletal system of the patient with haemophilia require clotting factor and immobilization of the affected body segment. Once the swelling has subsided, the patient may return to normal physical activity. In patients with inhibitors, however, this is not only inadequate but potentially dangerous, even though haemostasis has been accomplished. In such cases, a brusque change from complete inactivity to normal activity could provoke renewed bleeding episodes. Each stage in physiotherapy for patients with inhibitors must therefore be carefully programmed and include effective factor cover and a gradual return to normal activity under professional supervision.

An effective treatment for bleeding episodes used in physiotherapy is cryotherapy. This consists in reducing the temperature of the damaged body segment once it has been injured. Cryotherapy is also indicated after any exercise involving strenuous physical effort. The simplest method involves applying a plastic bag containing ice and wrapped in a cloth to the body segment. Disposable coldpacks are also available on the market, characteristically of small size, light weight (under 100 g) and low price (less than $1). They have been proven to keep the affected area at a temperature of 0–5° for approximately 10–15 min, affording relief from pain and effecting haemostasis. Coldpacks are an indispensable item in the first-aid box of every school and sports centre, and it is also advisable that every school-aged patient with haemophilia should carry one in his or her school bag.

Common problems and guidelines for diagnosis and management

Haematomas and haemarthroses are problems that affect patients with and without inhibitors alike. In both types of patient, recurrent haemarthrosis leads to synovitis and haemophilic arthropathy.

Haemarthrosis manifests itself through sudden swelling of the joint, which makes this disorder amenable to clinical diagnosis. Ultrasound examination is the most useful technique for confirming the presence of fluids in the joint cavity, although confirmation that the fluid detected is blood can only be obtained through arthrocentesis [9].

Haemarthrosis causes inflammation of the synovial membrane, which invariably requires replacement therapy with FVIII or FIX, or rFVIIa or aPCC. Physiotherapy is also indicated, consisting of immobilization, cryotherapy and progressive kinesiotherapy. Joints with haemarthrosis may benefit from orthotic devices that impede or facilitate mobility, or relieve the body segment of its weight-bearing function.

Recurrent haemarthrosis causes synovitis, which in patients with haemophilia are a result of physical and chemical processes. Chief among these are iron deposits that, because they cannot be eliminated, accumulate in the synovial cells, provoking inflammation, the first stage in joint destruction.

Together with replacement therapy and physiotherapy—the traditional mainstay of conservative treatment of synovitis—a new generation of non-steroidal anti-inflammatory drugs (NSAIDs) are now available, such as coxib (celecoxib and rofecoxib). Thanks to their selective action on COX-2, these NSAIDs do not affect platelet aggregation or have harmful effects on digestive mucus [10,11].

In addition to replacement therapy, muscle haematomas require immobilization in a painless posture with plaster of Paris splints and the application of cryotherapy for 10–15 min 3–4 times a day. After 24–48 h it is important to restore the affected segment to a functional position so as to prevent shortening of the muscle. Care must be taken in muscle mobilization. In the subacute phase, once it is established that passive mobility does not trigger pain, the affected muscle must be subjected to isometric exercise followed by progression from active mobility to normal activity (that is, a muscle balance of 5/5) through restoration of endurance capacity. The primary aim of muscle mobilization is to prevent, by means of ultrasound waves, the haematoma from forming a cyst, and to prevent muscle atrophy by means of electrotherapy and biofeedback.

Iliopsoas haematomas may be confused with hip haemarthrosis or even acute appendicitis, and therefore require differential diagnosis. The bleeding episodes that bring about any type of haematoma are usually so severe that factor cover requires large dosage administered over long periods of time. Absolute rest is prescribed, which in the case of patients with inhibitors can last as long as 2–3 weeks, with continuous infusion. It is imperative that hip flexion and ultrasonography monitoring be supervised.

Superficial haematomas caused by slight bruising to the skin and subcutaneous cell tissue are also extremely frequent during infancy. They are a frequent cause of visits to the surgery, but their treatment usually only requires the application of ice or heparinoid gels.

Haemophilic arthropathy

The physiopathological processes that bring about haemophilic arthropathy are described in Chapter 18. Immobilization, essential for the management of bleeds, conditions muscle hypotrophy and joint instability, which are factors in the vicious circle of haemorrhage, synovitis, loss of motion, instability and haemorrhage. Muscle hypotrophy benefits from kinesiotherapy and, particularly, electrotherapy [12]. Joint instability also requires kinesiotherapy, as well as orthotics adapted to the functional impediment. The physiopathological processes resulting in haemarthrosis may also cause osteoporosis following loss of motion. Osteoporosis requires exercise, but is also amenable to electrotherapy and phototherapy, as well as treatment with low frequency magnetic waves [13]. A side-effect of haemarthrosis is the enlargement of the epiphysis, a process caused by hyperaemia and alterations of cartilage characterized by irregular subchondral surface. In such cases use of laser is indicated on account of its biostimulant effects [13,14]. The evolution towards chronic arthropathy is characterized by narrowing of joint space, subchondral cysts, erosion of joint margins and, finally, joint abnormalities and deformities. All these features are measured on Pettersson's grading system developed for the radiological evaluation of damage caused by haemophilic arthropathy [15].

The clinical manifestations of haemophilic arthropathy can be summed up as pain, functional impediment and limited joint motion, all of which eventually cause disability and so require rehabilitation. The pain attendant on arthropathy can be dealt with in a number of ways: cryotherapy, ultrasound treatment, laser therapy, masotherapy and electrotherapy; of the latter, the best known is perhaps transcutaneous electrical nerve stimulation (TENS). Functional impediments and limited motion require kinesiotherapy, while disabilities call for orthotics to improve quality of life.

The principal aim in the management of haemophilic arthropathy is the prevention of haemarthrosis. The ideal way of accomplishing this end is prophylaxis with replacement therapy, which is unfortunately not possible in the case of patients with inhibitors, and physiotherapy with a view to obtaining and maintaining activities of daily living, including low-risk non-competitive sports. The beneficial effects of glucosamine sulphate on structural deformities and arthrosis have recently been reported [16]. In our experience with patients with haemophilia, this treatment has relieved pain and improved function, but in advanced arthropathy we have had to combine it with NSAIDs. We have also successfully used hyaluronic acid [17], although our experience with this medication in patients with haemophilia is limited to five cases.

Rehabilitation treatment of synovitis and arthropathy in patients with inhibitors

There are no substantial differences in the treatment of synovitis and haemophilic arthropathy in patients with and without inhibitors. However, in the former case, synoviorthesis is indicated in the event of recurrent joint bleeds or diagnosis of synovitis [18]. As regards kinesiotherapy as a means of restoring motion, it is desirable that restoration of full ROM and muscle strength should be accomplished by degrees. Other procedures used in physiotherapy do not require the special precautions in Table 25.5 with reference to the objectives that each one fulfils. The painkillers and anti-inflammatory medication used are indicated because of the security they afford to the gastrointestinal tract and haemostasis.

Problems affecting joints in patients with inhibitors

Rehabilitation of the shoulder

Function, examination and clinical evaluation

The shoulder consists of three synovial joints: the scapulohumeral, acromioclavicular and sternoclavicular. Besides these, we should add the so-called scapulothoracic joint, which aids shoulder elevation.

The examination of the shoulder involves defining its form, detecting swelling and ascertaining the state of the muscles and bones. The state of the sternoclavicular and acromiohumeral joints can be ascertained simply by feeling them. The aim of an examination is to establish joint and muscle balance (Fig. 25.1 and Table 25.4). The examination of the rotator muscles, consisting of the tendons of the supraspinous, infraspinous, lesser round and subscapular muscles, is especially important because of the stability they provide to the scapulohumeral joint.

Rehabilitation of shoulder haemarthrosis

The first step is to immobilize the shoulder in pain-free posture, which is generally with 10–15° abduction and slight flexion, 70–80° internal rotation, and 90° elbow flexion. Coldpacks should be applied for 10–15 min 3–5 times a day. During sleep this position should be kept by means of support straps or the orthotic devices described in Fig. 25.3.

Once bleeding has ceased kinesiotherapy can commence, starting with load-free pendular exercises. However, it is recommended that analytical exercises commence with abduction and flexion–extension, whereas rotation exercises are best avoided until the shoulder has reached a muscle balance of 3/5. Exercises using pulleys are also a feasible option, either with weights, or self-assisted, or with other forms of mechanotherapy. Kinesiohydrotherapy is equally indicated. It is fundamental that the physiotherapist inform the patient that the exercise should not be painful.

Even when muscle balance allows active and painless ROM of 90° for flexion and abduction, and 20–30° for extension, the patient should still wait until the strength of the deltoid muscle has recovered before attempting active rotation. Patients with inhibitors incapable of lifting weights of 2–3 kg by abducting or flexing the arm should not attempt active internal–external rotation of the shoulder, even though the motion does not cause the patient pain.

Once analytical muscle balance exceeds 3/5, muscle-strengthening exercises may commence following the guidelines described in the section on muscle strength and avoiding initial postures involving elbow extension and external shoulder rotation. Physiotherapy facilitating reabsorption following bleeds into the joint cavity are described in Table 25.5.

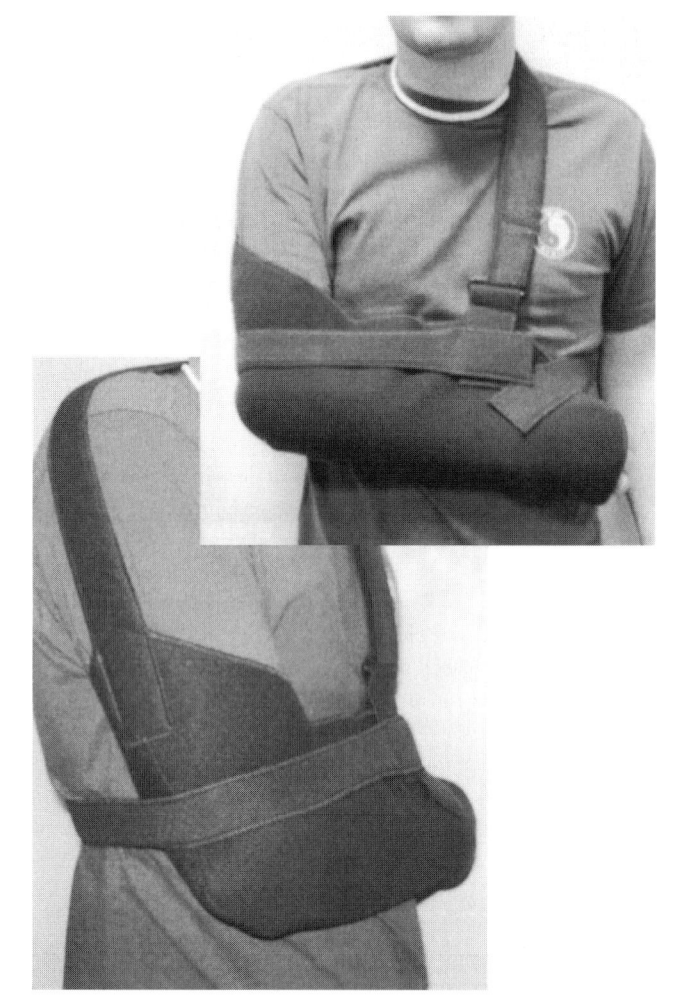

Fig. 25.3 Orthotic device for immobilization of the shoulder.

Rehabilitation of shoulder haematomas

In the acute phase, as with haemarthrosis, the shoulder joint should be immobilized in pain-free posture and cryotherapy utilized. However, if the haematoma is situated on the anterior surface and affects the long and/or short portion of the biceps, the elbow should be placed in neutral pronation–supination position with 90° flexion.

Ultrasound therapy in pulses may be indicated 24 h after cessation of bleeding has been established both clinically and by means of ultrasound. Active kinesiotherapy involving

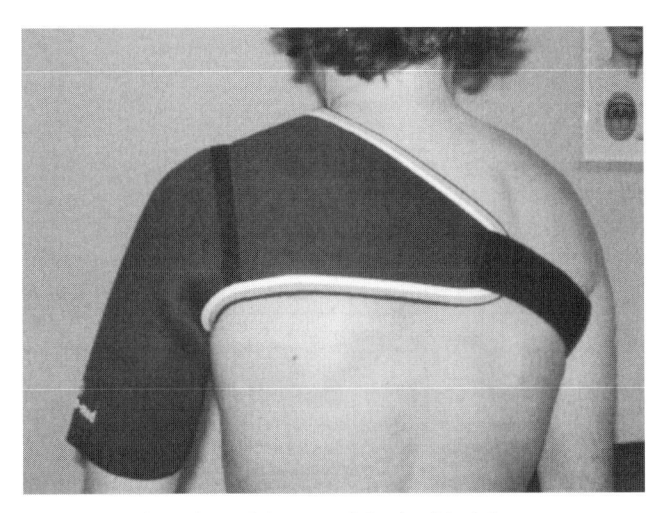

Fig. 25.4 Orthosis for stabilization of the shoulder joint.

gentle extension of the muscle may also be indicated. Each kinesiotherapy session should end with another of cryotherapy, a 10–15-min session if ice or coldpacks are applied, or a 3–5-min one if the affected area is exposed to cold air flows (−32°).

It is essential to ascertain that complete joint ROM has been restored, because the fibrosis caused by the haematoma may bring about muscle shortening, with the concomitant functional limitations.

Orthotics for the rehabilitation of the shoulder

The term 'orthotics' refers primarily to devices that 'straighten out' a deformity but nowadays the word is used in a broader sense. It is applied to any appliance or set of appliances that contribute not only to correcting but also to preventing deformities, as well as stabilizing biomechanical functions, assisting joint motion, immobilizing body segments, facilitating function and, in the case of tubular bandages, affording compression and heat for therapeutic purposes.

The orthoses most commonly used for immobilizing the shoulder are support straps. These set the shoulder in anteversion posture and internal rotation with the elbow in 90° flexion by strapping it to the trunk by means of a sling, allowing the arm to rest in a more or less fixed but comfortable position and thereby unburdening the shoulder joint (Fig. 25.3).

In the mobilization phase, shoulder pads with silicone cushions are used. These stabilize the joint and give the patient a 'safe' feeling while causing him or her to control joint motion (Fig. 25.4).

Rehabilitation of the elbow

Function, examination and clinical evaluation

The elbow is constituted by the union of the humerus with the radius and ulna, which allows flexion–extension and pronation–supination motion (Fig. 25.1). The inspection of a lesion affecting the elbow joint should include a tactile exploration of the area, especially the olecranal grooves and epicondyles, to detect a possible dislocation. The physician should observe the aspect of the skin, taking note of whether it is tense or shiny. He or she should also look for signs of haematoma, hyperaesthesia, pain, swelling and limitation of motion; all these can be signs of an acute bleeding episode. As with other joints, muscle atrophy, crepitus, instability and axial deformity can suggest the existence of a chronic injury.

Rehabilitation of elbow haemarthrosis

Besides factor cover, the elbow should be immobilized in pain-free posture with a splint or orthotic device during the acute phase. Generally, the elbow should be flexed in slight supination. If possible, ice or coldpacks are applied for 10–15 min 3–5 times a day. In patients with inhibitors the efficacy of the haemotological treatment is usually ascertained by improved motion in the first 24–48 h. After this period, a new splint may be placed with the elbow in functional position, that is, 90° flexion in neutral pronation–supination posture. The splint should be kept in place for at least a week. It may subsequently be removed by day because the patient will have regained control over active motion. At night the splint should be replaced to prevent bleeds caused by involuntary motion during sleep.

Once bleeding has ceased, kinesiotherapy may commence, starting with isometric exercise (contraction without joint motion) 3–5 times a day, followed by active and endurance exercises that do not cause pain. At the end of each session ice or coldpacks should always be applied. Passive motion should not be too strenuous, and combined flexion–extension and pronation–supination should be avoided during muscle strengthening. It is preferable that the recovery of each joint motion should be accomplished analytically.

In addition to active exercise, the patient also has recourse to exercise with pulleys, mechanotherapy and kinesiohydrotherapy. Other techniques that facilitate absorption of bleeds, relieve pain and strengthen muscles are described in Table 25.5.

Rehabilitation of muscle haematomas affecting the forearm

Haematomas affecting the forearm demand special care on account of the risk of compartment syndromes. Cryotherapy is essential from the outset. The bandages must also be adequately padded so as to control—'teach' both the patient and his or her relatives to control—possible changes in finger sensitivity and colouring, which would require instant medical attention. The specialist must define the extent of the lesion, which manifests itself in the first stage as tenseness of the skin. The perimeter of the forearm must be measured with a measuring tape and using anatomical features as reference points. If possible, ultrasonography should be performed as an aid for diagnosis and monitoring.

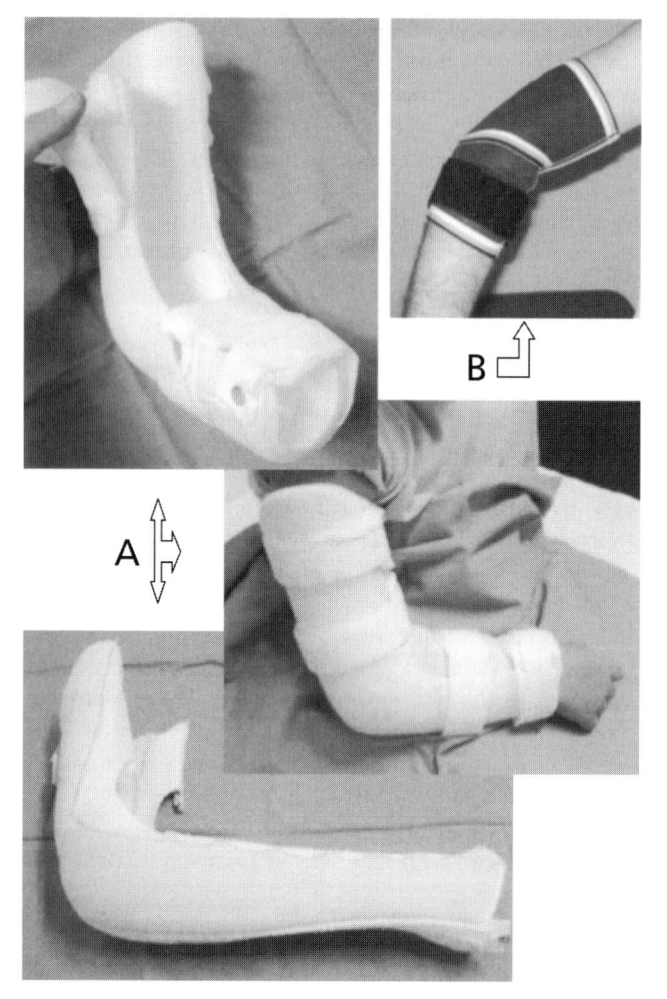

Fig. 25.5 Orthoses of termoplastic and resine (A) and splint (B) for stabilization of the elbow.

After 24–48 h, once bleeding has ceased, the haematoma tends to be absorbed or encapsulate. To prevent the latter, ultrasound pulses and kinesiotherapy may be utilized as these help accomplish normal muscle contraction–relaxation.

Elbow orthotics

So-called 'night-time' splints are indicated for children with haemophilia and inhibitors to prevent spontaneous haemarthroses. These splints set the elbow at 90° flexion in neutral pronation–supination or pronation posture so as to ensure function. Made of thermoplastic material or resin, they are both hardwearing and light (Fig. 25.5). They are used in acute haemarthrosis, either from the onset of the lesion when functional posture is possible, or 24–48 h afterwards, when the pressure caused by the bleed has subsided and motion is possible but joint stability must be ensured. Elbow pads with silicone cushions provide gentler stability and produce the same results as those described with shoulder orthotics. Adjustable elbow pads are indicated for cases of arthropathy in which motion within a prescribed range needs to be maintained.

Rehabilitation of the hip

Function, examination and clinical evaluation

The hip is a perfect example of enarthrosis. The femur fits into the cotyloid cavity of the hip bone in such a way as to allow flexion, extension, abduction and adduction, and rotation (Fig. 25.2). It is necessary to examine ROM and muscle balance to distinguish hip haemarthrosis from psoas haematoma. A clinical diagnosis of the former would include pain and functional limitation for all joint motions, whereas that of psoas haematoma would only be revealed by limitation of flexion–extension while rotation or abduction and adduction of the hip are not affected.

Rehabilitation of hip haemarthrosis

Kinesiotherapy for haemarthrosis consists of immobilization of the hip followed by restoration of motion and gait. The restoration of these functions is a gradual process beginning with complete load-relief with the aid of a walking frame or crutches, preferably equipped with devices that act as shock-absorbers, followed by partial weight-bearing function without walking aids before advancing to normal gait. Cryotherapy only has a slight effect on hip haemarthrosis, although its application is advisable in the early stages of kinesiotherapy and after exercise. The therapeutic benefits of exercise is enhanced when carried out in a swimming pool, given the reduced force of gravity in water.

Rehabilitation of muscle haematomas affecting the psoas and gluteus

Iliopsoas haematoma necessitates absolute rest that sometimes requires transcutaneous traction to prevent muscle shortening. The physiotherapist should forestall the risk of hip flexion by ensuring correct posture and supervising exercises to restore motion and strength, weight-bearing function not being allowed until motion is fully regained. As with haemarthrosis, ambulation assisted with walking aids should be encouraged so that normal gait can be restored. Kinesiohydrotherapy is also indicated, but exercise should be carried out at a gentle and leisurely pace. Gluteus haematoma lends itself to ultrasound treatment and phonophoresis, as well as standard kinesiotherapy.

Hip orthotics

Problems affecting the hip joint sometimes require limitation of rotary motion and relief from weight-bearing function. This is accomplished by means of a pelvipaedic device consisting of a pelvic belt corset connected to a thigh support. If total load-relief is necessary, an ischial weight-bearing brace is used. In our experience with patients with inhibitors, the latter type of orthotic device was resorted to on three occasions, in one case

REHABILITATION OF PATIENTS WITH HAEMOPHILIA

characteristics, the knee is one of the body segments most prone to bleeding. According to the data on the patients with inhibitors treated by us, knee bleeds ranked second. A large number of lesions are caused on account of weight-bearing flexion–extension combined with rotary movements, which result in excessive strain of the ligaments, excessive pressure on the menisci, and even torn cartilage.

An evaluation of the knee should take into account local temperature, presence of fluids and perimeter measurements, as well as ROM and muscle balance. Over 50 highly detailed functional tests have been described [19], which enable differential diagnoses of bone, joint and muscle disorders.

Rehabilitation of knee haemarthrosis

As with all joint bleeds, knee haemarthrosis calls for immobilization in pain-free posture, as well as factor cover and the application of ice. Subsequently, the knee is placed at 0° extension, and isometric exercise is prescribed followed by proprioceptive facilitation of the quadriceps. As soon as muscle balance has reached a value of 4/5, gait re-education exercises can commence. In the case of joint instability, the use of orthetic devices is recommended. Swimming or exercise in water are an excellent way of regaining knee function. Atrophy of the quadriceps is a significant feature of chronic synovitis of the knee, which requires NSAIDs, as well as factor cover and physiotherapy. If the outcome is unsuccessful, synoviorthesis must be considered.

Rehabilitation of muscle haematomas affecting the thigh

In our experience, thigh haematomas in patients with inhibitors are a result of bruising to the outer part of the abductors in the middle area of the thigh, and excessive stress on the quadriceps and adductors in the upper area of the thigh. The standard treatment for these lesions comprises rest, application of ice, gentle stretching, kinesiotherapy and a gradual return to weight-bearing function. In the case of quadriceps haematoma, splints affording partial weight-bearing relief may be required, or even ischiotibial support for complete weight-bearing relief until the muscle has regained sufficient strength for ambulation.

Knee orthotics

The knee joint is particularly amenable to orthotics. In patients with inhibitors the most commonly used devices are splints made of thermoplastic material, articularly aligned splints, and silicone knee pads with bars and straps to provide stability to the various body segments involved in moving the lower limbs.

As already indicated, ischiatic weight-bearing braces (Fig. 25.6) provide total relief from weight-bearing function so as to prevent renewed bleeds that may result from burdening the affected body segment. The use of walking aids is essential for the restoration of normal gait as it enables a gradual return to

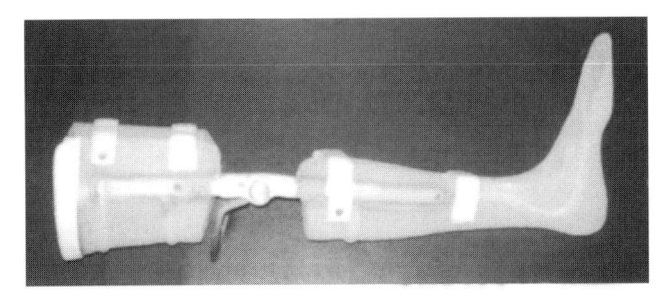

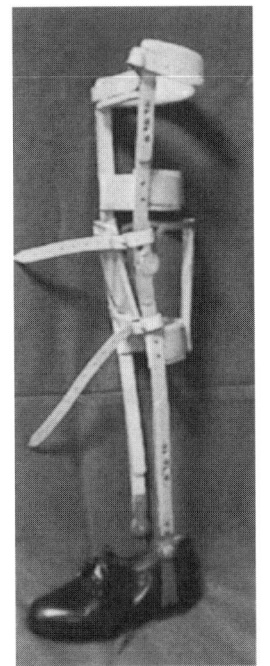

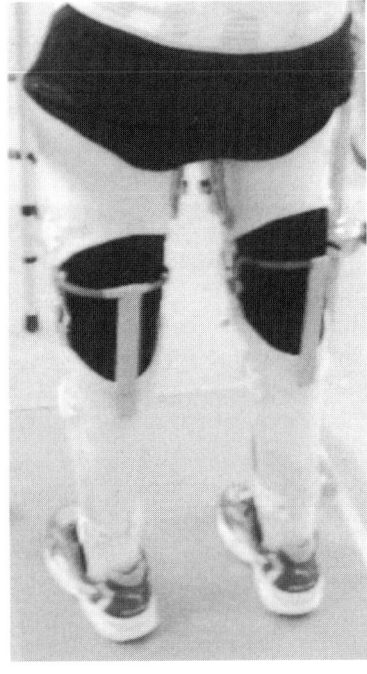

Fig. 25.6 Ischiopaedic orthosis.

owing to problems affecting both the hip and the knee (Fig. 25.6).

In all problems affecting the lower limbs, great care must be taken when restoring their weight-bearing function during deambulation. Gait re-education in patients with inhibitors, which involves the use of walking frames and crutches, can only be undertaken under the supervision of a physiotherapist. The orthotic devices used in gait re-education should be equipped with systems that reduce strain on the upper limbs. These systems include cushioned pads that act as shock absorbers and prevent the patient from losing his or her balance during deambulation. Springs may be placed inside the shaft to produce similar effects.

Rehabilitation of the knee

Function, examination and clinical evaluation

The knee consists of the femorotibial and femoropatellar joints. It is an extremely bulky and relatively unprotected joint with extensive synovial membrane full of blood vessels and continually under great biomechanical stress. Owing to these

Rehabilitation of the ankle

Function, examination and clinical evaluation

As the principal weight-bearing joint, the ankle is under continual stress, especially during sports activities. The tibia, fibula and astragalus make up the tibiotalal joint, which allows flexion–extension and is consequently prone to injury. The articulation of the astragalus with the calcaneal, scaphoid and cuboid bones allows eversion–inversion, together with movements combining abduction–adduction and pronation–supination. Ankle haemarthrosis generally causes slight equinus posture and inversion of the foot. The sensation of intra-articular bleed is more keenly felt on the front outer surface of the ankle, but confirmation can be obtained only by tactile exploration of the area, as well as evaluation of ROM and muscle balance and ultrasonography.

Rehabilitation of ankle haemarthrosis

Prophylactic treatment of lesions is also important in the rehabilitation of the ankle in patients with inhibitors. The first step is to perform annual studies of the foot in the course of the patient's regular visits, in particular in children and adolescents.

The action of the heel striking the ground during ambulation can result in lesions which are normally insignificant, but can sometimes cause haemarthrosis. In view of this, the patient should use insoles or special footwear to cushion the shock produced during ambulation [20].

The procedures to be followed in the management of the acute phase are the same as in other manifestations of haemarthrosis. First, the ankle joint is immobilized in pain-free posture, generally with sole flexion and inversion. After 24–48 h the ankle may be placed at 90° flexion in neutral eversion–inversion posture followed by gradual restoration of motion and strength. Both cryotherapy and hydrotherapy consisting of alternate baths have a beneficial effect on the oedema, while interferential electrostimulation or other forms of electrotherapy afford relief from pain and improve trophism. In our opinion, weight-bearing relief should be prolonged by means of suitable orthotic devices in the case of ankle haemarthrosis affecting patients with inhibitors. The return to total weight-bearing function should be carried out gradually and with greater care than in other joints, with recourse to walking aids and encouraging adequate gait patterns.

Rehabilitation of damaged ankle ligaments and muscle haematomas

Damage to the ligaments, or sprains, should be confirmed by means of ultrasonography so as to rule out the possibility of joint bleeds. The affected area is immobilized for 10–15 days and then stabilized by means of orthotic devices or functional bandages (taping or strapping). Weight-bearing function is restored gradually. Gastrocnemius haematomas severely limit gait and require standard treatment for haematomas. The orthotics used in these cases allow gradual modification of ankle dorsiflexion.

Ankle orthotics

The main purpose of orthopaedic footwear and insoles is to stabilize gait and cushion the heel to prevent the minor traumas that cause bleeds, or to rectify incorrect posture that brings about damage to the ligaments. Modern devices such as podobarometer or instrumented insoles help give accurate diagnosis of weight-bearing and tell the difference between weight-bearing with or without insoles. Patients generally report a sensation of increased comfort when walking with these orthotic devices.

In ankle haemarthrosis, the joint is generally immobilized with plaster of Paris splints. However, orthotic devices made of thermoplastic material or resin, or of the 'Airsplint™' type [21], are preferred because they allow a day-by-day monitoring of the problem and because they may be re-used in cases of recurrent haemarthrosis.

Walking is initiated during the subacute phase. The load-relieving orthotics that are, in our experience, best for this stage are the Air Cam Walker™ [3]. These devices are capable of unburdening the affected joint of up to 100% of its load by means of metal shafts and compression systems.

Silicone ankle pads equipped with anchorage and stabilizing systems to compensate or aid joint function are extremely well known because of their widespread use in sporting activities. They are especially useful for patients with inhibitors in that they forestall excessive stress in sports or activities of daily living when, in the final phase of management of a bleed, a return to normal living is permitted.

Sports

Sports and physical activity claim the interest of those involved in treating patients with haemophilia because they have a direct bearing on the patient's haematological, physical and psychosocial state [3,21–23]. All health programmes directed at the population at large emphasize the importance of sport as a means of keeping fit. The World Health Organization recommend non-competitive activities such as aerobics, jogging, walking and exercising on static bicycles and running belts. Exercise improves not only physical but also mental health, and for the patient with haemophilia it has therapeutic value because it prevents or restores lesions. However, with patients with inhibitors, the risks entailed by taking part in sport should be taken into account.

Gilbert *et al.* [23] classified sports into three categories. The classification was subsequently studied in an interesting survey on sports and haemophilia carried out by Heijnen et al. [24]. These authors found that not all sports are equally popular in different countries, and made a note of a large number of sports in which patients with haemophilia take part. The only kind of sport that patients with inhibitors may participate in are risk-free sports. Even so, these recommended activities should be performed taking all the precautions that ensure protection from injury. However, other factors should be considered when deciding on an activity. These include motivation and personal preferences, previous or recurrent joint problems, fitness and prior experience in sports, attitude of friends and relatives, and the incidence of injury during or following the chosen activity. Under no circumstances should competitive sports be recommended.

Swimming, snorkelling and rambling are probably the most suitable physical activities for patients with inhibitors. Two articles on t'ai chi ch'uan and haemophilia have been published recently [25,26], but in the West this kind of exercise is not very popular.

Conclusions

The development of inhibitors is the main complication in the treatment of haemophilia. Current therapy with rFVIIa and aPCC offers haemostatic coverage but control is difficult and it is necessary to increase the precautions to avoid musculoskeletal complications. In these patients, prophylactic treatment to avoid the development of haemophilic arthropathy is not feasible. In patients with inhibitors, rehabilitation should not differ, but be more careful and prolonged; moreover, it is necessary to increase the precautions to avoid musculoskeletal complications in other joints. The patient's physical condition is fundamental but it must be controlled by physiotherapy. There are only limited opportunities of taking part in sport for patients with inhibitors: the choice should be of a risk-free sport and, moreover, articular protection should be used.

References

1 Haya S, Aznar JA. Tratamiento del paciente hemofílico con inhibidores. In: Batlle J, Rocha E. *Guía Práctica de Coagulopatías Congénitas*. Madrid: Acción Médica, 2001: 123–38.
2 Di Michele DM. Inhibitors in hemophilia: a primer. *Haemophilia* 2000; **6** (Suppl. 1): 38–40.
3 Querol F, Almendariz A, López-Cabarcos C et al. *Guía de Rehabilitación en Hemofilia*. Barcelona, España: Baxter S.L., 2001.
4 Gilbert MS. Prophylaxis: musculoskeletal evaluation. *Semin Hematol* 1993; **30** (Suppl. 2): 3–6.
5 Manco-Johnson MJ, Nuss R, Funk S, Murphy J. Joint evaluation instruments for children and adults with haemophilia. *Haemophilia* 2000; **6**: 649–57.
6 Rodriguez-Merchan EC, Goddard NJ, Lee CA, eds. *Musculoskeletal Aspects of Haemophilia*. Oxford: Blackwell Science, 2000.
7 Buzzard BM. Proprioceptive training in haemophilia. *Haemophilia* 1998; **4**: 528–31.
8 Beeton K, Cornwell J, Alltree J. Muscle rehabilitation in haemophilia. *Haemophilia* 1998; **4**: 532–7.
9 Querol F, Herrero B, Lerma MA, Aznar JA. Artrocentesis en el hemartros agudo del paciente hemofílico. *Rev Iberoamer Hemostasia* 1997; **10**: 95–8.
10 Goldstein JL, Silverstein FE, Agrawal NM et al. Reduced risk of upper gastrointestinal ulcer complications with Celecoxib, a novel COX-2 inhibitor. *Am J Gastroenterol* 2000; **95**: 1681–90.
11 Leese PT, Hubbard RC, Karim A, Isakson PC, Yu SS, Gei GS. Effects of celecoxib, a novel cyclooxygenase-2 inhibitor, on platelet function in healthy adults: a randomized, controlled trial. *J Clin Pharmacol* 2000; **40**: 124–32.
12 Cometti G. *Los Métodos Modernos de Musculación*. Barcelona: Paidotribo, 2000: 109–41.
13 Rodríguez-Martín JM. *Electroterapia en Fisioterapia*. Madrid: Panamericana, 2000.
14 Redureau D. *Le Laser: Application en Physiotherapie*. Paris: Maloine, 1985.
15 Pettersson H, Ahlberg A, Nilsson IM. A radiologic classification of hemophilic arthropathy. *Clin Orthop* 1980; **149**: 153–9.
16 Reginster JY, Deroisy R, Rovati LC et al. Long-term effects of glucosamine sulphate on osteoarthritis progression: a randomised, placebo-controlled clinical trial. *Lancet* 2001; **357**: 251–6.
17 Wallny T, Brackmann HH, Semper H et al. Intra-articular hyaluronic acid for haemophilic arthropathy of the knee. In: Rodriguez-Merchan EC, Goddard NJ, Lee CA, eds. *Musculoskeletal Aspects of Haemophilia*. Oxford: Blackwell Science, 2000: 61–5.
18 Rodriguez-Merchan EC, Wiedel JD. General principles and indications of synoviorthesis (medical synovectomy) in haemophilia. *Haemophilia* 2001; **7** (Suppl. 2): 6–10.
19 Buckup K. *Pruebas Clínicas para Patología Ósea: Articular y Muscular*. Barcelona: Masson, 1997.
20 Heijnen L, Heim M. Orthotic principles and practice and shoe adaptations. In: Rodriguez-Merchan EC, Goddard NJ, Lee CA, eds. *Musculoskeletal Aspects of Haemophilia*. Oxford: Blackwell Science, 2000: 189–96.
21 Miller R, Beeton K, Goldman E, Ribbans WJ. Counselling guidelines for managing musculoskeletal problems in haemophilia in the 1990s. *Haemophilia* 1999; **3**: 9–13.
22 Arranz P, Costa M, Bayés R et al. *El Apoyo Emocional en Hemofilia*, 2nd edn. Madrid: Aventis Behring, 2000.
23 Gilbert MS, Schorr JB, Holbrook T, Tiberio D. *Hemophilia and Sports*. New York: National Hemophilia Foundation, 1994.
24 Heijnen L, Mause-Bunschoten EP, Roosendaal G. Participation in sports by Dutch persons with haemophilia. *Haemophilia* 2000; **6**: 537–46.
25 Beeton K. Tai Chi Chuan for persons with haemophilia: commentary. *Haemophilia* 2001; **7**: 437.
26 Danusantoso H, Heijnen L. Tai Chi Chuan for people with haemophilia. *Haemophilia* 2001; **7**: 437–9.

CHAPTER 26

Physiotherapy in the management of patients with inhibitors

K. Beeton and B. Buzzard

Patients with haemophilia who also have inhibitors present a particular challenge to the haemophilia team. All the members of the multidisciplinary team (doctors, scientists, laboratory technicians, nurses, physiotherapists and counsellors) have an important role in the management of these patients and experience in the management of haemophilia is essential. To date, the emphasis has been on the complex medical management of these patients and there is a wealth of literature available on this approach. However, there is very little published information available on the physiotherapy management of these patients. This chapter provides an overview of the physiotherapy assessment and management strategies that should be considered for a patient with inhibitors. Factors to be considered in the treatment of patients with acquired inhibitors are also reviewed. The principles are illustrated with three case studies.

The incidence of inhibitors in patients with haemophilia varies in the published literature with rates as low as <10% up to 52% reported in one study cited by Vermylen [1]. However, it is generally regarded that 20–33% of severely and moderately affected patients with haemophilia A develop inhibitors, although the incidence is less in patients with haemophilia B [2]. The variability in the reported incidence of inhibitors is thought to relate to the characteristics of the patient, treatment options and the diagnostic methods used [3,4]. The development of inhibitors is one of the major complications in the treatment of patients with haemophilia [2,5,6]. Indeed, Scharrer *et al.* [4] state that: 'the development of inhibitors is seen as the most serious complication of haemophilia A therapy, impacting on the efficacy of coagulation factor concentrates to treat bleeds, making prophylaxis impossible in many cases and creating a significant risk of life-threatening bleeds'.

Patients with inhibitors are generally considered to have a similar bleeding frequency as those without inhibitors, although it has been reported by some authorities that bleeding may be increased in the presence of inhibitors [7]. Rather than experiencing more bleeds, patients may experience more pain, disability and complications than those who do not have an inhibitor, because of the difficulty in managing the bleeding episodes [2,7,8]. Although surgery can be performed successfully for patients with inhibitors [9,10], for many patients surgery is not an option. For these patients, conservative management is the only treatment available and therefore the role of the physiotherapist is particularly important in assisting the patient to achieve maximal function and independence. Before any physiotherapy treatment is undertaken it is essential that the patient is fully assessed.

Initial examination

The initial examination by the physiotherapist includes both subjective and physical components. The components of the examination are described elsewhere [11,12] and it is not proposed to discuss the details of the initial examination in full. However, the particular issues relevant to patients with inhibitors are highlighted.

Subjective examination

The focus of the subjective examination is to establish baseline information regarding the patient's haemophilia status, obtain an understanding of the symptoms experienced, identify any factors which may be contributing to the problem and determine the history of the problem [13].

When assessing a patient with haemophilia and inhibitors, it is essential to obtain information about the inhibitor and the current medical management. The physiotherapist needs to have a clear understanding of the difficulties that can occur in managing a patient with inhibitors as inappropriate or vigorous physiotherapy management could cause or exacerbate bleeds.

Patients need to be questioned about the type of inhibitor they have. Patients may have a low or high titre inhibitor, measured in Bethesda units (BU). A low titre inhibitor is defined as <5 BU; a higher number indicates that there is more inhibitor present [2,14]. The patient may also be a high or low responder, depending on their response to factor replacement products [2]. Low responders have a less marked response to blood products and patients can be treated with large doses of factor replacement, preferably recombinant products [14]. High responders react very quickly and develop high levels of inhibitor following factor replacement [2]. It is often more difficult to control the bleeding in patients with a high titre inhibitor who are high responders and factor replacement may be reserved for life-threatening bleeds only. The type of the inhibitor can change over time [2] and therefore all physiotherapy for patients with inhibitors should be carried out judiciously in order to avoid any exacerbation of bleeding or rebleeding. It is also important to establish how long the patient has had the inhibitor, what treatment they may have had to try to eradicate the inhibitor and their current bleeding pattern and frequency.

Understandably, patients may be very anxious about having an inhibitor, or may be very depressed about the uncertainty or seriousness of their condition. This may impact on their ability to partake fully in any physiotherapy treatment programme. The child who develops an inhibitor poses additional problems to the parents and the multidisciplinary team. The physiotherapist may identify this problem during the examination or during the course of treatment. A further opinion on the management of this problem may need to be sought and the physiotherapist may refer the patient for counselling [15].

Some patients may present to the physiotherapist before it has been identified that the patient has an inhibitor. The physiotherapist must remain vigilant when assessing any patient with haemophilia. A lack of improvement following factor replacement, or repeated bleeds in spite of prophylaxis may be the first indication that an inhibitor has developed [16,17]. If the physiotherapist has any concerns, these should be discussed with the appropriate medical personnel prior to any intervention.

At the completion of the subjective examination, the physiotherapist should have a good understanding of the problem presented by the patient and be able to plan an appropriate physical examination.

Physical examination

The overall objective of the physical examination is to test the structures identified from the subjective examination as causing or contributing to the problem. Although care is always required when handling patients with haemophilia, in patients with inhibitors the scope of the physical examination may need to be more limited and particular attention given to careful and precise manual handling procedures in order to avoid causing or exacerbating bleeds.

For example, the examination of the joints usually includes assessment of active and passive physiological movements and passive accessory movements [18]. The aim of these tests is to identify particular movements which reproduce the patient's symptoms and to determine the quality of the movements. For patients with inhibitors, overpressures may be contraindicated even if the movements are pain-free and forceful end-of-range passive techniques must be avoided. The focus should be on assessing the active pain-free range of movement.

When testing the muscles, isometric muscle testing may be appropriate as there is no movement of the joint. Care should be taken if testing is to include muscle strength through the range of movement, as manual resistance by the physiotherapist involving long limb levers could trigger a bleed. In the same way, if the length of the muscle is to be tested, the limb should be positioned with care and careful and specific handling is required with respecting tissue quality and response to movement and the reproduction of any symptoms.

At the conclusion of the physical examination, the physiotherapist makes a decision based on the assessment findings. The patient may have a neuromusculoskeletal problem that requires physiotherapy. Alternatively he or she may just need advice and a self-management programme or may require further evaluation either from the orthopaedic surgeon or haemophilia consultant. The patient may be referred to a combined haemophilia/orthopaedic clinic when all members of the multidisciplinary team are present which may be very valuable in identifying the most suitable management approach for the patient with inhibitors. For those patients who require specific physiotherapy management, the findings from the examination can provide a baseline that can be used as a marker for the evaluation of treatment [11].

If physiotherapy is indicated, it is important to identify the treatment plan in conjunction with the patient, based on mutually agreed goals [19]. Goals should be SMART: specific, measurable, attainable, realistic and time-based. When managing the patient with an inhibitor it is essential that the goals are set taking into account the predisposition to bleeding and allowing for a slower recovery than when managing patients without inhibitors.

Specific issues in children with inhibitors

The development of an inhibitor is in itself a major complication for those with haemophilia, but when an inhibitor develops early in a child's life this can pose further social, economic and emotional problems for the family. Having a newly diagnosed child with haemophilia can be stressful even if there is prior knowledge of a family history of the condition. The need for comprehensive care is particularly important under these circumstances, as families need additional support to enable them to accept and adjust to the situation. An experienced team including a physiotherapist is essential as they will have intensive and prolonged involvement with the child and family throughout the early years. It is vitally important that the physiotherapist makes early contact with the child and family in order to promote understanding and trust [20]. Initially, physiotherapy may be furthest from the minds of the doctors, nurses and family; however, the physiotherapist has much to offer in these situations including advice, education and treatment. Physiotherapy intervention will be dependent upon the extent of the inhibitor present, whether the patient is a high or low responder and the site of the bleed. Joint and muscle bleeds must be treated with the same rapid response as in those individuals without inhibitors. Early recognition of bleeding by the parents will help to minimize the long-term pathological effects of these haemorrhages.

Venous access is often difficult [21] because of the poor vein development and the anxiety of the child. For the child with inhibitors, easy venous access for factor replacement is vital and a Port-A-Cath™ is now widely used in the developed world, despite the increased rate of infection [22] as the benefits outweigh the risks, with a view to immune tolerance programmes and the adequate treatment of bleeding episodes.

Children will often present in the early years of life more frequently with soft tissue bleeds, which are often difficult to diagnose and the presence of an inhibitor must always be suspected. The parent may observe that the child is fractious during bathing or dressing, or not using a limb normally. Advice and support from the comprehensive care team is essential and the physiotherapist is often able to identify the site of the bleeding episode through observation and knowledge of the child's developmental stages [23].

Annual reviews

All patients with haemophilia should be reviewed on a long-term basis [12,23,24]. This is particularly important for patients with inhibitors who may experience considerable symptoms from repeated and uncontrolled bleeding. Ideally, an annual or 6-monthly review should be undertaken in order to evaluate the musculoskeletal status and to identify any deterioration in joint range or muscle strength so that appropriate interventions can be undertaken as necessary. There are currently a number of assessment forms available, which can facilitate in the collection and monitoring of the musculoskeletal status [25–27].

Physiotherapy management for patients with inhibitors

This section reviews the management of acute joint bleeds, chronic synovitis and arthropathy that can occur in patients with inhibitors.

Management of acute joint and muscle bleeds

The joint or muscle should be rested until haemostasis has been achieved. Rest and/or immobilization of the affected part needs to be continued for these patients until it has been established that the bleeding has stopped in order to try and ensure that rebleeding does not occur. This may include bed rest, an element often not required for patients without inhibitors as a result of the effectiveness of factor replacement.

The main aims of physiotherapy are to:
- relieve pain;
- reduce swelling;
- maximize function with restoration of prebleed range of movement of joints and maximal muscle strength and length; and
- prevent recurrence of bleeding [12].

Modalities of treatment may include ice packs to reduce pain and swelling, and rest in a functional position. Splints may be required to provide appropriate support and protection for the affected limb and to prevent further bleeding. This may include the use of slings, plastic ready-made splints or plaster backslabs. For those patients with a high-responder inhibitor it is important that joint bleeds should be splinted before the administration of factor concentrates, which will maximize the effects of the treatment and minimize the risk of joint rebleeding [24]. The new generation of thermoplastic materials has made this task easier (Memory™, Johnston and Johnston). These splinting materials are easy to apply during an episode of acute bleeding. Parents are taught how to apply the splints at home, thereby ensuring that a joint is immobilized as soon as possible and so reducing the risk of further joint sequelae. The properties of some of these thermoplastic materials can be quite entertaining and can provide a distraction to the child during the application of the splint. Electrotherapy including pulsed short wave and ultrasound should be used with care in patients with inhibitors and should be avoided until it is established that any bleeding is under control.

Once bleeding has stopped and symptoms are improving, a graded programme of exercises can be commenced. Because of the severe pain that usually accompanies an acute bleed, there may be marked muscle inhibition, which can render the joint even more vulnerable to further trauma and rebleeding. Exercises should be started cautiously initially in non-weight-bearing positions with isometric exercises, progressing to closed and open chain exercises as recovery occurs. It is essential to avoid causing further bleeding by undertaking a vigorous programme of exercises [28]. Many patients have a good understanding of their condition and know how much they can safely do; others need to be counselled not to overdo activities and risk causing another bleed.

Any activities that are likely to provoke bleeds must be avoided until the bleed has completely resolved. If there was any specific event or activity that caused the bleed, this should be reviewed and strategies put in place to try to avoid repetition in the future [11]. Many of these patients will be unable to take part in sports or some of the more active pursuits enjoyed by patients with haemophilia who do not have inhibitors. It is important to try and identify other activities that the patient can do (such as swimming, snooker, t'ai chi ch'uan) [29]. This will help to increase confidence, improve self-esteem and promote social integration [30] so that patients do not feel isolated even if they are unable to take part in more active sports and team games.

Case report 1

JC was newly diagnosed at 2 years with severe haemophilia A and no previous family history.

January 1999

5 January 1999: increasing number of bleeds right upper limb noted.

8 January 1999: inhibitor detected 10 BU/mL human.

Inhibitor continued to rise to 40 BU/mL. The right upper limb remained the target area. The patient was treated with physiotherapy and modified splinting using a bandage. Physiotherapy intervention was difficult because of the age of the child, but with regular monitoring the child eventually began to use right upper limb normally. Full range of movement was restored and JC was using the arm normally in play activities.

April 1999

The patient was admitted for insertion of a Port-A-Cath™ for immune tolerance programme using daily factor VIII, as the right elbow remained the target joint (Fig. 26.1).

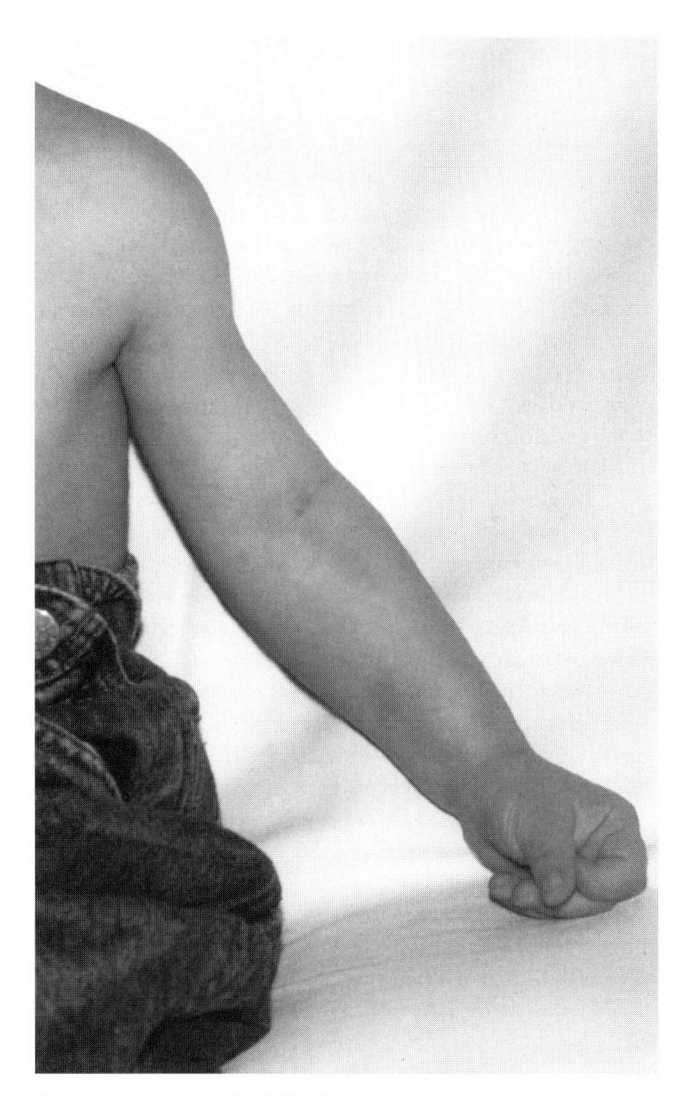

Fig. 26.1 Acute upper limb bleed in a child.

October 1999

The immune tolerance programme was working reasonably well until this month when the patient presented with generalized malaise and pyrexia. The Port-A-Cath™ was removed using activated factor VII. Blood cultures grew *Escherichia coli* and *Staphylococcus aureus*. The patient required a period of admission to the paediatric intensive care unit (PICU) because of additional complications of a left basal pneumonia.

Intensive physiotherapy was required during this period to assist respiratory function and clear residual secretions which was again carried out under the cover of factor VIIa and with close cooperation of the PICU staff and comprehensive care team.

November 1999

The patient was discharged from hospital and remained under review of the haemophilia centre staff.

2000

The patient continued to present with right upper limb bleeds and left lower limb bleeds, particularly the left ankle.

February–August 2000

A modified immune tolerance programme using factor VIII on alternate days was attempted, BU rose from 40 to 195 with continual bleeding into the right elbow, and as a result this programme was abandoned.

During this period, physiotherapy intervention consisted of early splinting using thermoplastic materials and low-dose pulsed short-wave diathermy (PSWD). The parents were taught how to apply and adjust splints for his right elbow and left ankle in order for them to be applied immediately if they observed a joint haemorrhage (Fig. 26.2).

2001

JC continues to present with numerous joint and muscle bleeds and is presently being treated with a factor VII protocol. He remains a very active 4-year-old with no clinical signs of arthropathy. He and his family will require continual support throughout his developmental years and regular physiotherapy input may help to diminish or prevent long-term disabilities.

Management of chronic synovitis

If uncontrolled bleeding continues, a chronic synovitis can develop. Patients with inhibitors are particularly vulnerable to developing chronic synovitis because of the likelihood of repeated bleeds into the same joint which may be difficult to control and slow to recover. In this situation there is often a large swollen warm joint with marked muscle weakness and

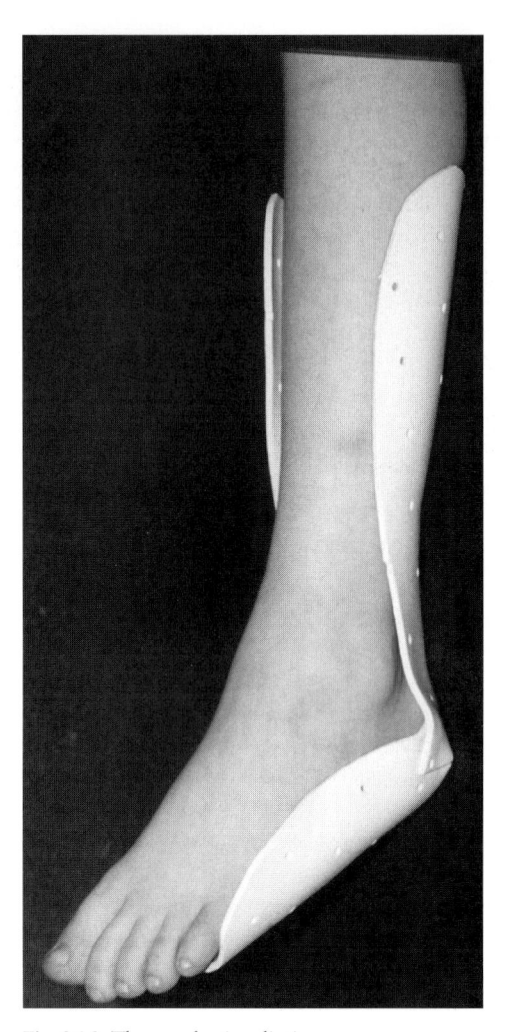

Fig. 26.2 Thermoplastic splinting.

atrophy, although pain is often not such a problem compared to an acute bleed.

The aims of physiotherapy treatment are to:
- reduce swelling;
- increase muscle strength;
- improve coordination and proprioception; and
- prevent recurrence of bleeding [12].

Splinting or braces may be required to rest the joint to prevent rebleeding and to provide additional support for the limb in the presence of muscle weakness. Crutches may be used to decrease weight-bearing. Shock-absorbing insoles, supportive footwear such as boots or trainers [31] and avoidance of activities which may lead to further bleeds are very important as prophylaxis may not be effective or indicated in these patients.

A progressive strengthening programme can be commenced with care when the bleeding is under control, with particular emphasis on joint coordination and proprioception training [32]. Hydrotherapy may be of particular benefit for patients with inhibitors as the buoyancy of the water can provide a supportive medium for the affected joints [33]. Ideally, the exercise programme should be continued until the joint has settled and the patient can return to functional activities [31]. This may be a prolonged process requiring months of rehabilitation [34].

Management of arthropathy

If acute bleeds and chronic synovitis are inadequately treated then eventually the joint will be destroyed. The presence of blood in the joint damages the articular cartilage leading to degenerative changes with loss of joint space, sclerosis and osteophytes evident on radiography [35]. The patient may complain of pain that can be severe. This may be associated with deformities, loss of range of movement, crepitus, decreased muscle strength, contractures and reduced function [35]. Although orthopaedic surgery may be a possibility for some patients with inhibitors, for many conservative treatment will be the only option and physiotherapy can play an important part in alleviating and improving the signs and symptoms which may arise as a result of arthropathy.

The main aims of physiotherapy are to:
- relieve pain;
- increase or maintain range of movement;
- increase or maintain muscle strength; and
- maximize function [12].

Modalities of treatment available for the treatment of arthropathy in patients with inhibitors are similar to those available for all patients with haemophilia. These include hydrotherapy, electrotherapy including transcutaneous electrical nerve stimulation (TENS), manual therapy techniques, braces, splints and advice and education which have been discussed in more detail elsewhere [12]. This section highlights some aspects particularly relevant to the patient with inhibitors.

Hydrotherapy involves exercises in warm water. These can be very useful to relieve pain and begin muscle strengthening without aggravating symptoms. However, it is still important to be aware that although the warmth of the water facilitates relief of pain, reduction of muscle spasm and promotes relaxation [32], overexertion and aggravation of symptoms is possible in the water. Treatment should begin with short sessions in the hydrotherapy pool, which can be lengthened once the effects of the treatment have been established.

Mobilization of joints can be very useful to reduce pain and restore range of movement [18]. These techniques can be performed extremely gently without reproducing any symptoms and yet still be very effective. The addition of gentle manual traction either in isolation or in conjunction with other manual techniques can also be beneficial [36].

Muscle imbalances often occur around the pelvis, trunk or shoulder girdle either because of habitual poor postures, inflexibility and poor patterns of movement [37] but also as a result of bleeds. Muscle re-education techniques involve the specific re-education of weak and lengthened muscles that provide a predominately stability function for joints and lengthening procedures for muscles which have a tendency to become overactive and tight and correction of the movement dysfunction

[38]. The techniques should always be pain-free, begin at low loads and be very precise. It may not be appropriate to progress to high load activities when rehabilitating patients with haemophilia [39] and for patients with inhibitors this may neither be indicated nor necessary.

Splints, braces, supportive footwear and custom-made insoles may all have a role in helping to alleviate symptoms affecting the lower limbs. Each modality should be carefully assessed, evaluated and their ongoing use monitored so that splints and braces are not used when no longer necessary. Advice and education are an essential part of any treatment and this aspect of management is particularly important for the patient with inhibitors. This can include advice on avoiding and minimizing aggravating factors, teaching the patient how to look after their joints to minimize stress, and a regular programme of exercises [40]. The principles for providing general mobilizing and strengthening exercises are similar to those outlined above for the acute bleed. Exercises should always be performed within pain-free range, and progressed slowly. Weight-bearing exercises should be undertaken with caution and high-impact vigorous exercises should be avoided in order to minimize the risks of precipitating bleeding episodes.

If patients have marked disability, referral to an occupational therapist may be appropriate and a visit arranged to the home so that adaptations can be made to minimize stress on joints and promote independence. This can include simple measures such as providing rails, bath aids and raised toilet seats, to more complex measures including the installation of showers and stair lifts. A wheelchair may also be required if walking distance is very limited. An electric wheelchair may need to be considered if there are problems with upper limb joints preventing functional use of the arms.

Case report 2

Patient SH is a 68-year-old patient with severe haemophilia A and high-responding inhibitor, diagnosed in the early 1970s following an allergic reaction after being treated with animal factor VIII preparations. His earliest recorded inhibitor titre was recorded in excess of 200 Oxford units, now recorded as BU. Since 1970 he has been treated successfully with non-activated prothrombin complex concentrates. The patient has suffered many joint bleeds over the years that have developed into severe haemophilic arthropathy affecting all joints (Fig. 26.3). He has limited range of movements in his elbows, hips, knees and ankles and also fixed flexion contractures of all these joints. The majority of these bleeding episodes have been treated conservatively with bed rest, splinting, ice and a graduated exercise programme. His inhibitor titre ranges between 23 and 40 BU and his main complaint at present is chronic pain in the left hip, right wrist and both ankles. Despite his severe arthropathy, SH has continued to lead an active life and remains mobile with the aid of two walking sticks.

His increasing joint arthropathy, which was compounded by a fracture of the left patella and right olecranon following a fall,

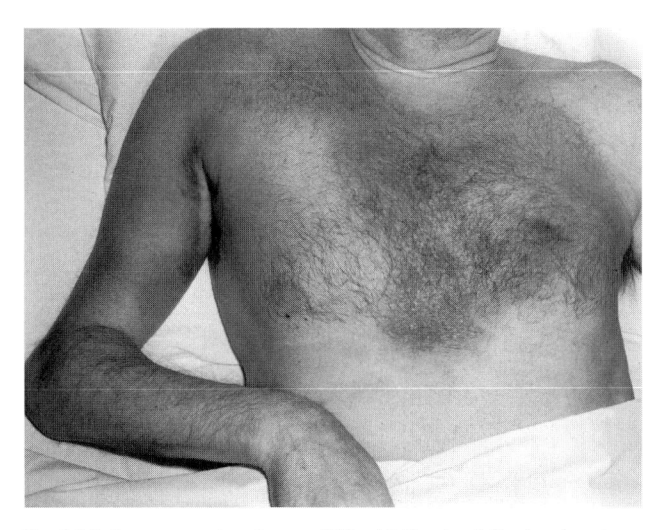

Fig. 26.3 Severe anterior chest wall bleed following fall, also showing fixed flexion contractures in elbow and wrist.

has necessitated the implementation of a daily living assessment and provision of the necessary equipment to ensure an independent and fruitful quality of life. Physiotherapy has provided a vital link to this patient's continued independent lifestyle through regular assessment and treatment. His current problem remains chronic pain, especially in his left hip for which operative intervention is not an option.

Acquired haemophilia/inhibitors

A thorough literature search has failed to provide any information on the physiotherapy management of this condition. It occurs in 0.2–1 in 1 million of the population per year, mainly affecting the elderly [41,42]. Despite the rarity of acquired haemophilia, those affected often suffer severe muscle haematomas [43], which may require intensive physiotherapy to aid resolution and restoration of function. Most patients present with a short history of easy bruising and haemorrhages [44]. Bleeding may at first be deemed inconsequential but nevertheless can be progressive, leading to massive haematomas which, if left untreated, can develop into compartment syndromes requiring immediate and often radical surgical intervention. This condition has a high mortality rate of 18–22%, usually occurring within the first few weeks from diagnosis [44].

Other conditions have been associated with the development of acquired haemophilia, such as rheumatoid arthritis and systemic lupus erythematosus. Pregnancy also appears to influence the susceptibility of autoantibody development in some cases [43]. It has been reported that the incidence of inhibitor development in elderly patients who have cancer may be purely coincidental [43].

The management of this condition is both complicated and costly and is therefore best administered within a recognized

comprehensive haemophilia care centre. The medical management of acquired haemophilia consists of:
1 treating the bleed; and
2 eliminating or 'switching off' the inhibitor.

Laboratory assays measure the level of inhibitor present using BU and treatment will be dependent upon these results. Replacement therapy is necessary to halt the bleeding episodes, which may include the use of factor VIII, Feiba, porcine and, more recently, recombinant activated factor VII. In addition to these products, steroid therapy using prednisolone has been shown to abolish the inhibitor in 30% of patients [44]. Those who fail to respond may require additional therapy using high-dose immunoglobulins and cyclophosphamides.

Those patients who present with massive muscle haemorrhages also require intensive physiotherapy input as would be administered to those with hereditary haemophilia A and B. The consequences of these bleeding episodes if left untreated would otherwise lead to contractures and deformity. Close liaison between the medical staff, nursing staff and the physiotherapist is essential. Physiotherapy treatment needs to be carefully progressed to avoid rebleeding. As these patients have no previous experience of bleeding episodes they may be particularly distressed and frightened. Management by an experienced physiotherapist with knowledge of haemophilia can help to alleviate the patient's anxiety. Many of the patients may also have underlying conditions as previously identified, such as rheumatoid arthritis or chronic obstructive pulmonary disease (COPD) or even cancer, which will need consideration when implementing a physiotherapy programme. Initial treatment will be rest, immobilization and possibly ice, in conjunction with the prescribed medical regimen. Progression of treatment will be dependent upon the site and extent of bleeding. Normal physiotherapeutic techniques can be used with care, based on haematological results. Physiotherapy intervention will often be prolonged and, because of the age of some of the patients, it may be necessary to continue with physiotherapy in the patient's home following discharge from hospital.

Case report 3

DF is a 76-year-old female diagnosed with acquired haemophilia, admitted in July 2001 with massive haematoma affecting the left lower limb (Fig. 26.4). She was transferred from another hospital for management following presentation of haematuria.
- Activated partial prothrombin time (APTT)
  Factor VIII:C 10%
  Inhibitor 2 BU/mL
  Haemoglobin 7.3
  Positive inhibitor screen.
- Previous medical history of COPD
  Cor pulmonale
  Total abdominal hysterectomy for endometrial cancer
  Hydronephrosis right kidney.

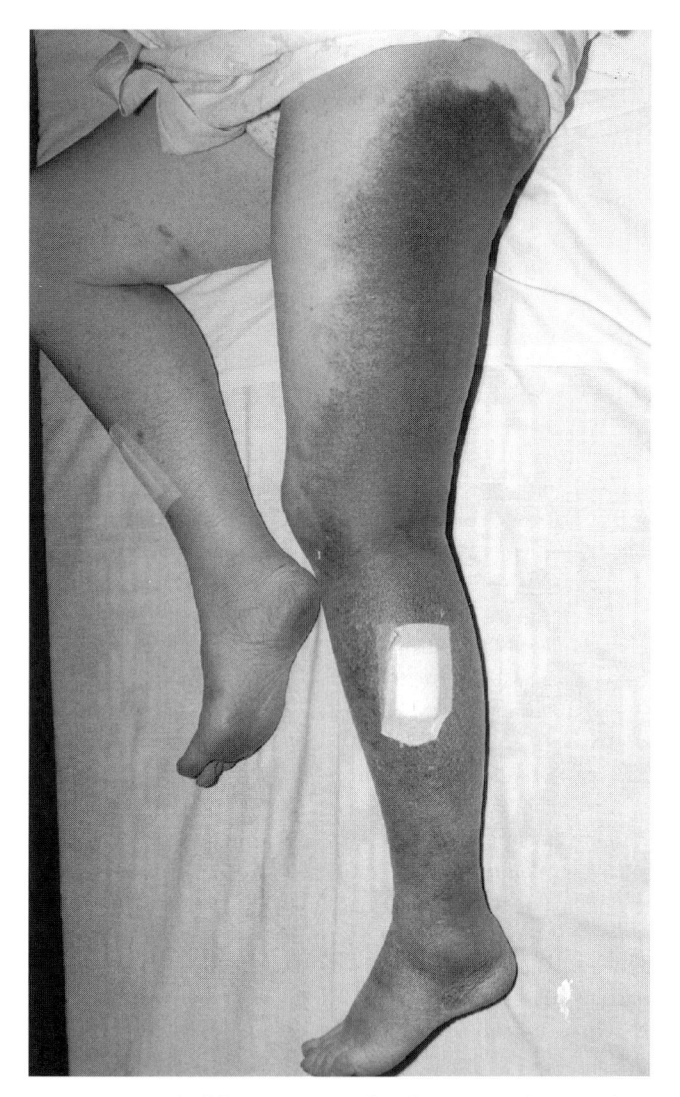

Fig. 26.4 Lower limb haematoma in a female patient with acquired inhibitor.

- Clinical findings
  Extensive bruising and swelling full extent of left lower limb consisting of 5 cm increased circumference throughout the length of the left lower limb
  Swelling is hard and tense with pooling of blood throughout the whole posterior aspect of the leg
  Range of movement is decreased in hip, knee and ankle because of swelling and pain
  Active range of movement left knee 0–80°
  Left ankle dorsiflexion −15°, plantar flexion 25°
  Sensation—intact
- Plan of treatment—medical
  Steroid therapy initiated using prednisolone and gradually reducing dose of intravenous gammaglobulin for 3 days
  Cyclophosphamides
  1 week later factor VIII level had risen to 23%
  Transfused 3 units of blood

- Physiotherapy management
  Full subjective and objective assessment carried out
  Strict bed rest
  High elevation of limb
  Gradual maintenance exercises consisting of foot and ankle exercises and static quadriceps exercises within the pain-free range
  Low dose PSWD
  Active assisted range of motion exercises to ankle, hip and knees.

Over the next 7–10 days the patient's condition improved with a positive reduction in swelling and increased range of motion of all affected joints. Active mobilization commenced 7 days postadmission, initially with a Zimmer frame then progressing to elbow crutches. The patient was discharged 14 days postadmission following an assessment by an occupational therapist. Community physiotherapy was implemented for further rehabilitation in the patient's home and she was reviewed every 2 weeks in the haemophilia centre.

The patient is currently fit and well but she continues to have persistent swelling and pain in her left ankle with residual bruising. She will continue to be followed up on a 2-weekly basis.

Outcome measures

Whichever modality of treatment or intervention is used it is important that it is evaluated using appropriate assessment tools [26]. Where possible 'a published, standardized, valid, reliable, and responsive outcome measure is used to evaluate the change in the patient's health status' [45]. There are a number of instruments available to evaluate outcome of physiotherapy including impairment measures, pain scales, functional scales, activity levels and quality of life measures [46,47]. This ensures that the effectiveness of the interventions is evaluated and the benefits of the treatment identified.

Conclusions

The medical management of the patient with inhibitors is complex and difficult. Fortunately, more effective methods of treatment are continually being identified and the future is optimistic [2]. The physiotherapist has much to offer these patients in helping to alleviate symptoms, advise on joint care and in enabling these patients to maintain their independence with an improved quality of life. Prevention of some of the bleeding problems associated with this condition should be emphasized. This can only be achieved by continually promoting patient education. The need for good musculoskeletal health and fitness by regular exercise and avoidance of activities that predispose to injury should also be encouraged, in conjunction with continued support and advice from those responsible for haemophilia care.

References

1 Vermylen J. How do some haemophiliacs develop inhibitors? *Haemophilia* 1998; **4**: 538–42.
2 Di Michele D. Inhibitors in haemophilia: a primer. *Haemophilia* 2000; **6** (Suppl. 1): 38–40.
3 Chang H, Sher G, Blanchester V, Title J. The impact of inhibitors on the cost of clotting factor replacement therapy in haemophilia A in Canada. *Haemophilia* 1999; **5**: 247–52.
4 Scharrer I, Bray G, Neutzling O. Incidence of inhibitors in haemophilia A patients: a review of recent studies of recombinant and plasma-derived factor VIII concentrates. *Haemophilia* 1999; **5**: 145–54.
5 Cahill M, Colvin B. Haemophilia. *Postgrad Med J* 1997; **73**: 201–6.
6 Freiburghaus C, Berntorp E, Ekman M *et al.* Tolerance induction using the Malmö treatment model 1982–95. *Haemophilia* 1999; **5**: 32–9.
7 Roberts H. Inhibitors and their management. In: Rizza C, Lowe G, eds. *Haemophilia and Other Inherited Bleeding Disorders*. London: W.B. Saunders, 1997.
8 Green D. Complications associated with the treatment of haemophiliacs with inhibitors. *Haemophilia* 1999; **5** (Suppl. 3): 11–17.
9 Penner J. Management of haemophilia in patients with high-titre inhibitors: focus on the evolution of activated prothrombin complex concentrate AUTOPLEX T. *Haemophilia* 1999; **5** (Suppl. 3): 1–9.
10 Faradji A, Bonnomet F, Lecocq J *et al.* Knee joint arthroplasty in a patient with haemophilia A and high inhibitor titre using recombinant factor VIIa (NovoSeven): a new case report and review of the literature. *Haemophilia* 2001; **7**: 321–6.
11 Beeton K, Ryder D. Principles of assessment in haemophilia. In: Buzzard B, Beeton K, eds. *Physiotherapy Management of Haemophilia*. Oxford: Blackwell Science, 2000: 1–13.
12 Beeton K. Physiotherapy for adult patients with haemophilia. In: Rodriguez-Merchan EC, Lee CA, Goddard N, eds. *Musculoskeletal Aspects of Haemophilia*. Oxford: Blackwell Science, 2000.
13 Jones M. Clinical reasoning process in manipulative therapy. In: Boyling J, Palastanga N. *Grieve's Modern Manual Therapy*, 2nd edn. Edinburgh: Churchill Livingstone, 1994: 471–89.
14 Inhibitor Subcommittee of the Association of Hemophilia Clinic Directors of Canada. *Haemophilia* 2000; **6** (Suppl. 1): 52–9.
15 Miller R, Beeton K, Goldman E, Ribbans W. Counselling guidelines for managing musculoskeletal problems in haemophilia in the 1990s. *Haemophilia* 1997; **3**: 9–13.
16 Brettler D. Inhibitors in congenital haemophilia. *Baillière's Best Pract Res Clin Haematol* 1996; **9**: 319–30.
17 Ribbans W, Giangrande P, Beeton K. Conservative treatment of haemarthrosis for prevention of hemophilic synovitis. *Clin Orthop* 1997; **343**: 12–18.
18 Maitland G. *Peripheral Manipulation*, 3rd edn. London: Butterworth–Heinemann, 1991.
19 Cott C, Finch E. Goal setting in physical therapy practice. *Physiother Can* 1991; **43**: 19–22.
20 Buzzard BM. Physiotherapy for prevention and treatment of chronic haemophilic synovitis. *Clin Orthop* 1997; **343**: 42–6.
21 Van den Burg HM, Fischer K, Roosendaal G, Mauser-Bunschoten EP. The use of the Port-A-Cath: a review. *Haemophilia* 1998; **4**: 418–25.
22 Westernberg F, Flacteh Janssen CW Jr. Central venous catheter with subcutaneous port (Port-A-Cath): eight years clinical follow up with children. *Pediatr Hemotol Oncol* 1993; **10**: 233–9.

23 Buzzard BM. Physiotherapy management of children with haemophilia. In: Rodriguez-Merchan EC, Goddard NJ, Lee CA, eds. *Musculoskeletal Aspects of Haemophilia*. Oxford: Blackwell Science, 2000: 169–76.
24 Buzzard BM, Jones PJ. Physiotherapy management of haemophilia: an update. *Physiotherapy* 1988; **74**: 221–6.
25 Pettersson H, Gilbert M. Hemophilic arthropathy. In: *Diagnostic Imaging in Haemophilia*. Berlin: Springer-Verlag, 1986: 23–68.
26 Chartered Society of Physiotherapy. *Standards for Haemophilia*. London: Powage Press, 1996.
27 Manco-Johnson M, Nuss R, Funk S, Murphy J. Joint evaluation instruments for children and adults with haemophilia. *Haemophilia* 2000; **6**: 649–57.
28 Heim M, Martinowitz U, Graif M, Ganel A, Horoszowski H. Case study: the treatment of soft tissue haemorrhages in a severe classical hemophiliac with an unusual antibody to factor VIII. *J Orthop Sports Phys Therapy* 1988; **10**: 138–41.
29 Danusantoso H, Heijnen L. Tai Chi chuan for people with haemophilia [Correspondence]. *Haemophilia* 2001; **7**: 437–9.
30 Jones P, Buzzard B, Heijnen L. *Go For It: Guidance on Physical Activity and Sports for People with Haemophilia and Related Disorders*. Montreal: World Federation of Hemophilia, 1998.
31 Heijnen L, Rosendale G, Heim M. Orthotics and rehabilitation for chronic haemophilic synovitis of the ankle. *Clin Orthop* 1997; **343**: 68–73.
32 Buzzard BM. Proprioceptive training in haemophilia. *Haemophilia: State of the Art* 1998; **4**: 528–31.
33 Titinsky R. Hydrotherapy and its use in haemophilia. In: Buzzard B, Beeton K, eds. *Physiotherapy Management of Haemophilia*. Oxford: Blackwell Science, 2000.
34 Buzzard BM. Physiotherapy for the prevention and treatment of chronic haemophilic synovitis. *Clin Orthop* 1997; **343**: 42–6.
35 Rodriguez-Merchan EC. Effects of hemophilia on articulations of children and adults. *Clin Orthop* 1996; **328**: 7–13.
36 De Kleijn, Beeton K. Management of shoulder. In: Heijnen L, ed. *Physiotherapy Management of Haemophilia*. East Sussex: Medical Education Network, 1995.
37 Comerford M, Mottram S. Functional stability re-training: principles and strategies for managing mechanical dysfunction. *Man Ther* 2001; **6**: 1–7.
38 Beeton K, Alltree J, Cornwall J. Rehabilitation of muscle dysfunction in haemophilia. *Treatment of Haemophilia Monograph* 2001: Series No. 24.
39 Padkin J. Muscle imbalance in haemophilia. In: Buzzard B, Beeton K, eds. *Physiotherapy Management of Haemophilia*. Oxford: Blackwell Science, 2000.
40 Haemophilia Chartered Physiotherapists Association. *Joint Care and Exercises*. Haemophilia Society, 1993.
41 Lottenberg R, Kentro TB, Kitchens CS. Acquired haemophilia: a natural history study of 16 patients with factor VIII inhibitors receiving little or no therapy. *Arch Intern Med* 1987; **147**: 1077–81.
42 Kessler CM. An introduction to factor VIII inhibitors: the detection and quantitation. *Am J Med* 1991; **91** (Suppl. 5a): 1S–5S.
43 Green D. Acquired factor VIII inhibitors and immunosupression of autoantibodies. In: *Acquired Haemophilia*, 2nd edn. 1995: 25–40.
44 Hay CRM, Negrier C, Ludlum CA. The treatment of bleeding in acquired haemophilia with recombinant FVIIa: a multi centre study. *Thromb Haemost* 1997; **78**: 1463–7.
45 Chartered Society of Physiotherapy. *Core Standards of Physiotherapy Practice*. London: Chartered Society of Physiotherapy, 2000.
46 Cornwall J. Disability and outcome measures. In: Buzzard B, Beeton K, eds. *Physiotherapy Management of Haemophilia*. Oxford: Blackwell Science, 2000: 101–10.
47 De Kleijn P, van Meeteren N. Use of a theoretical framework for health status in haemophilia care and research. In: Buzzard B, Beeton K, eds. *Physiotherapy Management of Haemophilia*. Oxford: Blackwell Science, 2000: 89–100.

CHAPTER 27

Elective orthopaedic surgery in haemophilia patients with high-responding inhibitors

I. Hvid, K. Soballe and J. Ingerslev

Repetitive intra-articular bleeding episodes eventually cause degenerative changes in joints because of enzymatic cartilage damage. Even with optimal control and prophylactic factor VIII (FVIII) treatment, haemarthrosis may occur often enough to initiate cartilage degeneration leading to arthropathy in some patients. Without these measures, arthropathy will occur in one or more of the more susceptible joints (knee, elbow, ankle, hip and shoulder) in almost all patients [1]. With the availability of FVIII, even major orthopaedic surgery can be safely undertaken although, in general, more complications should be expected in haemophiliac patients, some of which can be attributed to the increased risk of postoperative bleeding [2–5].

Nevertheless, average intra- and postoperative bleeding does not appear to be increased when comparing, for instance, total knee arthroplasty in patients with and without haemophilia [6–8]. In some reports, the frequency of postoperative infection is markedly increased in total joint replacement in haemophilia [2,4,5]. This might have been because of relatively poorer general health, including the presence of transmissible diseases such as HIV infection or chronic hepatitis. However, in our own experience with total joint replacement, the frequency of septic complications has not been significantly different than in non-haemophilic patients [3,7,9].

Historically, in haemophilia patients with inhibitors, and certainly those with high titres of inhibitors, surgical treatment was rarely undertaken, except for life-saving procedures. There were several options in dealing with inhibitors, including immune tolerance induction, high-dose FVIII or factor IX (FIX) administration, substitution therapy combined with immunosuppressive measures such as cyclophosphamide administration, and the use of activated products such as Autoplex™ and Feiba™. With the emergence of recombinant activated factor VIIa (rFVIIa), however, the ability to control haemostasis has become considerably more predictable [10]. At present our indications for performing elective orthopaedic procedures in patients with high titre inhibitors differ little from those we tend to apply in dealing with non-complicated haemophilia.

There has been a tendency to wait for more severe degenerative joint changes to occur in haemophiliac patients than in other patients before major procedures, such as total joint replacement, are performed. However, and this is particularly true for total knee replacement, waiting for too long will tend to make the operation more difficult (another source of postoperative complications), and obviously will adversely affect the quality of life of the patient. Once degenerative knee joint changes are established, malalignment and flexion contracture tend to develop, and progress with time. These conditions require soft tissue release to balance the collateral ligaments and restore extension. The extent of these releases typically increases with increasing deformity, and therefore adds to the possible sources of postoperative bleeding, thus increasing the risk of related complications. So increasing malalignment and flexion deformity of the knee occurring with significant arthropathy would indicate total knee replacement, even if pain were not a major problem. Eventually, patients with major knee flexion deformity will experience severe limitation of the ambulation, because in this situation walking is very energy-consuming.

Although a few major orthopaedic procedures in haemophiliac patients with high-responding inhibitors have been reported in the past [10,11], the advent of rFVIIa has resulted in increased confidence in the ability to adequately control the bleeding tendency [12–15]. We have invariably used repeated administration of rFVIIa rather than continuous infusion. For these surgeries, rFVIIa was given as an intravenous bolus injection of 74–125 μg/kg body weight just before surgery, then every 2 h for 10–30 h, then gradually prolonging the administration up to (3-, 4-) 6-hourly intervals [16]. This schedule was maintained until the sutures had been removed. Discrete daily doses were then given prior to physiotherapy sessions. If late active bleeding was encountered, administration intervals were again reduced until haemostasis was obtained. This regimen was supplemented with 6-hourly administration of an inhibitor of fibrinolysis, tranexamic acid, in doses of 25 mg/kg body weight. No serious adverse effects of this treatment regimen have been reported. Continuous infusion therapy has resulted in one reported case of disseminated intravascular coagulation [17].

Planning surgery

If the patient has not been treated with rFVIIa before the planned surgery (or if treatment has not been given for a longer period), the patient is given one or more test doses about 1 week prior to surgery in order to titrate the dose needed for haemostasis. Monitoring consists of platelet count, activated partial thromboplastin time (APTT), prothrombin time, factor VII:C, fibrinogen, D-dimers, antithrombin (III), and prothrombin fragment F1 + 2 [16]. These laboratory tests are repeated as necessary during and after surgery.

The surgeon should be intimately familiar with the intended surgery. In this day and age of increasing subspecialization, hardly any one orthopaedic surgeon will have mastered all aspects of orthopaedic surgery to perfection. On the other hand, the subspecialized surgeon may not be familiar with the particular problems related to surgical intervention in patients with haemophilia. Until such familiarity has been firmly established, it is recommended that the surgeon experienced in haemophiliac surgery join with the subspecialized surgeon to perform the surgery. In our centre, one orthopaedic surgeon acts as the coordinator responsible for haemophiliac patients. He will then call upon subspecialized colleagues to help evaluate the particular problem, and to plan and perform the surgery whenever necessary.

The pathology encountered in patients with haemophilia is often more severe than in those without; for instance, we have had to deal with bilateral knee flexion contractures of 50° or more in relatively young patients. Significant contracture may also be a problem in most other major joints including the hip, ankle and elbow. In the knee particularly, acute elimination of larger contractures by soft tissue release and musculotendineous lengthenings carries a risk for neuropraxia, and therefore may require a combination of soft tissue release, musculotendineous lengthening and tibial and femoral shortening (i.e. deeper terminal bone resections than otherwise required), and quite often a semiconstrained or even occasionally a constrained implant. Although we have seen very long-term survival of constrained implants, there is consensus among those performing knee surgery that the degree of constraint of an implant is inversely related to the expected time of service (implant survival). For that reason, unconstrained or semiconstrained implants should be used whenever possible in younger patients (<60 years of age). An alternative to one-stage surgery could be to deal with the flexion contracture first, making sure not to overstretch the tibial and peroneal nerves, followed by intensive physiotherapy, and then to perform the arthroplasty at a later date. However, this is obviously a more expensive option and, furthermore, it is not always possible because normal joint congruence is often lost in severely arthropathic joints, so that abnormal joint congruence would resist extension of the joint even after musculotendineous lengthening and capsular release (Fig. 27.1). Quadriceps weakening is a related problem because, with flexion contracture, the quadriceps muscle will not have been able to contract (shorten) normally for a long time. In moderate contracture (<20°), this may be overcome by intensive physiotherapy, but with more severe contracture, distalization of the tibial tubercle with the patellar tendon may be necessary. However, the benefit of increasing extension power in this way, may be countered by some loss of flexion. It is important that the patient understands these trade-offs that may have to be made, and that elimination of flexion contracture has priority over the ability to flex the knee.

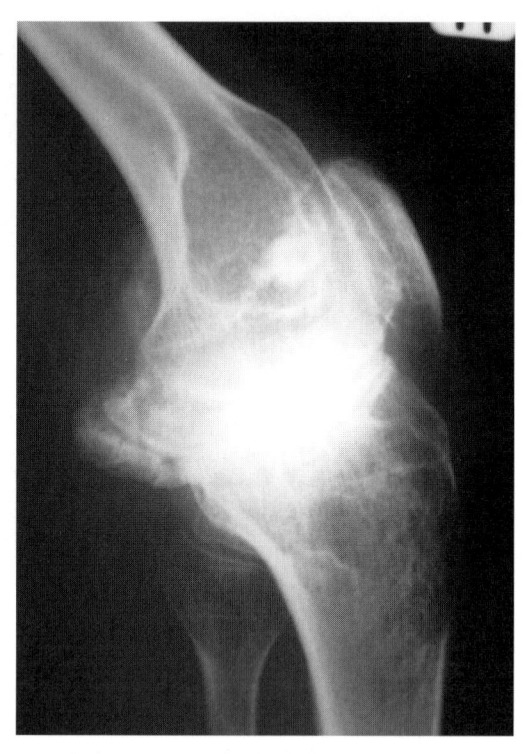

Fig. 27.1 In this lateral radiograph of a knee joint with severe haemophilic arthropathy and flexion contracture, abnormal joint congruence is seen. This joint would resist extension even if soft tissue problems were adequately addressed. Accordingly, soft tissue contractures and arthropathy must be dealt with simultaneously.

When planning total joint arthroplasty, the experienced surgeon will usually be able to choose the implant needed for the particular case. For the knee, we prefer a posteriorly stabilized (semiconstrained) implant because arthropathy is invariably quite severe in these patients. However, a more constrained device should always be at hand.

Surgical considerations, results and complications

The surgeon should be prepared to spend additional time performing the surgery. Bleeding should be dealt with as it occurs during dissection, also when a tourniquet is applied. When dissection is close to bone, there may be bleeding points directly from the bone. Any larger bony bleeding points should be sealed with bone wax.

Muscle can be a particularly worrisome source of bleeding. If not dealt with properly, postoperative haematoma will frequently occur, and tend to progress to form a pseudotumour (Fig. 27.2). Muscular dissection cannot always be avoided, in which case time should be taken to deal with every single

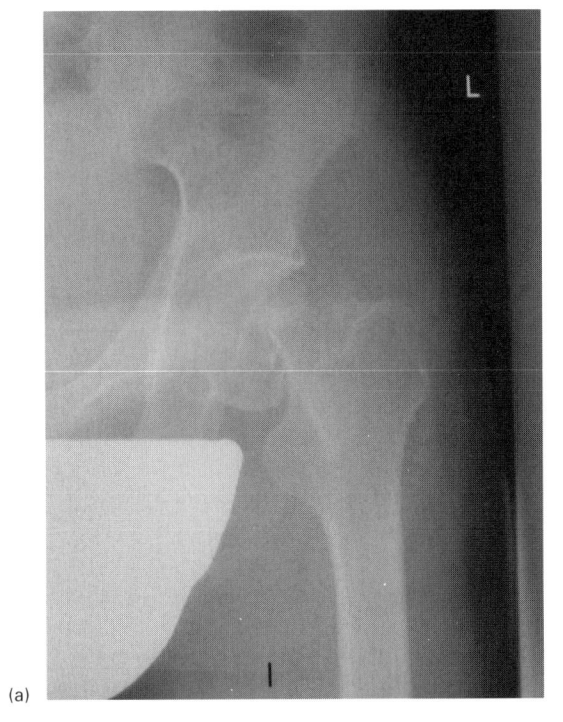

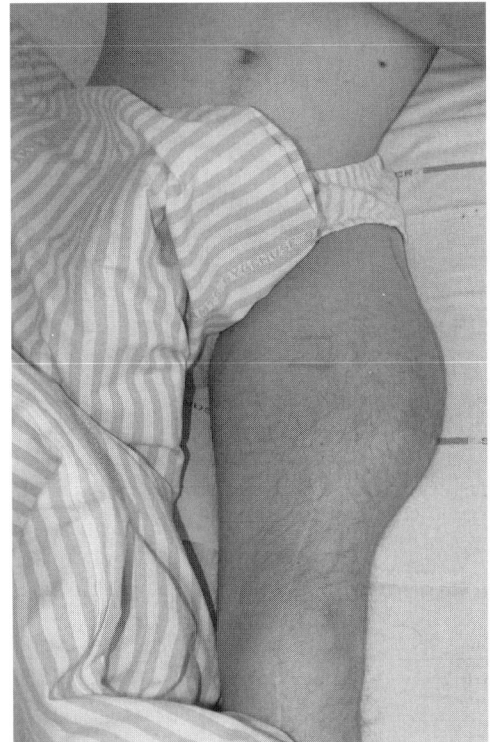

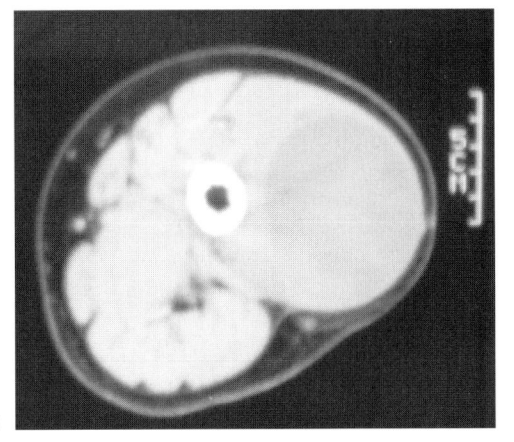

Fig. 27.2 Pseudotumour of the left thigh developing after closed reduction and internal fixation (CRIF) of a femoral neck fracture. (a) The patient woke up one morning with severe hip pain. Radiography showed this fracture. The patient had probably suffered an epileptic seizure. (b) Pseudotumour developed over several weeks. (c) Computed tomography scan showing cross-section of the mid thigh.

bleeding point. Use of a fibrin sealant should be considered to minimize surface oozing. This is also recommended after excision of a pseudotumour (Fig. 27.2) because oozing tends to be somewhat profuse, and dealing with the vast number of small bleeding points is not really possible. However, generally speaking, we do not recommend the use of a fibrin sealant. It has been claimed that this will reduce postoperative bleeding, but this has not been our experience, and certainly there is no documentation in the literature that this is the case.

Experience at any one centre of performing surgery in patients with haemophilia and high titre inhibitors will obviously be limited. Our centre serves a population of about 3 million people. Four patients with haemophilia A and inhibitors have undergone 11 orthopaedic procedures (Table 27.1). There were four complications in 11 surgeries (36%). Two surgeries (18%) were followed by reoperations because of haematomas: one evacuation of haemarthrosis after total knee arthroplasty (TKA); one pseudotumour of thigh after closed reduction and internal fixation (CRIF) of femoral neck fracture. The haemarthrosis arose because of a minor accident during postoperative radiography of the knee, the extremity being 'dropped' during transfer to the radiography table. One could therefore claim that this complication was not directly related to the surgery. There were no infections. The complication rate compares favourably to complication rates reported after surgeries in non-inhibitor patients [2–4].

Intra- and postoperative bleeding (Table 27.2) did not differ significantly from that expected in non-inhibitor haemophiliac patients [7], or even non-haemophiliac patients [6,8], except regarding the patient who developed pseudotumour after CRIF of a femoral neck fracture, and the patient who was treated for compartment syndrome. This last patient suffered from the effects of long-term chronic alcohol abuse, and died from multiple organ failure. It was felt that his alcoholism rather than his

Table 27.1 Orthopaedic procedures in four patients with haemophilia A and high titre inhibitors.

Patient	Procedure	Complications
1	ORIF left patellar fx	None
	Ankle arthrodesis	None
	Bilateral TKA	None
2	Left TKA	Haematoma (trauma)
	Elbow synovectomy	None
	CRIF left femoral neck fx	Pseudotumour
	Pseudotumour left thigh	None
3	Bilateral TKA	Arthrofibrosis (brissement)
	CR right supracondylar femoral fx	None
	THR	None
4	Forearm fasciotomy	Recurrent bleeding
		Died of unrelated cause

CR, closed reduction; CRIF, closed reduction and internal fixation; fx, fracture; ORIF, open reduction and internal fixation; THR, total hip replacement; TKA, total knee arthroplasty.

Table 27.2 Intraoperative bleeding (IOB) and postoperative bleeding (POB) related to the type of surgery.

Procedure	n	IOB (mean mL)	POB (mean mL)	Total range (mL)
TKA	5	500	770	365–2240
THR	1	250	450	700
CRIF Hip	1	400	100*	500*
Pseudotumour	1	1000	600	1600
Ankle arthrodesis	1	Negligible	Negligible	
Elbow synovectomy	1	210	0	210
Fasciotomy	1	Not measurable†		

CRIF, closed reduction and internal fixation; THR, total hip replacement; TKA, total knee arthroplasty.
*Recurrent intramuscular bleeding leading to pseudotumour formation.
†Recurrent bleeding not adequately quantified.

haemophilia was the underlying cause of death. It is apparent that in the elective surgeries, there was only one complication related to the bleeding tendency (one in eight surgeries), and this could have been prevented.

Our most frequent operation was TKA, which was performed in three patients (Tables 27.1 and 27.2); two patients received bilateral TKA. There is no reason to believe that bilateral knee surgery increases the risk of complications, and there is no indication that such was the case in this small series. From an economical point of view, bilateral surgery is obviously preferable because it reduces the consumption of factor. Figure 27.3 shows pre- and postoperative radiographs of one of the bilateral cases (CR, Table 27.1). The technical details of the operations have been described above, except that a synovectomy will usually need to be performed. Leaving an irregular vulnerable synovium would be a mistake, because this would pose a high risk of intra-articular haemorrhage later on. Again, even after synovectomy, we do not recommend the routine use of fibrin sealant. On the other hand, we see no contraindications to its use, although the risk of postoperative intra-articular adhesions may be increased.

Only one ankle arthrodesis was performed (Fig. 27.4). It was performed as an open procedure in order to secure haemostasis intraoperatively. In other patients with haemophilia, we have performed ankle arthrodesis using an arthroscopic and percutaneous approach, and this technique has not resulted in any problems so far [18], and would probably also be safe for patients with inhibitors.

In planning the surgery for the femoral neck fracture, it was considered that a strong internal fixation was necessary because there was a risk that the patient might experience more epileptic seizures. Conceivably, less invasive surgery, avoiding muscular dissection, would not have produced the postoperative pseudotumour of the thigh. Percutaneous insertion of three cannulated screws might have been a better treatment for this patient.

The three patients who have had multiple elective surgeries are generally satisfied with the results, and there is no doubt that their quality of life has been improved considerably.

ELECTIVE ORTHOPAEDIC SURGERY

(a)

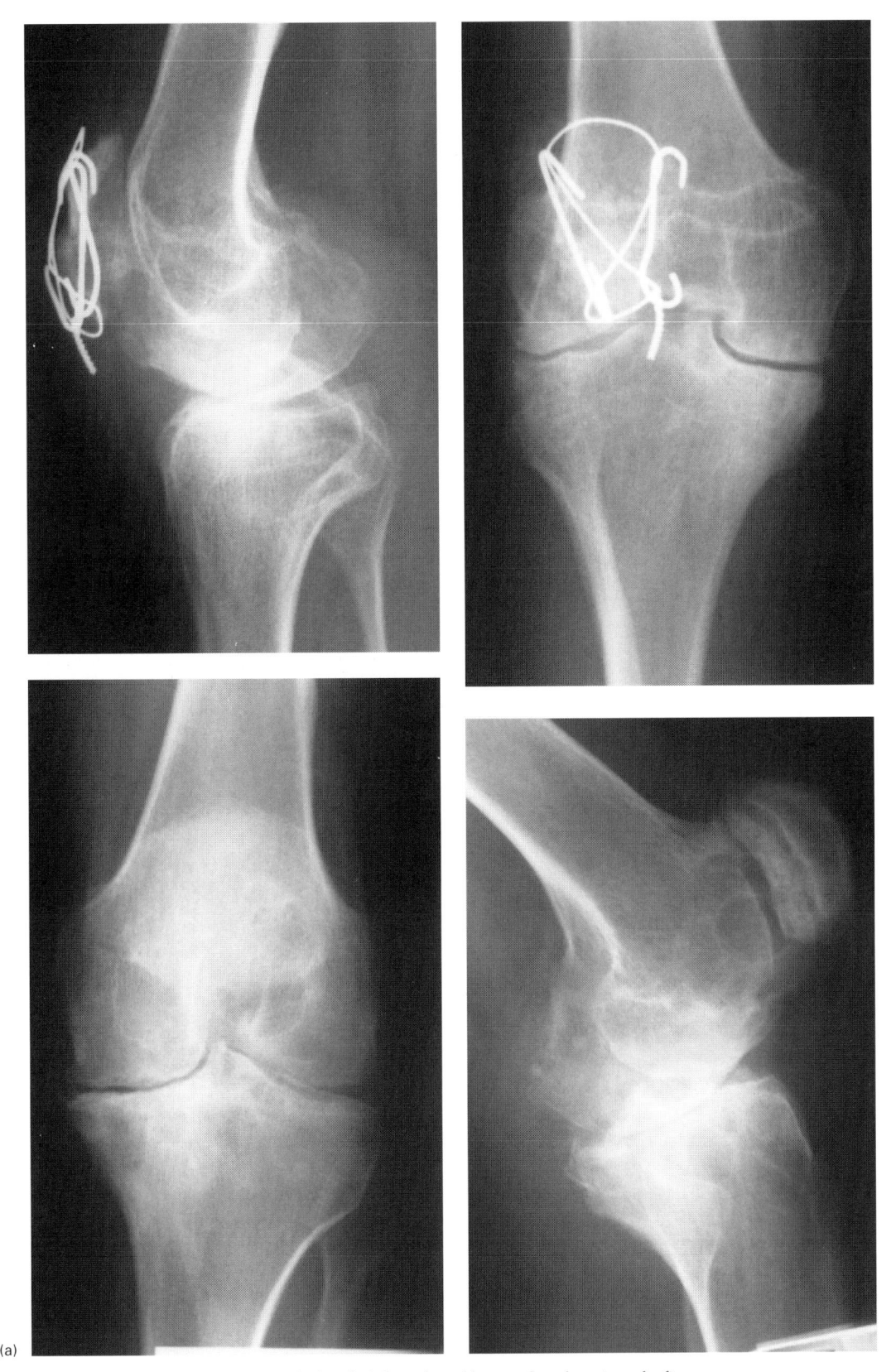

Fig. 27.3 Radiographs before (a) and after (b) bilateral total knee arthroplasty (*overleaf*).

(b)

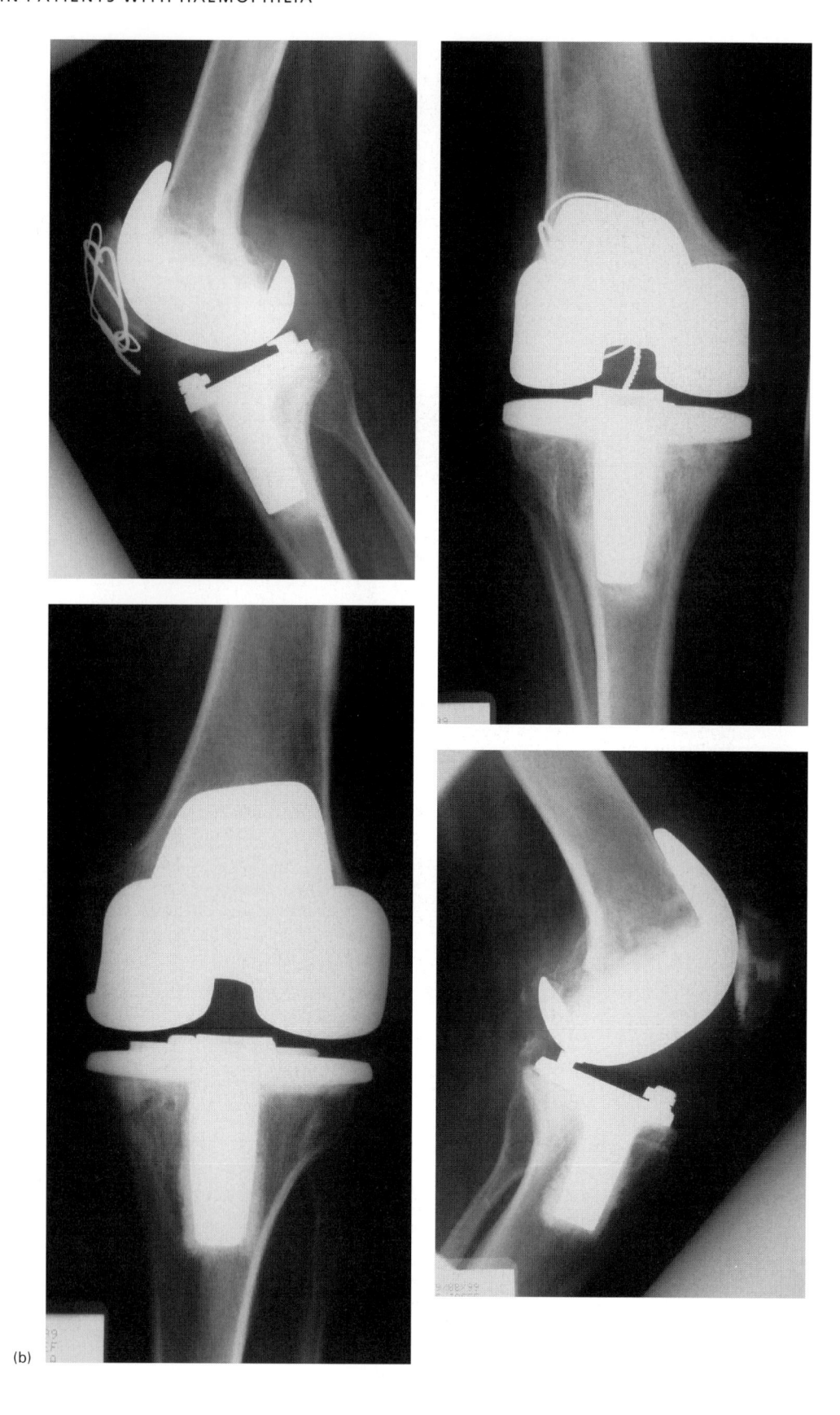

ELECTIVE ORTHOPAEDIC SURGERY

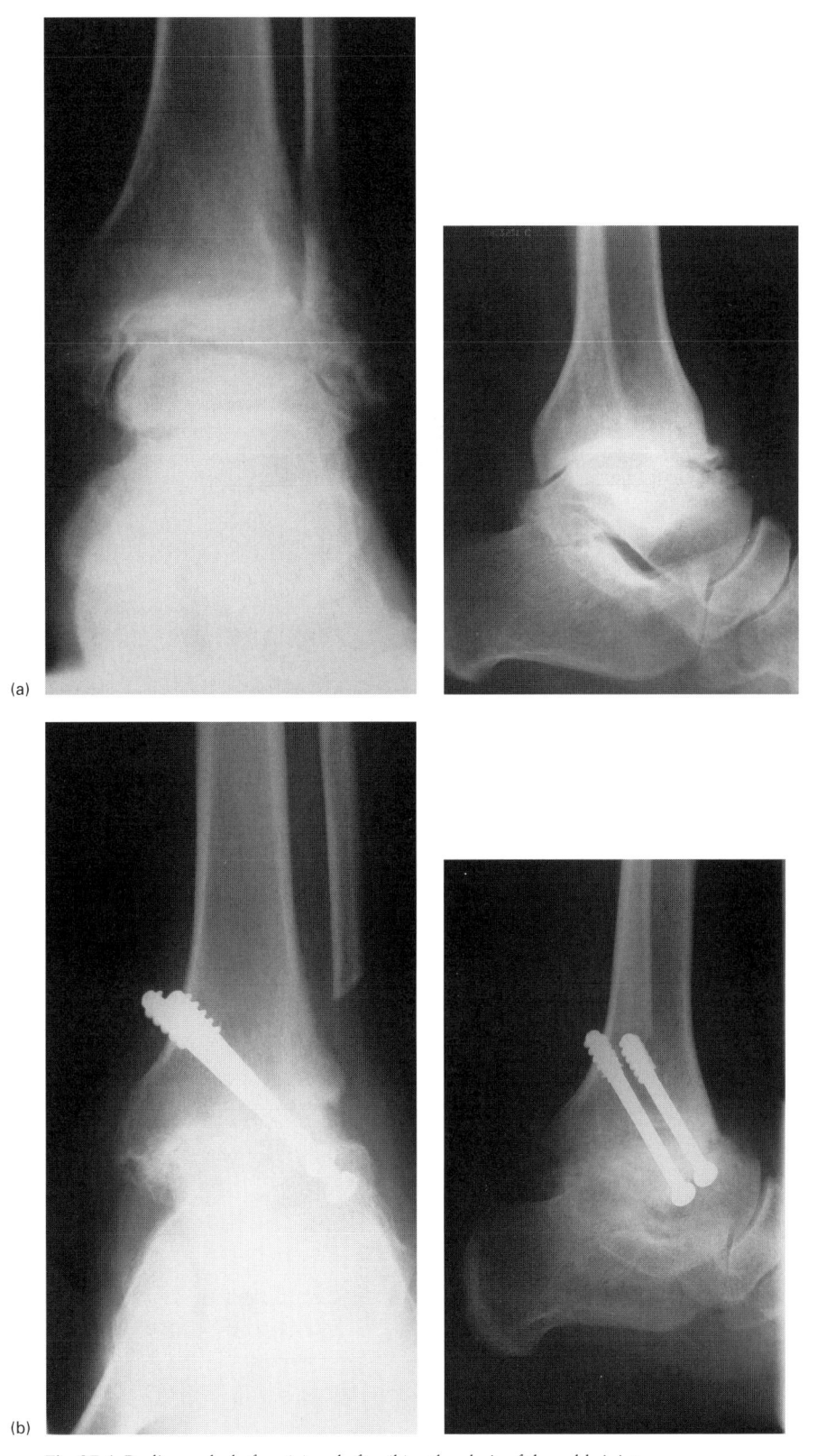

Fig. 27.4 Radiographs before (a) and after (b) arthrodesis of the ankle joint.

Conclusions

With proper planning and careful surgery, it appears that even major orthopaedic procedures can be performed safely in patients with haemophilia and high titre inhibitors using rFVIIa as described in this chapter.

References

1 Aznar JA, Magallon M, Querol F, Gorina E, Tusell JM. The orthopaedic status of severe haemophiliacs in Spain. *Haemophilia* 2000; **6**: 170–6.
2 Cohen I, Heim M, Martinowitz U, Chechick A. Orthopaedic outcome of total knee replacement in haemophilia A. *Haemophilia* 2000; **6**: 104–9.
3 Kjaersgaard-Andersen P, Christiansen SE, Ingerslev J, Sneppen O. Total knee arthroplasty in classic hemophilia. *Clin Orthop* 1990; **256**: 137–46.
4 Thomason HC III, Wilson FC, Lachiewicz PF, Kelley SS. Knee arthroplasty in hemophilic arthropathy. *Clin Orthop* 1999; **360**: 169–73.
5 Vastel L, Courpied JP, Sultan Y, Kerboull M. Knee replacement arthroplasty in hemophilia: results, complications and predictive elements of their occurrence. *Rev Chir Orthop Reparatrice Appar Mot* 1999; **85**: 458–65.
6 Adalberth G, Bystrom S, Kolstad K, Mallmin H, Milbrink J. Postoperative drainage of knee arthroplasty is not necessary: a randomized study of 90 patients. *Acta Orthop Scand* 1998; **69**: 475–8.
7 Crewe B, Hvid I, Ingerslev J. Forty-six total knee arthroplasties in haemophilia A. *Haemophilia* 2000; **6**: 377.
8 Widman J, Isacson J. Surgical hemostasis after tourniquet release does not reduce blood loss in knee replacement: a prospective randomized study of 81 patients. *Acta Orthop Scand* 1999; **70**: 268–70.
9 Crewe B, Soballe K, Ingerslev J, Hvid I. Total hip arthroplasty in congenital bleeding disorders. *Haemophilia* 2000; **6**: 377.
10 Rodriguez-Merchan EC, de la Corte H. Orthopaedic surgery in haemophiliac patients with inhibitors: a review of the literature. In: Rodriguez-Merchan EC, Goddard NJ, Lee A, eds. *Musculoskeletal Aspects of Haemophilia*. Oxford: Blackwell Science, 2000: 136–42.
11 Lauroua P, Barbier F, Dieu P, Dumora D, Moulinier J. Bilateral prosthesis of the knee in a hemophilia A patient with an inhibitor. *Ann Fr Anesth Reanim* 1986; **5**: 154–6.
12 Faradji A, Bonnomet F, Lecocq J *et al.* Knee joint arthroplasty in a patient with haemophilia A and high inhibitor titre using recombinant factor VIIa (NovoSeven): a new case report and review of the literature. *Haemophilia* 2001; **7**: 321–6.
13 Hvid I, Crewe B, Soballe K, Ingerslev J. Orthopaedic surgery in five patients with haemophilia A complicated by inhibitors. *Haemophilia* 2000; **6**: 382.
14 Ingerslev J, Freidman D, Gastineau D *et al.* Major surgery in haemophilic patients with inhibitors using recombinant factor VIIa. *Haemostasis* 1996; **26** (Suppl. 1): 118–23.
15 Tagariello G, De Biasi E, Gajo GB *et al.* Recombinant FVIIa (NovoSeven) continuous infusion and total hip replacement in patients with haemophilia and high titre of inhibitors to FVIII: experience of two cases. *Haemophilia* 2000; **6**: 581–3.
16 Ingerslev J, Sneppen O, Hvid I, Fredberg F, Kristensen HL. Treatment of acute bleeding episodes with rFVIIa. *Vox Sang* 1999; **77** (Suppl. 1): 42–6.
17 Schulman S, d'Oiron R, Martinowitz U *et al.* Experiences with continuous infusion of recombinant activated factor VII. *Blood Coag Fibrinolysis* 1998; **9** (Suppl. 1): S97–101.
18 Crewe BJ, Linde F, Ingerslev J, Hvid I. Talocrural arthrodesis in haemophilia. *Haemophilia* 2000; **6**: 382–3.

PART 7

General surgery

CHAPTER 28

General and emergency surgery in patients with high-responding inhibitors

B. White and O.P. Smith

Surgery in patients with inhibitors to factor VIII (FVIII) or factor IX (FIX) represents a considerable challenge and should only be undertaken as a matter of absolute necessity. The surgery should, whenever possible, be performed in centres with comprehensive care facilities which include access to medical, nursing and laboratory staff with coagulation expertise. In this chapter we will discuss the preoperative assessment and the treatment options available for the management of these patients.

Preoperative assessment

The preoperative assessment should include the following.
1 Definition of the current and peak historical inhibitor titre to human and porcine FVIII, or human FIX.
2 *Confirmation of the immunity to hepatitis A and B*. The immune status of patients is usually assessed on a yearly basis and vaccinations administered accordingly. The purpose of vaccination is to protect patients from viral infection secondary to contaminated red cell transfusions or plasma derived concentrates.
3 *Documentation as to whether patients are infected with HIV, hepatitis B or hepatitis C*. Surgical, anaesthetic and nursing staff should be aware of the viral status of patients. It is important to identify additional risk factors for bleeding associated with infection (e.g. thrombocytopenia or liver dysfunction) or antiviral therapy (protease inhibitors). Protease inhibitors are an integral component of current antiretroviral treatment strategies. These agents are associated with an increased risk of spontaneous and surgery-related bleeding complications [1]. As a result protease inhibitors should be discontinued prior to major surgery and recommenced after the completion of wound healing.
4 *Liaison with surgical, anaesthetic and nursing staff regarding postoperative analgesia*. Special consideration regarding analgesia is particularly important in patients with high-responding inhibitors because haemostasis cannot be guaranteed. Intramuscular injections and spinal or epidural analgesia are contraindicated due to the risk of bleeding. In addition non-steroidal anti-inflammatory agents (NSAIDs) should be avoided because of the risk of gastric erosion and platelet dysfunction.

5 *Documentation of management plan*. A treatment strategy should be outlined for each patient and should be available to the medical and nursing personnel who will be responsible for patient care. Adequate stocks of concentrates should be on site to cover the full duration of the postoperative period and a contingency supply should also be available for emergency procedures.

Therapeutic strategies

Immune tolerance

Immune tolerance may be effective in up to 80% of selected patients with severe FVIII deficiency who develop high-responding inhibitors [2–4]. If time permits, elective surgery should be postponed until a trial of immune tolerance has been completed.

Human FVIII or FIX concentrates

Patients with low titre inhibitors (<5 BU/mL) may respond to high doses of human FVIII (100 U/kg followed by 5–10 U/kg/h or 100 U/kg every 8–12 h). Similarly, a high dose of FIX (2–3 times the normal dose) may be effective in patients with a FIX inhibitor titre of <5 BU, although this treatment option is best avoided if there is a past history of allergic reactions to FIX-containing products, due to the risk of anaphylaxis. When inhibitor titres are in excess of 5 BU, human FVIII or FIX concentrate are unlikely to be effective unless treatment is preceded by plasmapheresis or protein A adsorption [3]. In patients with low titre high-responding inhibitors, the use of FVIII or FIX concentrates is likely to be associated with an anamnestic response at 6–7 days. The response to therapy should be carefully monitored and if anamnesis occurs in patients requiring continuing haemostatic cover, then alternative therapies will need to be considered [5].

Porcine FVIII

The use of porcine FVIII in the setting of high-responding inhibitors has considerable advantages. In patients with little or no cross-reactivity, porcine FVIII provides a highly effective treatment that can be monitored by a laboratory assay. In addition the risk of human viruses can be eliminated. However, the only commercially available porcine FVIII concentrate (HYATE:C™, Porton Speywood Ltd, UK) does not undergo viral inactivation and the exclusion of plasma from pigs with porcine parvovirus has resulted in a current scarcity of product.

Losier *et al.* reported a large survey of patients who were treated with porcine FVIII at the time of surgery [6]. Forty-five patients with a mean age of 25 years (range 1–67) underwent 54 surgical procedures (33 major and 19 minor). The mean maximum doses of porcine FVIII were 148 U/kg (range 40–665) and the duration of treatment was dependent on the type of procedure, with an average of 7.3 days (range 1–36). The haemostatic response was considered excellent or good in 43/54 (80%), fair in 6/54 (11%) and non-existent in 5/54 (9%) of patients. As expected, the post-treatment FVIII levels were predictive of response. However, some patients with a poor recovery still had a good or excellent clinical response. The pre-treatment porcine inhibitor titre was also predictive of clinical response. Only one patient with a porcine inhibitor titre >8 BU had a clinical response to therapy: he received 233 U/kg every 8 h for several days.

Adverse reactions occurred in 5/54 (9%) of episodes and usually consisted of fever, chills and diaphoresis. All reactions were self-limited or readily treated with antipyretics, antihistamines or corticosteroids. Thrombocytopenia occurred in 25% of patients but did not appear to have any deleterious effect on haemostasis.

With respect to anamnesis, human inhibitor titres decreased in 12/39 (31%) of evaluable episodes, remained unchanged in 5/39 (31%) of episodes and increased in 22/39 (56%) of episodes. Porcine inhibitor titres decreased during or after treatment in 3 episodes (8%), were unchanged in 9 episodes (23%) and increased in 27 episodes (69%). In 17 of these 27 episodes the inhibitor titre increased to >10 BU; however, a change in therapy was only required in 8/17 cases. Overall haemostasis was considered to be excellent or good in 14 of these 17 episodes and fair in the remaining 3 cases.

In summary porcine factor VIII provides a highly effective treatment option in patients with high-responding inhibitors who have porcine titres of <5 BU. Porcine FVIII can be administered as a bolus dose of 50–150 U/kg 8–12-hourly or by continuous infusion at a rate of 5–10 U/kg/h. The dose should be adjusted on the basis of the porcine inhibitor titre and the laboratory and clinical response. A poor laboratory response may still be associated with satisfactory haemostasis. Serious side-effects to porcine FVIII are rare, and although anamnesis is common it may be sufficiently delayed to limit the requirement for additional therapies.

Activated prothrombin complexes

There are limited published data on the use of activated prothrombin complexes in patients with high-responding inhibitors undergoing surgical procedures. Lusher *et al.* initially demonstrated that non-activated prothrombin complexes were more effective in controlling haemarthrosis than an albumin placebo (50% vs. 25%) [7]. Subsequent studies suggested that activated prothrombin complexes such as Feiba were slightly but significantly more effective than non-activated prothrombin complex concentrates (PCCs) [8,9]. The published data on the use of the activated PCC called Autoplex as surgical prophylaxis is limited to two small case series which reported no excess bleeding in 3/3 and 3/3 patients, respectively [10,11]. The largest study on the use of Feiba at time of surgery reported on 23 patients who underwent minor ($n = 19$) or major ($n = 4$) procedures. No excess bleeding was noted in the patients undergoing minor surgery. One of the four patients who underwent major surgery had excess blood loss at the time of surgery. However, all four cases had a successful outcome. The doses of Feiba used were 100–210 U/kg/day and 210 U/kg/day for the minor and major procedures, respectively. The duration of treatment was similar to non-inhibitor patients. An additional seven patients in this study were commenced on Feiba after developing an anamnestic response to human ($n = 5$) or porcine ($n = 2$) FVIII which had been used to cover the perioperative period. Only two of these patients, both of whom had acquired haemophilia, had excess bleeding. In two smaller studies Feiba was effective in 3/4 and 6/7 patients undergoing surgery [9,12].

Anamnesis has been reported in 4.9–31.5% of patients receiving Feiba [5,9,12]. This wide variation reflects differences in the definition of anamnesis and may also have been influenced by the prior administration of FVIII and/or blood transfusion. Despite the continuing use of Feiba the inhibitor fell in the majority of cases and even in those patients with persistently elevated titres the response rate to Feiba was undiminished. Similarly, anamnestic responses have been associated with the use of Autoplex. However, as with Feiba the rise in inhibitor did not appear to impair the efficacy of subsequent treatment with Autoplex [13].

Allergic reactions, transient hypertension, disseminated intravascular coagulation (DIC) and thrombosis are well recognized but rare complications of activated PCCs. The DIC and thrombotic complications are likely to be minimized by adhering to the recommended maximum single dose of 100 U/kg and the maximum daily dose of 200 U/kg. The risk of anaphylaxis to activated PCCs should be considered in the presence of FIX inhibitors and FVIIa is probably the treatment of choice for these patients.

The published data suggests that Feiba is effective in approximately 90% of surgical procedures undertaken in patients with inhibitors. The dose, treatment interval and duration of therapy depend on the type of surgery. However, for major surgery a total daily dose of 200 µ/kg administered at 6–12-hourly intervals and continued for approximately 14–21 days appears reasonable. The concomitant use of antifibrinolytic therapy is generally avoided because of the theoretical risk of thrombosis. However, the combination of Feiba and antifibrinolytic therapy has been successfully used in some patients within the published data [5].

Recombinant FVIIa

The haemostatic efficacy of FVIIa in the setting of high-responding inhibitors was first reported in 1988 in a patient undergoing synovectomy [14]. In a subsequent case series of 12/13 evaluable surgical procedures (2 minor and 11 major), FVIIa was associated with an excellent response in 11/12 patients and an efficient response in the remaining patient [15]. Postoperative haemostasis and a satisfactory surgical outcome were obtained in all cases. The dose of FVIIa used was 90 µg/kg 2–3-hourly for two days, after which the interval was prolonged with a total duration of treatment ranging from 6 to 43 days. Lusher reported the use of FVIIa in 103 surgical procedures (21 major, 57 minor and 25 dental) with an excellent or effective response rate of 81%, 86% and 92%, respectively [16]. The majority of patients received an initial dose of 90 µg/kg (range 60–120). However, further information on the dosing interval and duration of therapy was not provided. Scharrrer reported the use of FVIIa in 22 surgical procedures (17 minor and 5 major) [17]. The dose of FVIIa was 90 µg/kg at 2–3-hourly intervals for 1–2 days. For major surgery dosing could be continued at 2–4 hourly intervals for 6–7 days increasing to 6–8 h for a further two weeks. Postoperative bleeding was assessed as absent or minimal in 88% of minor and all major procedures. Shapiro et al. reported on a prospective randomized trial comparing two doses of FVIIa (35 µg/kg vs. 90 µg/kg) in haemophilia patients with inhibitors undergoing surgery (11 major and 18 minor procedures) [18]. The treatment interval was 2-hourly for two days then 2–6-hourly for three days. Thereafter all patients requiring further treatment received 90 µg/kg. In patients undergoing minor surgery satisfactory homeostasis was achieved in 7/10 and 8/8 of the low-dose and high-dose arms, respectively. In major surgery the response rate was 2/5 for the low-dose arm and 5/6 for the high-dose arm. The overall response rate for patients receiving 90 µg/kg was statistically significantly better than the 35 µg/kg arm.

The insertion and removal of central venous access devices (CVADs) have been successfully undertaken in children with high-responding inhibitors. In one study four CVADs were inserted without any excess bleeding [19]. A further study reported on the insertion or removal of 19 CVADs [20]. Two out of 19 children developed minor haematomas at 72 h, which responded to further therapy. There was no excess bleeding in the remaining 17 cases. The treatment regimens were: 90 µg/kg 2-hourly for 24 h, followed by 90 µg/kg every 3–4 h for a further 24–48 h. All patients received tranexamic acid (15 mg/kg 6-hourly) for 1 week.

A number of recent studies have reported on the administration of FVIIa as a continuous infusion (CI). Eight patients (two with acquired haemophilia and six with severe congenital FVIII deficiency) were infused with a bolus dose of 90 µg/kg followed by a CI at a fixed rate of 16.5 µg/kg/h [21]. Two minor and six major surgical procedures were performed and the median duration of treatment was 13.5 days (range 1–26). There was effective haemostasis in only 2/6 major operations and 1/2 minor procedures. Three patients experienced excessive bleeding despite a plasma FVII:C level of 10 IU/mL. The two other treatment failures developed serious bleeding as a result of procedural errors unrelated to the FVIIa. The authors concluded that a 16.5 µg/kg/h infusion reliably achieves plasma FVII:C levels of 10 IU/mL, but that this level does not provide reliable haemostasis for surgery. A follow-up study assessed the efficacy of 90 µg/kg followed by a CI of 50 µg/kg/h designed to achieve FVII:C levels of 30–50 IU/mL [22]. FVII:C levels were >30 IU/mL in 90% of plasma samples. This treatment regimen was effective in 7/9 patients undergoing major surgery. Of the two remaining cases, one had minor bleeding and one had pump failure. However, the overall outcome was satisfactory in all nine cases. Santagostino et al. reported a response rate of 9/11 for major surgery and 12/14 for minor surgery in patients receiving a low-dose CI [23]. An initial bolus of 90–150 µg/kg was followed by a median CI rate of 20 µg/kg/h (range 15–50) and 17 µg/kg/h (range 10–20) for major and minor procedures, respectively. The conclusion of the authors was that low-dose CI (15–20 µg/kg) of FVIIa was effective in inhibitor patients undergoing surgery. The discrepancy between this study and the study by Smith et al. is difficult to explain but may relate to treatment intensity, the concomitant use of antifibrinolytic therapy and differences in patient-specific parameters. The median dose of FVIIa and the resultant FVII:C levels were similar in both studies. However, while all patients reported by Smith et al. received a fixed CI of 16 µg/kg/h, some patients in the Santagostino study received doses as high as 50 µg/kg/h and three patients undergoing knee arthroplasty were given an additional bolus dose of 60 µg/kg intraoperatively, prior to the removal of the tourniquet. It is possible that the concomitant administration of tranexamic acid, which was permitted by Santagostino et al. but not by Smith et al., may have had a role in the differing outcome. Nine of the 25 patients in the Santagostino study were children in comparison to 0/8 in the Smith study. There was also an unexpected increase in the FVII:C clearance over the duration of treatment in the Smith paper, which was not seen in the Santagostino study. This may reflect as of yet unexplained but important patient-specific variables.

The administration of FVIIa by CI does not appear to convey the same advantages over bolus dosing as has been demonstrated with FVIII or FIX concentrates. The concerns about the efficacy of the low-dose CI (15–20 µg/kg/h) means that the requirement for a CI rate of 50 µg/kg/h limits any saving on concentrate over bolus dosing. The availability of pumps to deliver 2-hourly boluses means that CI is no longer a more convenient method of administration. Furthermore there is no evidence that the steady state of FVIIa is more effective than the peaks and troughs associated with bolus dosing. On the contrary, the enhanced thrombin burst associated with peak levels in patients receiving the bolus dosing may prove to be more effective [16]. Consequently, it is the authors' opinion that bolus dosing is probably the preferable method of administration of FVIIa. However, if CI is selected, the higher infusion rate of 50 µg/kg/h should be used until further data is available on the efficacy of the low-dose regimen.

Although thrombosis has been reported in association with FVIIa, the side-effect profile is good and it does not cause an anamnestic increase in inhibitor titre. Tranexamic acid has been successfully used in combination with FVIIa, and one study suggested that this combination is superior to FVIIa alone [24]. However, there is no consensus on the use of antifibrinolytic therapy in patients receiving FVIIa.

In conclusion, FVIIa appears to be effective in 80–90% of patients with high-responding inhibitors undergoing surgery. The published data suggest that 90 µg/kg is more effective than 35 µg/kg. The initial recommended treatment interval is 2–3-hourly for the first 24–48 h. Thereafter the treatment interval and duration of therapy will depend on the type of surgery. In the case of major surgery, it would seem reasonable to administer 90 µg/kg every 2–3 h for the first 48 h and then continue therapy at 4–6-hourly intervals for approximately 2–3 weeks.

Conclusion

Surgery should be undertaken with caution in patients with high-responding inhibitors, and elective procedures require strong justification as no product can guarantee haemostasis. If the inhibitor titres are low, human or porcine concentrates may be effective. However, bypassing agents such as recombinant FVIIa and Feiba are the treatment of choice for the majority of patients. Adjunctive therapies such as antifibrinolytics or tropical fibrin may provide additional benefit, although their role remains unproven. Finally, the restriction of surgery to centres with coagulation expertise is likely to minimize the risk of major morbidity and mortality.

References

1 Wilde JT, Lee CA, Collins P *et al*. Increased bleeding associated with protease inhibitor therapy in HIV- positive patients with bleeding disorders. *Br J Haematol* 1999; **107**: 556–9.
2 DiMichele DM, Kroner BL. Analysis of the North American Immune Tolerance Registry (NAITR) 1993–97: current practice implications. ISTH Factor VIII/IX Subcommittee Members. *Vox Sang* 1999; **77** (Suppl. 1): 31–2.
3 Nilsson IM, Berntorp E, Zettervall O. Induction of immune tolerance in patients with hemophilia and antibodies to factor VIII by combined treatment with intravenous IgG, cyclophosphamide, and factor VIII. *N Engl J Med* 1988; **318**: 947–50.
4 Brackmann HH, Oldenburg J, Schwaab R. Immune tolerance for the treatment of factor VIII inhibitors—twenty years' 'bonn protocol'. *Vox Sang* 1996; **70** (Suppl. 1): 30–5.
5 Negrier C, Goudemand J, Sultan Y *et al*. Multicenter retrospective study on the utilization of FEIBA in France in patients with factor VIII and factor IX inhibitors. French FEIBA Study Group. Factor Eight Bypassing Activity. *Thromb Haemost* 1997; **77**: 1113–19.
6 Lozier JN, Santagostino E, Kasper CK, Teitel JM, Hay CR. Use of porcine factor VIII for surgical procedures in hemophilia A patients with inhibitors. *Semin Hematol* 1993; **30**: 10–21.
7 Lusher JM, Shapiro SS, Palascak JE *et al*. Efficacy of prothrombin-complex concentrates in hemophiliacs with antibodies to factor VIII: a multicenter therapeutic trial. *N Engl J Med* 1980; **303**: 421–5.
8 Sjamsoedin LJ, Heijnen L, Mauser-Bunschoten EP *et al*. The effect of activated prothrombin-complex concentrate (FEIBA) on joint and muscle bleeding in patients with hemophilia A and antibodies to factor VIII. A double-blind clinical trial. *N Engl J Med* 1981; **305**: 717–21.
9 Hilgartner MW, Knatterud GL. The use of factor eight inhibitor bypassing activity (FEIBA immuno) product for treatment of bleeding episodes in hemophiliacs with inhibitors. *Blood* 1983; **61**: 36–40.
10 Hanna WT, Madigan RR, Miles MA, Lange RD. Activated factor IX complex in treatment of surgical cases of hemophilia A with inhibitors. *Thromb Haemost* 1981; **46**: 638–41.
11 White GC. Seventeen years' experience with Autoplex/Autoplex T. evaluation of inpatients with severe haemophilia A and factor VIII inhibitors at a major haemophilia centre. *Haemophilia* 2000; **6**: 508–12.
12 Hilgartner M, Aledort L, Andes A, Gill J. Efficacy and safety of vapor-heated anti-inhibitor coagulant complex in hemophilia patients. FEIBA Study Group. *Transfusion* 1990; **30**: 626–30.
13 Green D. Complications associated with the treatment of haemophiliacs with inhibitors. *Haemophilia* 1999; **5** (Suppl. 3): 11–17.
14 Hedner U, Glazer S, Pingel K *et al*. Successful use of recombinant factor VIIa in patient with severe haemophilia A during synovectomy. *Lancet* 1988; **2**: 1193.
15 Ingerslev J, Freidman D, Gastineau D *et al*. Major surgery in haemophilic patients with inhibitors using recombinant factor VIIa. *Haemostasis* 1996; **26** (Suppl. 1): 118–23.
16 Lusher J, Ingerslev J, Roberts H, Hedner U. Clinical experience with recombinant factor VIIa. *Blood Coagul Fibrinolysis* 1998; **9**: 119–28.
17 Scharrer I. Recombinant factor VIIa for patients with inhibitors to factor VIII or IX or factor VII deficiency. *Haemophilia* 1999; **5**: 253–9.
18 Shapiro AD, Gilchrist GS, Hoots WK, Cooper HA, Gastineau DA. Prospective, randomised trial of two doses of rFVIIa (NovoSeven) in haemophilia patients with inhibitors undergoing surgery. *Thromb Haemost* 1998; **80**: 773–8.
19 Smith OP, Hann IM. rVIIa therapy to secure haemostasis during central line insertion in children with high-responding FVIII inhibitors. *Br J Haematol* 1996; **92**: 1002–4.
20 O'Connell N, McMahon C, Smith J *et al*. Recombinant factor VIIa in the management of surgery and acute bleeding episodes in children with haemophilia and high responding inhibitors. *Br J Haematol* 2002, in press.
21 Smith MP, Ludlam CA, Collins PW *et al*. Elective surgery on factor VIII inhibitor patients using continuous infusion of recombinant activated factor VII. plasma factor VII activity of 10 IU/ml is associated with an increased incidence of bleeding. *Thromb Haemost* 2001; **86**: 949–53.
22 Ludlam CA. *Thromb Haemost* 2002, in press.
23 Santagostino E, Morfini M, Rocino A *et al*. Relationship between factor VII activity and clinical efficacy of recombinant factor VIIa given by continuous infusion to patients with factor VIII inhibitors. *Thromb Haemost* 2001; **86**: 954–8.
24 Schulman S. Safety, efficacy and lessons from continuous infusion with rFVIIa. rFVIIa-CI Group. *Haemophilia* 1998; **4**: 564–7.

CHAPTER 29

Dental extraction in patients with haemophilia and inhibitors

E.A. Rey, S.A. Puia and W. Castillo

Clotting inhibitors represent a significant complication in the management of patients with haemophilia. Treating bleeding disorders in the presence of an inhibitor as well as neutralizing the inhibitor are real challenges in most cases. The inhibitors are antibodies to factors VIII and IX. They develop in patients with haemophilia A or B after the transfusion of blood products which contain the abovementioned clotting factors, and they eventually neutralize the therapeutic transfusion of factors VIII and IX. Between 10 and 30% of patients with severe haemophilia A, and 2–5% of patients with severe haemophilia B or mild–moderate haemophilia A develop an inhibitor against factors VIII or IX after treatment with either plasma-derived or recombinant products.

The importance of early detection of inhibitors prior to a surgical procedure is obvious. It is well known that haemophilia patients presenting with inhibitors are challenging to manage when they require dental extraction. The surgical technique is no different from a conventional one, but haemophilia is a very severe pathology within the blood disorders. Historically, in order to obtain control of the bleeding, in patients with inhibitors it is necessary to provide coadjuvant substitute therapy.

Surgical strategy

Surgical strategy and surgical tactical planning should be followed carefully. Every single surgical detail should be considered, from the preoperative periodontal preparation of the tooth to be extracted to the final discharge of the patient. The experience and confidence of the surgeon is often paramount because the postoperative course is unpredictable. Some patients with clotting inhibitors present with bleeding which may start several days after the surgical procedure. Before the availability of activated recombinant factor VII (rFVIIa), surgical procedures in patients with inhibitors were rarely performed because of the difficulty in obtaining effective haemostasis so that, in most cases, only emergency procedures were carried out.

In the patients treated by us there were not only extractions, but also a case of a pseudotumour in the lower jaw. This chapter presents an evaluation of patients with haemophilia and inhibitors who underwent teeth extractions. Our experience strongly suggests the importance of local haemostatic measures as a fundamental and basic means to be combined with the coadjuvant haematological treatment. To treat this kind of patient, the relationship with the haematologist is very important, because the haematologist should lead the general treatment, as he or she knows how to judge the best moment for the intervention.

The administration of local anaesthesia to patients with haemophilia has been fraught with controversy. Some authors reported a case of fatal subcutaneous haemorrhage after two inferior alveolar blocks in a haemophilic patient's airway. It has been recommended that conservative procedures could be carried out without local anaesthesia and it has also been suggested that deep nerve block injections be avoided and that only infiltration anaesthesia be used.

Some authors have suggested infiltration or pericemental injections of anaesthetic agent for dental extractions after adequate factor VIII replacement therapy is given. Some have also recommended infiltration anaesthesia for extraction of teeth. Others prefer to avoid inferior alveolar nerve blocks, even if replacement therapy has been given. Many authors do not think that haemophilia is a contraindication to deep nerve block injections, but if they are used, they should be performed with great caution, using an aspirating syringe, and only after adequate factor VIII replacement therapy has been administered. If a bloody aspirate is detected during a nerve block injection, additional factor VIII should be administered for 24–36 h postoperatively, or longer if necessary, at a level sufficient for surgical haemostasis [1].

The most recent literature tends to indicate that, to provide profound local anaesthesia for oral surgical procedures, deep nerve block injections can be safely used for patients with haemophilia if certain safeguards are observed, such as maintenance of adequate factor VIII replacement, caution during injection, and the use of an aspirating syringe [2].

Surgical technique

Regarding surgical technique, it is imperative to carefully design incisions and flaps, avoiding any type of injury which may translate into postoperative haemorrhage. The surgical assistant should be advised on the need to be extremely gentle with tissues. Sometimes the injury caused by aggressive retraction leaves worst sequelae than the surgical procedure itself.

When osteotomies are required, the surgeon should take into consideration the advantage of keeping a cavity with bony walls, which may eventually contain haemostatic material. The wound toilette is no less important than the previous operative considerations. Leaving behind granulation tissue in the extraction socket may become a bleeding source. To conclude, the suture constitutes the most significant final step of the surgical procedure, adding to proper haemostasis. The data suggest that a standard local therapy with fibrin glue and tranexamic acid can prevent bleeding complications during oral surgery in most patients [3–6].

The fibrin glue is prepared by mixing in two syringes: one containing thrombin and calcium chloride and the second containing fibrinogen, factor XIII and aprotinin. Thrombin converts fibrinogen to unstable fibrin clot, factor XIII stabilizes the clot and aprotinin prevents its degradation. Usually, 0.5 mL of fibrin glue is enough to apply in the wall of the socket. Gauze saturated with antifibrinolytic agent is kept in place for 120 min and an ice pack is placed on the cheek for 60 min [7].

In all patients with blood disorders undergoing dental treatment, ε-aminocaproic acid or tranexamic acid should be prescribed. Tranexamic acid binds to lysine-binding sites on plasminogen and plasmin. This mechanism blocks the binding of plasmin to fibrin, thus acting as a potent inhibitor of fibrinolysis. The adult dose is 500 mg four times a day, to be taken for 10 days postoperatively; for children, this dose must be modified for weight and age [2].

Mouthwash (e.g. 10 mL 5% tranexamic acid rinse four times daily for 7–10 days) can also be used for dental bleeding. The greater clinical efficacy of tranexamic acid mouthwash vs. placebo has been previously documented by Sindet-Peddersen et al. [7–10] who suggested that the concentration of tranexamic acid in the saliva after mouth rinsing four times a day is sufficient to reduce the incidence of postoperative bleeding complications, in accordance with previously reported results of treatment of patients with haemophilia.

Protective celluloid splint is used as an adjuvant therapy after dental extraction. The patient is instructed to consume only liquids on the days after surgery and to contact the department if postoperative bleeding occurs that cannot be controlled by compression with gauze for 20 min. Paracetamol is used as a postoperative analgesic; aspirin and other non-steroidal anti-inflammatory drugs are not used because they interfere with the coagulative cascade [11].

The authors of this chapter have treated 12 patients with haemophilia and inhibitors who received a tooth extraction. The age range was between 6 and 50 years and the indication for extraction was several dental caries. Patients had been prepared by their haematologists, using high-dose factor VIII, prothrombin complex concentrate or rFVIIa [12]. The results, in general, were very good. However, four patients presented with light bleeds 5 days later, and were treated by local therapy with trichloroacetic acid 30%.

Conclusions

It is possible to treat patients with low titre inhibitors under adequate local and systemic conditions, without the major complications that patients without inhibitors could cause, although extreme care must be taken. Dental decay prevention programmes are now possible, leading to less dental extractions being necessary in the future.

References

1 Larson CE, Chang JL, Bleyaert A. Anesthetic considerations for the oral surgery patient with hemophilia. *J Oral Surg* 1980; **38**: 516–19.
2 Harrington B. Primary dental care of patients with haemophilia. *Haemophilia* 2000; **6** (Suppl. 1): 7–12.
3 Federici AB, Sacco R, Carpenedo E et al. Optimising local therapy during oral surgery in patients with von Willebrand disease: effective results from a retrospective analysis of 63 cases. *Haemophilia* 2000; **6**: 71–7.
4 Tock B, Drohan W, Hess J et al. Haemophilia and advanced fibrin sealant technologies. *Haemophilia* 1998; **4**: 449–55.
5 Martinowitz U, Varon D, Heim M. The role of fibrin tissue adhesives in surgery of haemophilia patients. *Haemophilia* 1998; **4**: 443–8.
6 Suwannuraks M, Chuansumrit A, Sriudomporn N. The use of fibrin glue as an operative sealant in dental extraction in bleeding disorder patients. *Haemophilia* 1999; **5**: 106–8.
7 Sindet-Pedersen S, Ramstrom G, Bernivel S et al. Hemostatic effect of tranexamic acid mouthwash in anticoagulant treated patient undergoing oral surgery. *N Engl J Med* 1989; **320**: 840–3.
8 Sindet-Pedersen S. Distribution of tranexamic acid and plasma and saliva after oral administration and mouth rinsing: a pharmacokinetic study. *J Clin Pharmacol* 1987; **27**: 1005–8.
9 Sindet-Pedersen S, Stenbjerg S. Effect of local antifibrinolytic treatment with tranexamic acid in hemophiliacs undergoing oral surgery. *J Oral Maxillofac Surg* 1986; **44**: 703–7.
10 Sindet-Pedersen S, Stenbjerg S, Ingerslev J. Control of gingival hemorrhage in hemophilic patients by inhibition of fibrinolysis with tranexamic acid. *J Periodontal Res* 1988; **23**: 72–4.
11 Zanon E, Martinelli F, Bacci C et al. Proposal of a standard approach to dental extraction in haemophilia patients: a case–control study with good results. *Haemophilia* 2000; **6**: 533–6.
12 Rodriguez-Merchan EC, de la Corte H. Orthopedic surgery in haemophilic patients with inhibitors: a review of the literature. In: Rodriguez-Merchan EC, Goddard NJ, Lee CA, eds. *Musculoskeletal Aspects of Haemophilia*. Oxford: Blackwell Science, 2000: 136–42.

PART 8
Psychosocial issues

CHAPTER 30

Psychosocial impact of inhibitors on haemophilia patients' quality of life

E. Remor, P. Arranz and R. Miller

Significant increases in survival have been reported for a wide range of chronic diseases of childhood. These improvements have generally been achieved through the use of increasingly aggressive treatment protocols. This raises questions in healthcare settings about the quality of life in relation to the quantity of survival where there is no cure for children with chronic disease [1]. Frequent lengthy hospitalizations, painful treatments and lack of certainty about the future may all impair the quality of life of the child and family, and continue into adult life. During the 1990s there has been an upsurge in developing tools to assess health-related quality of life in children [2,3] and adults [4–7] with chronic disease.

Haemophilia, like other chronic diseases, has experienced advances in treatment provision, including blood products, comprehensive care from a team of specialists (physicians, nurses, psychologists, social workers, physiotherapists, dentists) and corrective orthopaedic surgery [8,9]. All these have contributed to the prognosis and quality of life for patients with haemophilia and their families. Recently, research addressing quality of life issues for those with haemophilia has increased, with an emphasis on the importance of evidence-based practice in health care [5–7,9–13].

However, limited information exists in the literature regarding the impact of inhibitors on the quality of life of people with haemophilia and other related bleeding disorders. One study from Australia by Brewin *et al.* [10] assessed the quality of life of six children with inhibitors who underwent transition from ineffective treatment for inhibitors to treatment with on-demand NovoSeven™. These authors hypothesized that lack of effective treatment for children with long-standing inhibitors results in crippling disease with poor quality of life, poor self-esteem and a sense of helplessness. The results were obtained from quality of life (QoL) questionnaires (Child Health Questionnaire—child and parent form, EuroQoL questionnaire and an additional questionnaire specific to life issues relevant to haemophilia). Major improvements noted were a reduction in time between bleeds and treatment, duration of pain, days when immobile, and hospital visits. Improvements in participation in everyday activities, future expectations and self-esteem were reported for the group on NovoSeven™.

The authors concluded that treatment improvements make it possible for these children to approach adulthood with similar expectations as children without inhibitors [10].

Children with inhibitors are at greatly increased risk of developing damaged joints which can deteriorate in adulthood. Damaged painful joints lead to immobility which has a significant effect on daily living activities, family, social and work life. The ability to offer orthopaedic surgery to alleviate pain and immobility has contributed considerably to the quality of life of many patients [14].

The importance of the quality of life for those with haemophilia, especially for those with complications such as inhibitors, hepatitis C and HIV, is now a target issue and intensive research is needed to address this aspect effectively.

In this chapter the authors study some of the psychological, social, emotional and physical issues facing those with inhibitors. We use a small descriptive study of children with inhibitors carried out in 2001 at the La Paz University Hospital, Madrid, to highlight the serious complications of inhibitors and to consider how these affect personal life, health care and the future outlook of these children when they become adults. Compared to patients on treatment regimen, those with inhibitors cannot benefit as readily from remedial surgery, have more damaged joints, have to be more cautious about trauma and face bigger risks if invasive procedures are required.

We conclude by suggesting some guidelines of how support might be offered to children and young adults with inhibitors regarding their perceived control or improvement of their well-being and quality of life. Dealing with these issues at a young age might help to improve the chances of managing better in later life.

The children's inhibitor study

Aim of study

The main objective of this study, carried out at the La Paz University Hospital Haemophilia Centre from March to May 2001, was to verify differences in the style and quality of life for children with haemophilia, comparing those with and without inhibitors.

Method

The subjects of the study were eight children with severe haemophilia A: four with and four without inhibitors. The children were grouped into two age groups of 8–12 years and 13–16 years. All the children live in the city, are treated at the same Comprehensive Haemophilia Care Centre, attend a state school, and have similar socioeconomic status.

Measures

The patients' QoL was assessed by questions selected from the pool of items of the Haemo-QoL questionnaire [13], using a likert-scale response format throughout, including 'never' to 'always'. The description of the measure, details of the number of questions in each domain and reliability data of the sample are shown in Table 30.1.

Results

The results were classified under the domains of psychosocial (self-esteem, family attitude, relationship with friends), haemophilia-related and physical status (bleeds status, physical health, health distress, perceived health) and school-related (school and sport activities). A quality of life global indicator was calculated, scoring across the nine domains described in Table 30.1. Of the nine specific domains, differences were identified in all dimensions between the children with inhibitor from those who did not have this condition. Age groups were take into account, and the same pattern of differences was observed in both groups. Results are shown in Table 30.2.

Table 30.2 shows that children with an inhibitor presented less self-esteem, fewer relationships with friends, experienced more difficulties in school activities, reported a greater number and frequency of bleeds associated with pain, worse physical health status, less participation in sport activities and worse perceptions of health. Family attitude (communication and

Table 30.1 Measures and reliability.

Measures	Number of items	Reliability (α cronbach)
Self-esteem Feelings of pride and being pleased with oneself, feeling on top of the world, with good ideas and able to cope with haemophilia	5	0.82
Family attitudes Feelings of getting on well with parents, feeling fine at home, quarrelling at home, feeling of being restricted by parents, of being treated like a baby because of the haemophilia	5	0.82
Relationships with friends Feelings of 'success' with friends, getting along well with friends, of being different from others, avoiding others fearing haemophilia will be noticed	5	0.50
School activity Doing school work, finding school interesting, worrying about getting bad marks or grades, missing school days because of haemophilia, difficulty in paying attention to school activities because of pain	5	0.51
Bleeds status Frequency of bleeds last month, troubled by bleeds, severity of bleeds, lying in bed, strange sensations in joints before bleed	6	0.74
Physical health Presence of bruises, pain in joints, stiff joints, difficulties in mobility of arms, legs or feet, afraid of bleeds	9	0.89
Health distress Bad mood, sadness, anger, worried and lonely because of haemophilia	6	0.68
Sports activity Had to refrain from sports, indoor activities, doing as much sports as other kids, feeling sports are dangerous	11	0.73
Perceived health In general, how is your health?	1	–

Table 30.2 Impact of inhibitor on quality of life of children with haemophilia.

QoL domain	Age group 8–12 years		Age group 13–16 years		Age group 8–16 years	
	Without inhibitor (mean)	With inhibitor (mean)	Without inhibitor (mean)	With inhibitor (mean)	Without inhibitor (mean)	With inhibitor (mean)
Self-esteem[a]	16	12	19.5	16.5	17.7	14.2
Family attitude[b]	2.5	0	3	4.5	2.7	2.2
Relationship with friends[c]	19.5	16	17	16	18.2	16
School activity[d]	18.5	16	14	10.5	16.2	13.2
Bleeds status[e]	10	7	6.5	10.5	8.2	8.7
Physical health[f]	14	9.5	3	11	7	10.7
Health distress[g]	2	0	2	3	2	1.5
Sports activity[h]	7.5	23	6	17	6.7	20
Perceived health[i]	0.5	2.5	0	2	0.2	2.2
Overall HR-QoL[j]	54.5	70	42	77	48.2	73.5

Notes: (a) high score, high self-esteem; (b) high score, difficulties in family relationship; (c) high score, good relationship with friends; (d) high score, normal scholar activity; (e) high score, high number of bleeds and troubles related to bleeds; (f) high score, worst physical health; (g) high score, high health distress; (h) high score, less participation in sports activities; (i) high score, perception of self as an ill person; (j) high score, worst quality of life.

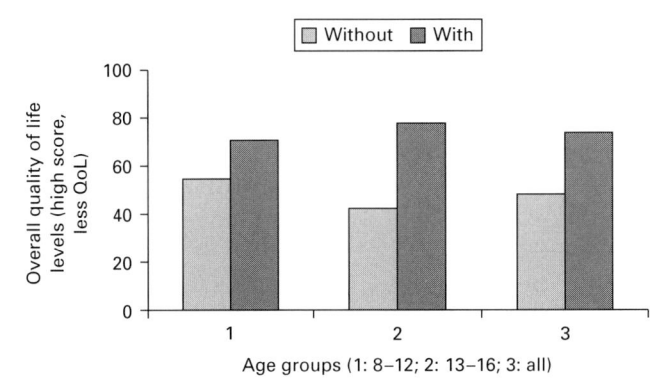

Fig. 30.1 Comparison between children with and without inhibitors of health-related quality of life.

Discussion: comparison between children with and without inhibitors

The presence of inhibitors (or lack of effective treatment) in children with haemophilia results in crippling disease with poor quality of life. Moreover, many other life dimensions may be affected for those with this complication, such as self-esteem, relationships with friends, school activities, physical health (including frequency of bleeds), sports activities and perceived health. Brewin et al. [10] found that even for patients with inhibitors a personalized treatment approach can help achieve an improved quality of life. Currently, there is great interest in the problem of inhibitor development and much research is focusing on this problem. Quality of life measures need to be more routinely included in the research that is being carried out in this field and in the evaluation of the global status and alternative treatments. Only from cooperative approaches will we be able to understand the total burden of the disease and its treatment experienced by those affected. The value of quality of life work remains in the balance until research finds appropriate measures to capture the difficult psychosocial dimensions and evaluate them.

overprotective attitude) and health distress (sadness, anger and worry because of haemophilia) was not less in children with inhibitors compared to children without.

Finally, to represent graphically differences between children with and without inhibitors, a global category called overall health-related quality of life was calculated including all quality of life domains. These differences are presented in Fig. 30.1, and means are detailed in Table 30.2. Results show a worse overall quality of life in children with inhibitors.

In summary, it is now known that inhibitors are likely to decrease the well-being and quality of life of those affected by this condition. However, if personalized interdisciplinary and comprehensive care is carried out, some of these consequences can be reduced and in some instances overcome. Current medical treatment of inhibitors is extremely costly and thus not available in all haemophilia care services. On the other hand, psychosocial help is less costly and, appropriately used, can help patients deal with their situation more effectively.

Guidelines for psychological management of haemophilia patients with inhibitors

From the limited research described here it is evident that the development of inhibitors is a serious complication of haemophilia. It adds new challenges to treatment and is often a source of frustration and suffering to the patient, his or her family and the care team. Individuals who develop inhibitors have to learn to cope with this additional problem [15–18] which is not always easy, but some treatment options are available [19,20]. Psychosocial support may help to manage this challenge as effectively as possible in each individual case.

In this chapter some suggestions are made about how this can be achieved. The major component of the delivery of medical care is good communication in the context of a supportive relationship; this cannot be improvised, but it can be learned. The central principle of effective therapeutic dialogue is that the patient and parents should feel that their emotions have been understood and acknowledged by the professional, which reduces distress.

Two important steps are first to give information about inhibitors to patients and their families and, secondly, to stimulate their personal resources, adaptive coping and a positive view of themselves. This strategy can help people to diminish uncertainty, maximize their perceived control and minimize associated emotional distress [21].

Step 1: giving and eliciting information

How information is given influences the resulting understanding. It is usually more effective first to ask people about their knowledge, beliefs and concerns before giving information. In this way misconceptions and misunderstandings are revealed, people's beliefs are identified and their concerns are brought into the open. Gaps in information can be filled in and this is done at the patient's pace rather than overwhelming them with information that may not be absorbed [21,22].

Information about inhibitors can address the frequency and nature of inhibitors, who may develop an inhibitor and why, and information about the treatment of bleeding in the presence of inhibitors [23].

Information about the frequency and nature of inhibitors

The formation of inhibitors cannot be prevented, but the risk of developing an inhibitor does not remain the same during the lifetime of a person with haemophilia. The literature indicates that most inhibitors are known to develop during childhood. One study shows that the age at first infusion seems to be an important risk factor, which is inversely correlated with inhibitor incidence [24]. Other studies conducted on persons with haemophilia A who received only recombinant factor VIII, show increased inhibitor development compared to those who previously received plasma-derived products. Similarly, inhibitor development after fewer treatments with recombinant factor VIII than with plasma-derived products is observed [25]. There is less information on the nature of factor IX inhibitors because of the low prevalence of inhibitors in the haemophilia B populations [19].

Information about who may develop an inhibitor and why

Inhibitor development occurs more frequently in individuals with certain inherited conditions. There is less frequent inhibitor development in haemophilia B than in haemophilia A. The incidence of inhibitors is highest among those with severe or moderately severe haemophilia and is less common among persons with mild haemophilia [19,25].

Information about the treatment of bleeding in the presence of inhibitors

The treatment of bleeding in those with haemophilia and inhibitors can be a challenging experience for both patient, family and physician [21]. As conventional factor VIII replacement is ineffective, inhibitors to treatment significantly complicates the management of these patients.

Although there are several therapeutic options for bleeding in patients with haemophilia and inhibitors, there are none as yet that guarantee the same effective outcome as specific factor VIII or IX treatment. Consequently, those affected by inhibitors frequently suffer from many more infections (e.g. Port-A-Caths™), orthopaedic problems, and life-threatening complications from haemophilia, and subsequently have greater disability in their day-to-day lives than those who do not develop an inhibitor. In addition, they may not be able to benefit from future advances in gene therapy aimed at better treatment and even a cure for the condition. For these reasons, the eradication of the inhibitor is the best option for most of these individuals [19,26,27]. This is done by a process called immune tolerance, consisting of regular infusions of factors VIII or IX administered for a period of weeks or years, with or without drugs that dampen the immune system's reaction to the factor.

Providing information about the options of palliative treatments, future therapeutic options and their consequences, over the short or long term, may be one way to help patients with inhibitors to cope with their problems. This information addresses the certainties and uncertainties in a way that balances hope with reality.

Step 2: empowerment of personal resources

The second step is to help people deal with the reality of unforeseen haemorrhages, the difficulties in controlling the bleeding process and its consequences, in a way that least disrupts personal, social and emotional life. This is a major challenge to the creativity and resources of patients, families and health carers.

The patient with an inhibitor is in a situation of greater physical and psychological vulnerability. The consequent social implications include threats to their self-perception and self-confidence, endurance of pain, fears for present and future health and well-being, and absenteeism from school and work.

Feelings of helplessness are present to a greater or lesser extent [14,22,28–30].

These risk factors have to be considered in dealing with both children and adults in order to achieve optimal comprehensive management and care. How can we decrease those risk factors to enhance well-being for these patients? The problem is not the difficulties themselves but the way in which we cope with them. In other words, patients with inhibitors depend on their own and their families' skills and resources to cope with stressful situations which arise largely from difficulties in dealing with uncertainty [31].

An approach to psychosocial issues

The following points are suggested as a way to approach the psychosocial issues related to inhibitors.

1 To identify difficulties recognizing that they pose problems for individuals and families: clarify the difficulties, reappraise the situation, and adopt a realistic point of view, as well as introducing a more optimistic perspective which may make the current situation more tolerable; for example, that there is on-going research and treatments will evolve in time.

2 To decrease the risk factors (physical and psychological): the target is to identify and manage all those circumstances and conditions that increase susceptibility and reduce the well-being of the person with haemophilia.

3 To increase protection factors: the target is to enhance all the conditions that will help the child or teenager with haemophilia to be stronger and more resistant when coping on a daily basis with the risk linked to his or her condition. The protection factors that can facilitate stress management associated with inhibitors include:

- facilitate parents' acceptance of haemophilia and its consequences;
- information that highlights the importance of early treatment;
- avoid blaming related to accidents;
- direct and open communication between parents and children—the importance of listening to the child;
- allow children to explore their own limits, to promote results from their own actions, autonomy and self-esteem;
- train children in techniques that favour self-regulation: identifying and changing irrational thoughts, relaxation techniques, promotion of self-care habits, development of activities that can enhance self-esteem;
- promote good communication with physicians and other professionals;
- enhance autonomy;
- increase techniques to reduce frustration;
- facilitate a relationship of trust and openness among the family members; and
- avoid the tendency to social isolation and guilty feelings.

The overall aim is to normalize the routine daily implications that this clinical condition requires, so that patients are not overwhelmed by the distressing aspects and can be helped to balance hope with reality.

Conclusions

Advances in treatment for those with haemophilia have reduced the frequency of hospitalizations and the accompanying disruption of normal development and complications of haemophilia (e.g. joint damage, social or psychological problems) and of its treatments (e.g. HIV or hepatitis infections). Despite these notable advances, those with the condition battle throughout their lives against complications of both the disease and its treatment. Inhibitor development is one of the most serious of these complications, which can decrease well-being and quality of life and is associated with a high level of psychological stress. A personalized, multidimensional and interdisciplinary approach in a supportive relationship, in which patients perceive acknowledgement of their concerns, can minimize inhibitor-associated emotional stress. To help children with problem-solving, coping with stress, self-care and decision-making allows the capacity to adapt to changes and to recover from life's stings without resentment. Paying attention to this must be a central issue in the Comprehensive Health Care Centre.

It is the combination of medical treatment and appropriate psychosocial support that is needed to address the complex problems of inhibitors. Delivery of effective medical care is not possible without effective communication. The best way to motivate people with inhibitors to adhere to complex treatment and to implement behaviour changes in a way that will enhance their quality of life is for the clinician to demonstrate effective listening by developing open and clear communication in the context of a trusting mutual relationship between the team, the child and both parents.

Acknowledgements

The authors wish to thank the psychologist José Luis Ramos for the collection of data and Drs A. Villar and M. Morado for their suggestions on the first version of this manuscript.

References

1 Eiser C. Children's quality of life measures. *Arch Dis Child* 1997; **77**: 350–4.
2 Eiser C, Morse R. A review of measure of quality of life for children with chronic illness. *Arch Dis Child* 2001; **84**: 205–11.
3 Varni JW, Seid M, Rode CA. The PedsQL: measurement model for the pediatric quality of life inventory. *Med Care* 1999; **37**: 126–39.
4 Herdman M, Fox-Rushby J, Badia X. 'Equivalence' and the translation and adaptation of health-related quality of life questionnaires. *Quality Life Res* 1997; **6**: 237–47.

5 Miners AH, Sabin CA, Tolley KH *et al*. Assessing health-related quality-of-life in patients with severe haemophilia A and B. *Psychol Health Med* 1999; **4**: 5–15.
6 Solovieva S. Clinical severity of disease, functional disability and health-related quality of life: three-year follow-up study of 150 Finnish patients with coagulation disorders. *Haemophilia* 2001; **7**: 53–63.
7 Aznar JA, Magallón M, Querol F, Gorina E, Tusell J. The orthopaedic status of severe haemophiliacs in Spain. *Haemophilia* 2000; **6**: 170–6.
8 Miller K, Buchanan GR, Zappa S *et al*. Implantable venous access devices in children with hemophilia: a report of low infection rates. *J Pediatr* 1998; **132**: 934–8.
9 Rodriguez-Merchan EC, de la Corte H. Orthopaedic surgery in haemophiliac patients with inhibitors: a review of the literature. In: Rodriguez-Merchan EC, Goddard NJ, Lee CA, eds. *Musculoskeletal Aspects of Haemophilia*. Oxford: Blackwell Science, 2000: 136–42.
10 Brewin T, Ekert H, Davey P. Recombinant VIIa (NovoSeven) treatment of six children with long-standing inhibitors improves quality of life. *Haemophilia* 2000; **6**: 414.
11 Khair K. Quality of life in children with haemophilia. *Haemophilia* 2000; **6**: 419.
12 Sek J, Saleh M, Furlong W *et al*. Health-related quality of life in people living with haemophilia or von Willebrand disease in a geographical population. *Haemophilia* 2000; **6**: 426–7.
13 von Makenzie S, Bullinger M, Ravens-Sieberer U, Negrini C. Health-related quality of life in children and adolescents with haemophilia: development of the new Haemo-QoL questionnaire. *Haemophilia* 2000; **6**: 427.
14 Miller R, Beeton K, Miners A, Padkin J, Goddard N. Quality of life issues: patients with haemophilia and joint replacements. Abstracts of the World Federation of Haemophilia 5th Musculoskeletal Congress, Sydney, 14–16 April 1999.
15 Rivard GE. Inhibitors: a complication of hemophilia. *Hemophilia Today* 2000; **Fall**: 8.
16 Santavirta N, Bjorvell H, Koivumaki E *et al*. The factor structure of coping strategies in hemophilia. *J Psychosom Res* 1996; **40**: 617–24.
17 Miller R, Telfer P. HCV counselling in haemophilia care. *Haemophilia* 1996; **2**: 1–4.
18 Bor R, Miller R, Latz M, Salt H. *Counselling in Health Care Settings*. London: Cassel, 1998.
19 Di Michele DM. Inhibitors in haemophilia: a primer. *Haemophilia* 2000; **6** (Suppl. 1): 38–40.
20 Inhibitors Subcommittee of the Association of Hemophilia Clinic Directors of Canada. Suggestions for the management of factor VIII inhibitors. *Haemophilia* 2000; **6** (Suppl. 1): 52–9.
21 Arranz P, Costa M, Bayés R *et al*. *Emotional Support in Hemophilia*. Montreal: WFH, 2000.
22 Miller R. Some discussion essential for musculoskeletal care and prior to joint replacement in patients with haemophilia. In: Rodriguez-Merchan EC, Goddard NJ, Lee CA, eds. *Musculoskeletal Aspects of Haemophilia*. Oxford: Blackwell Science, 2000: 218–26.
23 Miller R, Beeton K, Goldman E, Ribbans WJ. Counselling guidelines for managing musculoskeletal problems in haemophilia in the 1990s. *Haemophilia* 1997; **3**: 9–13.
24 Lorenzo JI, López A, Altisent C, Aznar JA. Incidence of factor VIII inhibitors in severe haemophilia A: the importance of age at start of therapy. *Haemophilia* 2000; **6**: 293–4.
25 Carcao MD, Stain AM, Sparling C, Blanchette VS. High-titre inhibitors to factor VIII occurring after the first exposure to recombinant factor VIII in mild haemophilia A. *Haemophilia* 2000; **6**: 291–2.
26 Buchanan GR. Factor concentrate prophylaxis for neonates with hemophilia. *J Pediatr Hematol Oncol* 1999; **21**: 254–6.
27 Gruppo RA. Prophylaxis for hemophilia: state of the art or state of confusion? *J Pediatr* 1998; **132**: 915–17.
28 Haemophilia Foundation Australia. *The Pain Management Book for People with Haemophilia and Related Bleeding Disorders*. Treatment of Hemophilia Monograph Series. No. 22, WFH, Montreal, 2000.
29 Varni JW, Gilbert A, Ditrich SL. Behavioral medicine in pain and analgesia management for the hemophiliac child with factor VIII inhibitor. *Pain* 1981; **11**: 121–6.
30 Kasper CK. Hereditary plasma clotting factor disorders and their management. *Haemophilia* 2000; **6** (Suppl. 1): 13–27.
31 Arranz P, Rodriguez-Merchan EC, Hernández-Navarro F. Burnout syndrome in haemophilia staff. In: Rodriguez-Merchan EC, Goddard NJ, Lee CA, eds. *Muculoskeletal Aspects of Haemophilia*. Oxford: Blackwell Science, 2000: 212–17.

PART 9

General strategy for management of inhibitor patients

CHAPTER 31
General strategy for management of inhibitor patients (summary)
I. Scharrer

The most important complication of treatment of haemophilic patients nowadays is the development of inhibitors that neutralize the effect of factor VIII (FVIII) concentrates and may lead to life-threatening conditions, recurrent bleeding episodes and progressive joint damage. Inhibitors occur in about 20–50% of individuals with severe haemophilia A. The aim of this volume is to elucidate the best management of haemophilic patients with inhibitors, with discussion of the diagnosis, likely contributory factors, especially genetic disposition, clinical treatment and complications. Lee and Rodriguez-Merchan deserve great credit for encouraging leading experts in haemophilia to contribute to this very important 'hot' topic: inhibitors in patients with haemophilia.

J. Lusher (*Natural history of inhibitors in severe haemophilia A and B: incidence and prevalence*; p. 3) points out that patient factors such as severity of haemophilia, underlying gene defects, race and family history of inhibitors play an important role in inhibitor development. The percentage of inhibitors in severely affected previously untreated patients (PUPs) with haemophilia A developing inhibitors over a period of 5 years or >100 ED has been 28–30% in most series (regardless of the product received) with a prevalence at the end of these studies between 11.1 and 19%. She also points out that in various cohorts only 0–3.8% of patients with haemophilia B develop inhibitors. However, in approximately 50% of these patients anaphylaxis or severe allergic manifestations may occur after exposure to any factor IX (FIX)-containing product.

K. Peerlinck and M. Jacquemin (*Characterization of inhibitors in congenital haemophilia*; p. 9) state that the first occurrence of an inhibitor was reported by Lawrence in 1941. Most of the patients have antibodies of the IgG type with a preponderance of IgG4 subclass. The Oxford or Bethesda method may be used for quantification of inhibitors. The Bethesda assay is used most often. An incubation mixture consisting of one part of patient plasma diluted with imidazole buffer and one part of pooled plasma (as a source of FVIII) is left at 37°C for 2 h. Also, a control of one part of normal plasma and one part of buffer is incubated in the same manner. Residual FVIII activity is assayed.

A Bethesda unit (BU) is then defined as the amount of inhibitor which would inactivate half the FVIII in the incubation mixture. For increasing the sensitivity and specificity of the Bethesda assay it is important to substitute FVIII deficient plasma for imidazole buffer in the control incubation mixture and to buffer that substrate plasma to a pH of 7.4, as described by Verbruggen and coworkers. ELISA-based techniques may be used but for patient management the functional FVIII inhibitor methods seem to be more important.

The mechanisms of FVIII inhibition are not yet fully elucidated. Anti-A_2 antibodies reduce the catalytic activity of FX activating complex. Anti-A_3 antibodies prevent FIXa interaction with FVIIIa. Anti-C_2 antibodies prevent FVIII binding to phospholipids and to vWF. Binding sites for FVIII inhibitors have been found in the A_2 and C2 domains and in the A_3–C_1 domains. Anti-FVIII antibodies inactivate FVIII by proteolysis. It is difficult to establish a relationship between epitope specificity and mechanisms of FVIII inactivation, even when using polyclonal antibodies in these investigations. Many different human monoclonal antibodies directed towards FVIII and other pathogenic antibodies have been produced (BO2C11 or LE2E9).

T.T. Yee and C.A. Lee (*Incidence and prevalence of inhibitors and type of blood product in haemophilia A*; p. 14) note that multiple factors contribute to the risk of development of inhibitors to FVIII in haemophilics. Besides genetic factors, FVIII concentrates may occasionally play an important role. FVIII products can become more immunogenic by the preparation procedure or by methods used for viral inactivation. With the introduction of recombinant FVIII concentrates there was also concern that these products could induce more inhibitors. However, in six recombinant PUP studies of 360 severely affected PUPs the incidence was comparable with that of plasma derived products: 22–31%. Switching from one product to another may also increase the inhibitor risk.

Regarding the issue of purity of FVIII products as a further risk for inhibitor formation, some studies with intermediate purity von Willebrand factor-containing FVIII concentrates had a lower rate of inhibitor development. Yee and Lee remind the reader that there are still a number of unresolved issues regarding the type of FVIII concentrate as a potential risk for the development of FVIII antibodies. More studies investigating pharmacoviligance with new FVIII concentrates are needed both in PUPs and in previously treated patients (PTPs).

J. Oldenburg and E. Tuddenham (*Genetic basis of inhibitor development in severe haemophilia A and B*; p. 21) emphasize that several studies have demonstrated that the type of gene mutation is the most decisive risk factor for inhibitor development. Large deletions, nonsense mutations, and in haemophilia A the intron inversions, are associated with a high risk of inhibitor development. These mutations prevent endogenous synthesis of FVIII/IX protein. The major histocompatibility complex (MHC) may be another (but less conclusive) risk factor for inhibitor development.

Data from the HAMSTeRS mutation register and from the Bonn Centre revealed that patients with large deletions are at the highest risk (88%) for inhibitor formation. The mutation type of small deletions/insertions shows a low risk of inhibitor development in haemophilia A. The inhibitor prevalence in haemophilia B is much less, being 3% vs. 20–30% in haemophilia A. One reason for this difference may be the different proportions of null mutations, which are less than 20% in haemophilia B but more than 70% in haemophilia A.

The influence of race on inhibitor or inhibitor formation could be shown in three US studies and one from France. In African-Americans the inhibitor incidence in severe haemophilia patients was doubled. A high incidence of inhibitors was found in siblings (i.e. if a boy developed an inhibitor, his haemophilic brother(s) was more likely to have one as well).

Regarding the management of treatment E. Berntorp describes *Plasmapheresis and protein A immunoadsorption* (p. 29). Immunoadsorption using Citem 10/Immunosorba-system has been performed in Malmö since 1980. The Malmö protocol comprises immunomodulation, the administration of cyclophosphamide and intravenous gammaglobulin. If the inhibitor titre is below 10 BU, immune tolerance treatment (ITT) is started without protein A adsorption. If the inhibitor titre is above 10 BU the patient has extracorporeal protein A adsorption. In Malmö 16 patients with haemophilia A (13 HR >300 BU) have been treated with this protocol a total of 21 times, 62.5% (10) of haemophilia A patients with success. Seven patients with haemophilia B were treated with this protocol and 86% of them became tolerant. The Malmö protocol may be useful in haemophilia A patients with longstanding inhibitors but is seldom used in very young patients with inhibitors of recent onset. However, Berntorp suggests that in haemophilia B, the Malmö protocol may be appropriate as a first-line treatment.

M. von Depka and A. Huth-Kuehne (*Immunoadsorption with anti-immunoglobulin antibodies using Ig-TheraSorb columns*; p. 199) state that the antibody-based adsorption of immunoglobulins (Ig-TheraSorb system) utilizes the principle of affinity chromatography. All subfractions of IgG (1–4) are eliminated to the same extent by this system. Ig-TheraSorb may be used to reduce the titre of FVIII inhibitor like the Malmö protocol.

In Germany, in Hannover and Heidelberg, 14 patients have been treated with this system (12 acquired inhibitors and 2 inhibitors in congenital haemophilia A). All patients had severe life-threatening bleeding episodes. Between 4 and 44 apheresis sessions were performed. Out of 14 patients, 13 are in remission and one is in partial remission. The use of Ig-TheraSorb seems to be an effective, specific and safe tool for the elimination of FVIII inhibitors. However, this must be proven in a multicentre randomized clinical trial.

The management of *venous access in children with inhibitors* is often very difficult, as R.C.R. Ljung points out (p. 36). A peripheral vein is always the first choice in a patient with an inhibitor but in many children a central venous access device is necessary, especially for ITT.

Infection is the most frequent complication when using a central venous line. Various studies have shown that 50–83% of patients with inhibitors and central venous lines may acquire an infection. The most frequent pathogens were *Staphylococcus epidermidis* or *S. aureus*. Other less common complications include thrombosis or leakage. The final decision for using a central venous access has to be a compromise between the medical goal and the risk of complications.

H.H. Brackmann and T. Wallny describe *immune tolerance with high-dose regimen* (p. 45). Brackmann is well qualified to discuss this since the first curative treatment protocol for inhibitor patients—known as the 'Bonn protocol'—was published by Brackmann and Lormsen in 1977. ITT has been performed in a total of 60 haemophilia A patients with inhibitors (36 high responders, 24 low responders). The overall success rate is 87% (52 patients). Nowadays the complete ITT is performed using 150 IU FVIII/kg twice a day. Feiba is only used in those patients with a high bleeding frequency.

Brackmann and Wallny point out that although many successes have been achieved with high-dose ITT, little is known about the underlying immune mechanism. It would be helpful for future immunologically based studies to study the pathomechanisms of ITT and help define parameters to predict success.

E.P. Mauser-Bunschoten describes *immune tolerance with low dose-regimen* (p. 49). Low-dose ITT was developed because it is less demanding for patients and staff as patients are infused only two to three times a week. This regimen was introduced in 1981 in The Netherlands. The dosage used was 25–50 U FVIII/kg bw with some inevitable dose adjustments. A group of 27 patients with severe haemophilia A was treated, and 19 of them had inhibitor titres below 40 BU before the start of treatment.

Until 1997 success was defined as a decrease of inhibitor to <2 BU/mL with FVIII recovery of at least 50% of normal and half-life of 6 h. In evaluation clinical success was chosen as an endpoint. In 23 patients (85%) success was obtained with low dose. The main disadvantage of low-dose ITT may be the longer period of time to achieve success in comparison with the ITT regimens with high daily FVIII doses.

W. Kreuz emphasizes in the chapter *Immune tolerance and choice of concentrates* (p. 55) that the best management of inhibitor patients is rapid ITT. Various regimens such as administration of high doses of FVIII twice daily or lower doses three times weekly have been attempted.

The success of ITT depends on the inhibitor titre at start of ITT, the maximum titre during ITT, the age of the patient, inflammation during ITT and the number of exposure days from detection of the inhibitor until the start of ITT. Efficacy rates range from 62 to 90% in retrospective studies. Up to now no prospective randomized trial has been performed focusing on optimal FVIII dosage and choice of concentrate. However the Bonn and Frankfurt centre has demonstrated that high dosages of FVIII (100–150 U FVIII/kg bw every 12 h) and FVIII concentrates with high vWF content may influence the success rate significantly (100% for low responders and 88% for high responders). In patients with inhibitors directed against the light chain of FVIII, FVIII concentrates with high amounts of vWF are preferable.

D. DiMichele and B. Kroner describe *The North American Immune Tolerance Registry* (p. 59). This registry is a retrospective data-gathering tool of 180 patients who underwent ITT. It was initiated in 1992 as a project of the ISTH (Factor VIII/IX Scientific SubCommittee). The data were collected between March 1993 and December 1999 with respect to therapeutic regimens, therapeutic outcomes, potential predictors of success and complications of therapy.

Haemophilia A patients who completed ITT were treated for a mean of 16.8 months, and haemophilia B patients for a mean of 11.6 months. Successful tolerance was achieved in 70% of patients. The definition of ITT success was a negative inhibitor titre, normal FVIII recovery, normal factor survival and conversion from high to low responder status. Only 25% of the patients were successfully tolerized on the basis of hard endpoints such as negative inhibitor titre plus normal factor recovery and survival. The three most common causes for failure were a late rise in inhibitor titre after an initial decline during ITT, the patient's desire for terminating ITT and central venous access device complications. FVIII product purity had no impact as a predictor of success of ITT in high-responder haemophilia A patients, in contrast to the results of Kreuz.

In a multivariate analysis, the peak titre during ITT and the preinduction titre were the most significant predictors of success. In contrast to the data from the international registry higher doses of FVIII were not associated with a higher rate of success. In haemophilia A patients 6% of ITTs were complicated by 14 adverse reactions, mostly allergic reactions. In contrast in haemophilia B patients adverse events occurred in 65% (11 courses out of 17 ITT courses).

Regarding the management of bleeding episodes, S. Schulman and the rFVIIa CI Group describe the special requirements of *life- and limb-threatening episodes and intracranial bleeds* (p. 78). The treatment for life- or limb-threatening bleeds in patients with inhibitors has evolved from none to activated prothrombin complex concentrates (aPCC) to rFVIIa. rFVIIa provides excellent haemostatic efficacy especially when it is used very early as first-line therapy. Treatment options for bleeding in patients with inhibitors include PCC, aPCC, porcine FVIII and rFVIIa.

In the 1970s, the efficacy of PCC could be demonstrated in a randomized study against an albumin placebo in patients with joint bleeds. The overall effectiveness of aPCC was assessed as about 60%. In the 1980s, the efficacy of porcine FVIII could be demonstrated in a survey of hospitalized patients with 491 haemorrhagic episodes in over 90% of the cases. The effect was reduced in the presence of high antiporcine inhibitor titres. rVIIa was effective in 79% of haemarthrosis and 65% of muscle bleeds, depending on the early onset of treatment.

In general, intracranial haemorrhage has been associated with fatal outcome especially in inhibitor patients. Feiba, porcine FVIII and rFVIIa were used for central nervous system (CNS) bleeds. Comparing these therapeutic possibilities, treatment with rFVIIa appears to be the best option for inhibitor patients with CNS bleeds. Addition of tranexamic acid may be considered.

Regarding life-threatening internal bleeding rVIIa has often been life saving according to published results. All products may be given as bolus injections or by continuous infusion (CI) with the exception of PCC (only bolus injections); however, the dosing and monitoring of rFVIIa, when given by CI has not been well established. While the overall results with rFVIIa appear not to be influenced by method of injection, some patients will need bursts of thrombin generation. Precautions in using the infusion pump also have to be considered. Side-effects of aPCC are thrombosis; porcine FVIII may result in allergy and transient thrombocytopenia, and thrombotic complications have been described following use of rFVIIa.

G. Kenet and U. Martinowitz compared three different treatment regimens (*Standard, high and mega bolus doses of NovoSeven: comparison of three different treatment regimens*; p. 206) in children with inhibitors. They have used an initial bolus dose of 90 µg/kg followed by 15 µg/kg/h with CI for treating bleeding episodes and for prevention of peri- and postoperative bleeding. They achieved high FVII:C steady state levels with an augmented regimen (160–180 µg/kg bolus followed by 30 µg/kg/h). A mega bolus dose of rFVIIa (300 µg/kg) resulted in higher efficacy (75% response) and faster response (30–40 min) in treating bleeding episodes. No thrombotic complications were documented. Based on their results they recommend the use of rVIIa 'mega dose' for bleeding episodes in young patients with inhibitors.

M. Heim and coworkers remind the reader (*Treatment of iliopsoas haematomas and compartment syndromes in patients with haemophilia who have a circulating antibody*; p. 139) that a compartment syndrome in haemophilic patients with inhibitors is an orthopaedic emergency and requires prompt decompression. The diagnosis of a bleed into the iliopsoas or of a compartment syndrome should be made by ultrasound. In case of surgical treatment the patient should be covered by rVIIa for relatively long periods until wound healing.

Regarding the treatment of patients with *Factor IX inhibitors and anaphylaxis* (p. 87), I. Warrier points out that anaphylaxis and the nephrotic syndrome may occur in treatment of haemophilia B inhibitor patients with FIX preparations. This problem is seen primarily in patients with FIX gene deletions or stop codon abnormalities. In view of the risk of nephrotic syndrome in haemophilia B patients with inhibitors, anaphylaxis and ITT, ITT should be undertaken only when there are no other treatment options. The period of ITT should be shortened as much as possible. rFVIIa is probably the treatment of choice in these patients.

The problems of *inhibitors in mild and moderate haemophilia A* (page XX) is addressed by C.R.M. Hay and C.A. Lee. The development of FVIII inhibitors in patients with mild haemophilia may be more frequent than hitherto expected. Hay *et al.* reported the annual incidence of inhibitors in the UK with 3.5 per 1000 patients with severe haemophilia and 0.84 per 1000 patients with mild and moderately severe haemophilia. Inhibitors in the latter patients tend to occur at a median age of 33 years, later in life than those observed in severe haemophilia A.

66% of these patients demonstrate a bleeding pattern similar to that of acquired haemophilia. Both type I and type II kinetic reactions of the inhibitor have been described. The causes for inhibitor development in mild haemophilics are not fully elucidated. Hay found a familial predisposition. Most of the patients with mild or moderate haemophilia A have missense mutations. It has been speculated that the mutation may lead to conformational changes in the circulating FVIII molecule. Uncommon mutation defects have also been reported. Intensive studies in these patients have revealed much about the structural/functional relationship of the FVIII molecule.

Most of these inhibitors develop after an intensive period of replacement therapy, such as for surgical coverage, or continuous infusion for a serious bleeding episode. These patients can sometimes successfully be treated with 1-deamino-8-D-arginine vasopressin (DDAVP), and ITT has also been used. In patients with a risk mutation a combination of rFVIIa, DDAVP and antifibrinolytics may be used for surgery. It is recommended that intensive exposure to FVIII should be avoided in these patients if possible.

The general strategy for management of inhibitor patients should focus on several important issues: identification of patients at highest risk, characterization of the antibodies, laboratory diagnosis of inhibitors, clinical features, immediate treatment of acute bleeding episodes and the reduction and complete and permanent eradication of inhibitors.

CHAPTER 32

Immunoadsorption with anti-immunoglobulin antibodies using Ig-TheraSorb columns

M. von Depka and A. Huth-Kuehne

The coagulation system, together with endothelial cells, ensures and maintains the integrity of the vascular system through a complex system of interactions where pro- and anticoagulatory effects are well balanced. Disorders of this system can result in severe and eventually life-threatening disease. Pathological interactions of the immune system with the coagulation system, e.g. the formation of antibodies against clotting factors, are among the most threatening of these diseases.

Immunopathology of coagulation inhibitors

Reaction of the immune system to circulating factor VIII (FVIII) molecules involves both the humoral and the cellular immune response. The humoral response comprises production of immunoglobulins, namely IgG and/or IgM, which are taken up by antigen presenting cells (APCs), T and B lymphocytes after binding of FVIII. Soluble protein antigen can also be picked up directly from the circulation by major histocompatibility complex (MHC) Class II molecules on APCs and presented to CD4+ T helper cells, which are necessary for the production of antibodies by FVIII-specific B lymphocytes [1]. Via their B-cell receptor (BCR), B-cells can also internalize FVIII, process the molecule and present it with MHC Class II molecules to CD4+ T helper cells.

Antibodies to coagulation FVIII can be directed towards functional, non-functional and conformational epitopes. Some domains and regions of the molecule seem to be more immunogenic than others. The domains A2, A3 and C2 are the main targets of immunoglobulins, while A1 and B seem to be poorly or not at all immunogenic. Inhibitors interfere with the binding of FVIII to FIX, factor X, phospholipids and von Willebrand factor (vWF) [2]. Within one patient, normally a polyclonal immune reaction is involved which may shift over time. Thus, inhibitors may interfere with more than one of the interactions between the four molecules mentioned.

Antibodies to FVIII are found in three clinical settings:
1 patients with haemophilia A receiving substitution therapy with FVIII preparations [3];
2 acquired inhibitors in autoaggressive diseases; and
3 natural antibodies occurring normally in healthy individuals. Natural antibodies in healthy individuals are controlled by anti-idiotypic antibodies as shown in various neutralization experiments [4].

In patients with haemophilia A, substitution with exogenous FVIII may lead to the formation of alloantibodies against the 'foreign' protein. It has been shown that the risk for developing inhibitors is higher in patients with severe haemophilia A, e.g. with FVIII activity <2 U/dL [5]. It has also been shown that the type of the gene defect may influence the likelihood of inhibitor induction. Patients with nonsense point mutations, inversions or large gene deletion are more prone to develop inhibitors [6]. Such gene defects may eventually result in no detectable FVIII molecules, probably causing a lack of immune tolerance induction. Recently, Oldenburg *et al.* [7] demonstrated that severe molecular defects within the FVIII gene, such as intron 22 inversions, nonsense mutations and large deletions are associated with a 7–10 times higher inhibitor prevalence compared to other gene defects.

The type of the administrated factor—the antigen—is also important for the development of inhibitors. Vermylen [1] points out that production processes, e.g. (i) virus reduction and inactivation steps; or (ii) purity of products, and repeated switching of products may increase immunogenicity.

It has also been shown that the age at initiation of the replacement therapy is important. Lorenzo *et al.* [8] report a cumulative incidence of inhibitor development of 41% in patients starting therapy before the age of 6 months, 29% in patients starting therapy between 6 and 12 months and 12% in those starting replacement therapy above 1 year of age. However, this may just reflect the type of genetic defect and the resulting low or absent concentration of functional FVIII molecules and thus the need for early start of replacement therapy.

Several investigators have analysed the association of inhibitor formation with the HLA haplotype. The MHC Class II is of special interest because of their involvement in the presentation of antigens. The MHC Class II alleles *DQA0102*, *DQB0602* and *DR15* seem to occur more frequently in patients with than without inhibitors (relative risk 1.9–4.0). It is also speculated that coincidence of inflammatory processes, e.g. infections in early childhood, with exposure to FVIII during replacement therapy may increase the risk of developing an inhibitor [7].

The fact that some patients develop transient low titre inhibitors supports the concept that FVIII substitution stimulates the formation of antibodies and that suppressing anti-idiotypic antibody formation is delayed. Once anti-idiotype formation has caught up, idiotypic suppression controls the inhibitors [3].

Kazatchkine et al. [4] suggest that in congenital and acquired haemophilia A two distinct populations of anti-FVIII antibodies are found. One population results from clonal selection and affinity maturation and results in antibodies with high affinity to FVIII. The other population has properties similar to those of anti-FVIII antibodies present in healthy individuals. Different strategies seem to be necessary to regain control over both clones.

Rationale for immunoglobulin depletion

Treatment of coagulation inhibitors involves several aspects. In many cases, immediate treatment of acute bleeding episodes is necessary. However, eradication or suppression of the inhibitor-forming cells using immune tolerance therapy (ITT) is helpful on a long-term basis. Both can be achieved by the use of immune adsorption.

Treatment of acute bleeding episodes

Immunoadsorption

Since Nilsson et al. [9] first reported successful eradication of FVIII inhibitors, several groups have adapted this concept, especially in patients with high titres. This treatment concept combines plasmapheresis or immune adsorption with immunosuppressive agents such as steroids or cyclophosphamide, the administration of FVIII concentrates and intravenous immunoglobulin.

Knöbl et al. [10] applied antibody-based immune adsorption successfully to deplete coagulation inhibitors. With the elimination of the inhibitor, a dramatically reduced dose of FVIII was sufficient to provide a measurable and haemostatically effective FVIII activity. Consequently, this results in an enormous cost-saving potential, and bleeding episodes can then be treated with FVIII concentrate. However, if recombinant activated factor VII (rFVIIa) is available, this is the treatment of first choice because of its very rapid haemostatic effect. The combination of both immune adsorption and administration of rFVIIa usually allows considerable saving of rFVIIa.

Immune tolerance therapy to eradicate inhibitor antibodies

Immune tolerance therapies aim to eradicate or at least suppress the inhibitors in the long term. Different strategies have been developed mainly for haemophilic patients. However, these have also been applied to patients with autoantibodies.

In the 1970s, Brackmann and Gormsen [11] started to treat patients with inhibitors with large doses of FVIII and activated prothrombin complex concentrates (aPCC). This protocol which has evolved with the availability of new factor concentrates is highly successful in inducing tolerance. In a series of 60 patients, 52 (87%) were found to have developed immune tolerance [7].

Because of the high costs related to the large quantities of factor concentrates, other groups have tried modifications with either lower doses of FVIII [12,13] or with lowering of the inhibitors prior to the start of ITT using extracorporeal adsorption of immunoglobulins [14–16].

The Ig-TheraSorb system

The antibody-based adsorption of immunoglobulins (Ig-TheraSorb system) utilizes the principle of affinity chromatography. The stationary phase, the immunoglobulin columns, consists of a glass housing containing sepharose CL4B. Sepharose is a material with good biotolerability and inert to plasma proteins. By coupling specific antibodies to the sepharose in a covalent manner, selective removal of plasma constituents—depending on the specificity of the antibodies—is performed.

The principle used in the immunoglobulin-based adsorption columns was developed in the late 1970s at the University of Cologne [17] for the selective binding of apolipoprotein B-100 containing lipoprotein particles. Stoffel et al. [18] were the first to describe the clinical application of the system in patients suffering from familial hypercholesterolaemia. The anti-apolipoprotein B-100 antibody was generated by immunizing sheep with human low density lipoprotein (LDL) particles. The system was commercialized in the mid 1980s by Baxter Germany (Unterschleissheim, Germany).

Immunizing sheep with a commercially available polyclonal intraveneous immunoglobulin (Gammagard™, Baxter Hyland, Unterschleissheim, Germany) resulted in polyclonal sheep anti-human immunoglobulin antibodies.

These antibodies bind all classes and subclasses of human immunoglobulins, fragments of immunoglobulins and immune complexes [19]. Other plasma constituents do not bind to the column.

The Ig-TheraSorb system consists of a pair of immunoadsorption columns, a tubing set connecting the patient with the extracorporeal system, and the instrument directing the fluids (Ig-TheraSorb column, TheraLine, Life 18™ instrument, respectively; PlasmaSelect AG, Teterow, Germany).

During therapy, the patient's circulation is connected with the system via venous access. A small amount of the patient's blood volume is continuously processed outside the body. Anticoagulant (mainly citrate) is added to the blood, which is then separated into blood cells and plasma. The plasma is then directed to one of the two columns. While plasma passes the column, the immunoglobulins become in contact with the sheep antihuman immunoglobulin antibodies and thus selectively

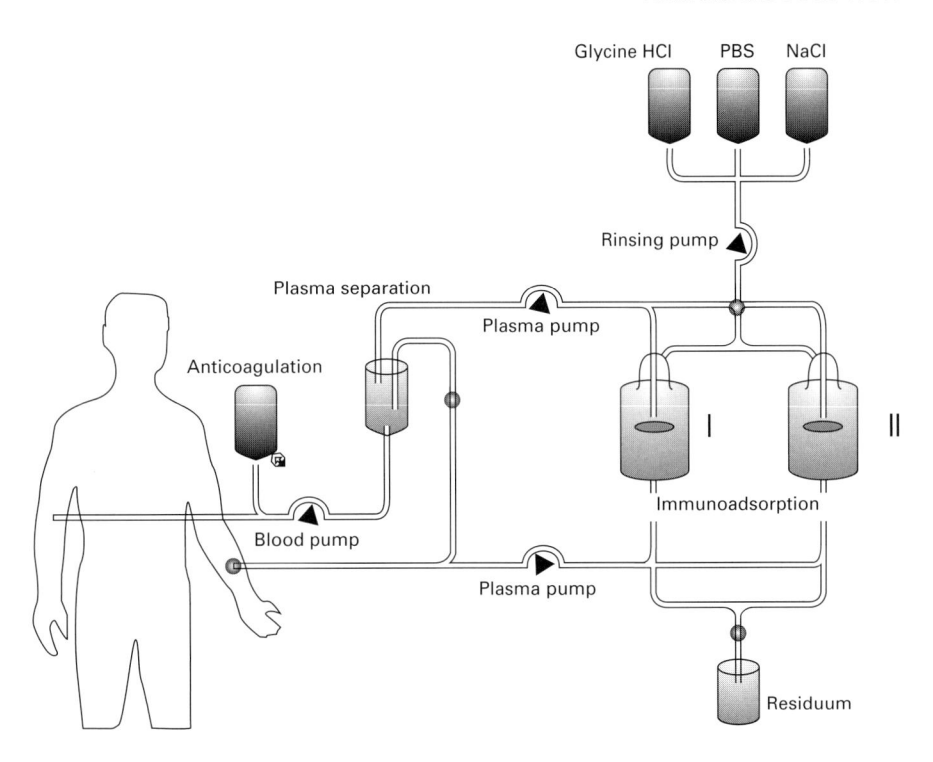

Fig. 32.1 Principle of Ig-TheraSorb™ immunoadsorption.

bind to the adsorption column. After a predetermined amount of plasma has been processed, the plasma is directed to the second column. While the second column is loaded, the first column is regenerated by the use of regeneration buffers with low pH (mainly glycine hydrochloride), to break the antibody–antigen bond and saline (Fig. 32.1). Loading and regenerating can be repeated as often as is necessary to reach the target immuno-globulin level.

The safety of the Therasorb procedure has been extensively tested in LDL-TheraSorb™ immunoadsorption. Richter et al. [20] reported data on safety parameters in Ig-TheraSorb™ immunoadsorption. No short- or long-term changes in electrolytes, creatinine, haemoglobin, liver function tests or fibrinogen levels have been observed. Post immunoadsorption, a transient increase in leucocytes was found. Human antisheep IgG antibodies (hASA) were within the normal range at all times. A moderate decrease of complement C3 and C4 and a slight increase in C3a occurred, while C5a remained unchanged (Table 32.1). No column-related side-effects were observed. Only a few moderate and transient side-effects related to the extracorporeal circulation and the anticoagulation were observed.

Elimination of immunoglobulins depends on the amount of plasma treated. By adjusting the plasma volume, any desired reduction in immunoglobulins can be achieved. Treatment time depends on the patient's plasma volume, immunoglobulin concentration, and the plasma viscosity and plasma flow of the patient. Typically, in one session 1.5–2.5 plasma volumes are processed, resulting in an approximately 70% reduction of immunoglobulins.

All subfractions of IgG (1–4) are eliminated to the same extent. IgM and IgA were reduced, as well as circulating immune complexes and fragments of immunoglobulins (κ- and λ-light chains). In *in vitro* experiments with plasma from patients with atopy, elevated IgE was eliminated with a kinetic comparable to IgG reduction.

The columns are produced according to international standards (EC-GMP-Guidelines). The sheep used for antibody production are kept in a closed flock for more than 11 years and are subject to strict safety controls.

As biological material is used in the production of the LDL-Therasorb™ columns, the respective European guidelines for viral safety evaluation have been applied. In a downscale experimental setting, production material from different production steps has been spiked with a set of different model viruses (e.g. lipid-enveloped DNA and RNA and non-lipid-enveloped DNA and RNA viruses). Reduction and inactivation occurs in several independent steps and exceeds the requirements of the abovementioned guidelines.

Immunoadsorption with anti-immunoglobulin antibodies

Ig-TheraSorb in the management of acute bleeding

The first paper on the use of Ig-TheraSorb in patients with FVIII inhibitors was published in 1995 by Knöbl et al. [10]. In a recent update by Jansen et al. [16] the group reports on their experience in the treatment of different subsets of patients with coagulation inhibitors.

Five patients were treated for acute bleeding (one patient with haemophilia A, four patients with acquired FVIII

Table 32.1 Effect of long-term low-density lipoprotein-TheraSorb™ immunoadsorption on safety parameters.

Safety parameters					
Total protein	No change	Factor II	No change	AST	No change
Transferrin	No change	Factor V	No change	ALT	No change
Haptoglobin	No change	Factor VII	No change	AST	No change
Ceruloplasmin	No change	Factor VIII	No change	LDH	No change
Retinol-binding protein	No change	Factor IX	No change	Calcium	No change
Complement C3	No change	Factor X	No change	Platelets	No change
Complement C4	No change	Factor XI	No change	Leucocytes	No change
IgG	No change	Factor XII	No change	RBC count	Reduction*
IgA	No change	AT	No change	Hematocrit	Reduction*
IgM	No change	Protein C	No change	Hemoglobin	Reduction*
IgE	No change	TAT	No change	Ferritin	No change
PT	No change	PAI	No change	Triiodothyronine	No change
PTT	No change	TPA	No change	Thyroxin	No change
Fibrinogen†	No change	Creatinine	No change	Thyrotropin	No change
Plasminogen	No change	Urea	No change	Vitamin A, E	No change

*Presumably because of frequent blood sampling required by the study protocol.
†If within normal range at entry.

inhibitors). The inhibitor titres ranged from 18 to 540 BU/mL. In four of five patients, inhibitors disappeared after 4–14 Ig-TheraSorb immunoadsorptions (mean 6.5 immunoadsorptions). In the remaining patient, the inhibitor titre decreased from 540 to 20 BU/mL and the bleeding stopped. FVIII levels increased to 0.42 BU/mL and the aPTT became normal. Two patients received human FVIII, two patients rFVIIa, and one patient Feiba. Excluding the haemophilia A patient, all were treated in combination with immunosuppressive therapy. The authors conclude that the combination of Ig-TheraSorb with antihaemophilic drugs (human or recombinant FVIII, rFVIIa, porcine FVIII, prothrombin complex concentrates (PCC), aPCC) offers the possibility to substantially reduce the amount of drugs necessary to achieve measurable FVIII activity. Moreover, it may shorten the time to normalize FVIII concentration.

Toepfer et al. [21] reported the treatment of a patient with an acquired FVIII antibody with Ig-TheraSorb as monotherapy. Within five sessions of immunoadsorption, FVIII inhibitors decreased and effective haemostasis was re-established.

In 1995, the Vienna group reported the rapid elimination of a high titre spontaneous factor V antibody by Ig-TheraSorb immunoadsorption in combination with immunosuppression [22]. In this rare condition, control of bleeding is difficult to manage as no factor V concentrate is available. PCC or aPCC have only a low and unpredictable efficacy and did not work in the patient reported by the group. Other sources of factor V, such as fresh frozen plasma or platelet concentrates, may be used but the low concentrations are unlikely to overcome the effects of the inhibitor.

Originally developed by Nilsson et al. [9] in the 1970s, the combination of depletion of FVIII inhibitors with immunosuppressive therapy (prednisone and cyclophosphamide), FVIII (the antigen) and immunoglobulin allows successful elimination of high titre FVIII inhibitors.

Groups in Bonn, Heidelberg, Vienna and Hannover have modified this protocol using Ig-TheraSorb for depleting inhibitors. The protocol used by the Bonn group, also referred to as the 'modified Bonn protocol', is depicted in Fig. 32.1 [23].

Ig-TheraSorb in elective surgery

Ig-TheraSorb has also been used in haemostatic regimens for haemophilia patients undergoing elective surgery. Again, elimination or reduction of inhibitor antibodies allows the treatment of patients with lower concentrations of antihaemophilic drugs (human, recombinant, or porcine FVIII) [16].

Ig-TheraSorb in pregnancy

The control of bleeding during pregnancy is another potential field for the application of Ig-TheraSorb, especially in patients with autoantibodies, where the use of immunosuppressive drugs is naturally limited. Julius et al. [24] and Swoboda et al. [25] have shown that the TheraSorb technology can safely be applied during pregnancy.

Hannover and Heidelberg experience with immunoadsorption with anti-immunoglobulin antibodies

In our two centres, so far 14 patients have been treated with the 'Hannover–Heidelberg protocol' which is a modification of the Malmö protocol (Fig. 32.2).

Twelve of the patients had acquired and two patients had congenital haemophilia A. Patients were treated with long-term immunoadsorption treatment cycles (each cycle consisting of

Ig-TheraSorb immunoadsorption 1.5–2.5 × plasma volume

Intravenous immunoglobulin (0.3 mg/kg BW) on day 5, 6, 7 every week

| Week 1 | Week 2 | Week 3 | Week 4 |

FVIII substitution (initially 100 U/kg BW bid) reduction, when factor VIII activity >1 U/ml

Mycophenolatmofetil 10 mg/kg BW bid OR cyclophosphamide (2 mg/kg/d) until remission

Prednisolone (1 mg/kg/d) until remission

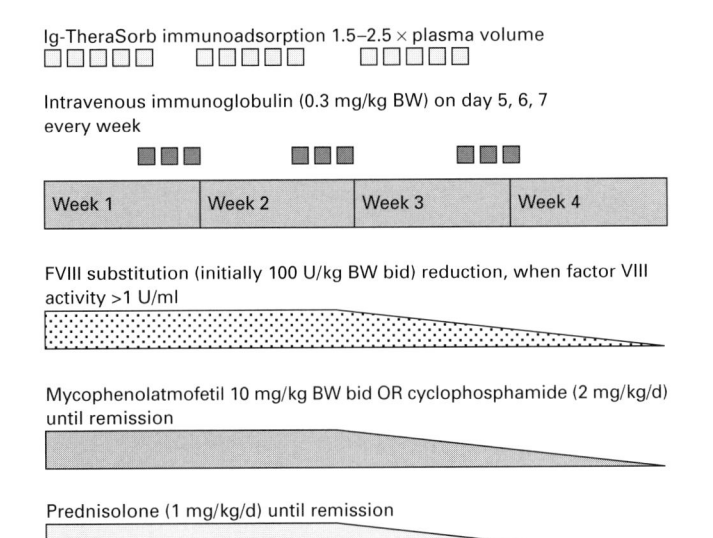

Fig. 32.2 The Hannover–Heidelberg immunoadsorption treatment protocol.

5 sessions/week) until patients reached normal FVIII:C levels above 1 U/L. Patients received a high-dose FVIII substitution (bolus of 200 U/kg body weight followed by continuous FVIII infusion 200 U/kg/day or 100 U/kg body weight twice daily as boluses). As soon as patients reached normal FVIII:C plasma levels, the dose was subsequently reduced. All patients received rFVIIa to control bleeding.

Concomitantly, patients received immunosuppressive therapy. In Heidelberg, patients were treated with prednisolone (1 mg/kg body weight) and cyclophosphamide (2 mg/kg body weight), while in Hannover patients received mycophenolate mofetil (10 mg/kg body weight twice daily) instead of cyclophosphamide. On days 5–7 of each cycle, intravenous immunoglobulin (0.3 g/kg body weight) was given.

Patients had a mean maximum of 79.6 BU (± 109; range 0.9–394) and a mean titre prior to apheresis of 27.4 BU (± 49.6; range 8.8–196). The mean FVIII activity prior to the first immunoadsorption was 0.37 U/L (± 4.4; Tables 32.2 and 32.3).

All patients had various severe life-threatening bleeding episodes, which did not correlate to either FVIII activity or inhibitor titres. Between four and 44 apheresis sessions (median 8) were performed. The FVIII activity reached 30 U/dL after 2–21 immunoadsorption sessions (median 4). FVIII replacement therapy was performed on a mean of 28 days (± 21.1; median 23, range 9–66).

As of October 2001 the mean follow-up time of patients was 29.5 months (± 16.3; median 21.5, range 11–52 months). Thirteen of 14 patients are in remission, one patient is in partial remission. No adverse events have been observed.

In another small group of three patients in Heidelberg the same protocol was applied without FVIII replacement therapy. Two of three patients had a complete remission after 11 and 26 apheresis sessions, respectively. This raises the question of the necessity of FVIII substitution in this protocol. This question will be addressed in a multicentre randomized clinical trial conducted by the German Haemophilia Association (principal investigator: M. von Depka).

Table 32.2 Laboratory data of patients with factor VIII inhibitors treated with TheraSorb™ immunoadsorption.

Patient	Max. inhibitor titre (BU)	Inhibitor titre prior to IA (BU)	FVIII activity prior to IA (%)
1	19.2	4.5	0.7
2	51.0	44.8	2.0
3	10.4	5.8	6.0
4	71.7	5.6	2.0
5	27.5	27.5	0.7
6	57.6	57.6	1.0
7	150	10.6	2.0
8	246	196	11.0
9*	64.0	4.0	<1.0
10	5.4	4.8	8.0
11	13.0	13.0	1.0
12	3.2	2.6	1.0
13*	394	6.0	<1.0
14	0.9	0.8	15.0
Mean	79.6	27.4	3.7
SD	109	49.6	4.4

BU, Bethesda units; IA, immunoadsorption, SD, standard deviation.
Patients 1–8, Heidelberg; patients 9–14, Hannover.
*Patients with haemophilia A.

Table 32.3 Number of immunoadsorptions and outcome in the Hannover–Heidelberg cohort.

Patient	No. IAs until FVIII >30%	Total No. IA	FVIII replacement (days)	Follow-up (months)	Outcome
1	5	20	23	52	Remission
2	2	7	12	51	Remission
3	4	17	16	51	Remission
4	10	31	32	48	Remission
5	3	10	12	46	Remission
6	21	44	66	35	Remission
7	2	6	9	18	Partial remission
8	2	15	22	18	Remission
9*	2	7	Prophylaxis	18	Remission
10	4	4	28	17	Remission
11	5	8	22	24	Remisson
12	6	5	60	11	Remission
13*	7	8	56	5	Remission
14	3	8	35	19	Remission
Mean	5.4	13.6	30.2	29.5	
SD	4.9	11	18.3	16.3	

IA, immunoadsorption; SD, standard deviation.

Economical aspects

While patients with coagulation inhibitors, specifically FVIII inhibitors, comprise only a small number, their therapy has a high impact on the health-care system [26].

The main factors contributing to high costs are those for factor (VIII, VIIa) concentrates. Additionally, significant expenses are related to the treatment of complications (e.g. haemarthrosis) and for the hospitalization of patients.

Using ITT, although costly in the short term, lifetime costs of patients can be reduced substantially. Colowick et al. [27] calculated in a decision analysis model, that ITT will save ~ US$ 1.7 million in total lifetime cost.

The use of Ig-TheraSorb results in demonstrable cost savings in the acute management of patients with bleeding episodes as well as in immune tolerance protocols. According to Jansen et al. [16], Ig-TheraSorb treatment is economically justified, as the reduction of high-dose therapy by only 1 or 2 days replaces the cost of immunoadsorption therapy. Relevant costs were calculated as follows (per day): aPCC: ~ €12.0, FVIII concentrates ~ €15.7 and FVIIa ~ €36.5. The average total costs of the immunoadsorption including 15 cycles is ~ €20.0.

Discussion and conclusions

Patients with haemophilia and inhibitors to FVIII or FIX (alloantibodies) or with acquired autoantibodies to FVIII or FIX constitute a serious medical as well as health economical challenge.

Several regimens to manage (acute) bleeding as well as to induce immune tolerance have been developed. They have been successfully applied in many patients. However, the associated costs are among the highest in health care.

Strategies have been introduced to eliminate or substantially reduce the concentration of circulating inhibitors, thus allowing the reduction of the amount of antihaemophilic drugs (human, recombinant, porcine FVIII, FVIIa, PCC, aPCC) necessary for haemostatic control and in immune tolerance induction.

Ig-TheraSorb is an effective and specific tool for elimination of the complete spectrum of immunoglobulins in patients. There are almost 10 years of clinical experience with the application of Ig-TheraSorb for various autoimmune diseases. Its application has been demonstrated to be safe and efficient. Particularly, experience has now accumulated showing that Ig-TheraSorb is effective in various challenges associated with the management of patients with coagulation inhibitors.

Removal of immunoglobulins can influence the autoimmune processes in many ways. With the Ig-TheraSorb system all immunoglobulins, including IgM, and all immune complexes are effectively reduced. Thus, the concentration of immune complexes consisting of FVIII and IgG or IgM is decreased. This may result in a decreased uptake by APCs and a reduced stimulatory effect for T helper cells. Elimination of circulating IgG and IgM also lowers the formation of new immune complexes. In this way it can also be assumed that free antigen (FVIII) will become available. The increase in not unopsonized antigen could facilitate the induction of anergy in T cells as well as in B cells. Free antigen is required for silencing high affinity clones during affinity maturation [28,29]. Activated T and B cells have a limited life cycle. Elimination of immunoglobulin could interrupt the continuous stimulation and recruitment of new antigen-specific cells.

Anti-idiotypic control of low affinity (naturally occurring) inhibitory antibodies could also be re-established through immunoadsorption regimens. As the balance has shifted towards the inhibitory antibodies, e.g. they are present in much higher concentration than the controlling anti-idiotypic antibodies [4], these antibodies have a higher statistical chance for elimination, thus outbalancing this ratio. Administration of intravenous immunoglobulin might have a synergistic effect, as it should provide more control over anti-idiotypic anti-bodies [30].

Immunoadsorption may thus influence the formation of inhibitors on two levels:
1 anergy of high affinity clones; and
2 re-establishment of the idiotypic control of low affinity clones.
Furthermore, regimens including Ig-TheraSorb immunoadsorption can substantially reduce the costs of inhibitor therapy.

References

1 Vermylen J. How do some haemophiliacs develop inhibitors? *Haemophilia* 1998; **4**: 538–42.
2 Scandella DH. Properties of anti-FVIII inhibitor antibodies in haemophilia A patients. *Semin Thromb Hemost* 2000; **26**: 137–42.
3 Brackmann HH, Schwaab R, Effenberger W *et al*. Antibodies to FVIII in haemophilia A patients. *Vox Sang* 2000; **78**: 187–90.
4 Kazatchkine MD, Lacroix-Desmazes S, Moreau A, Kaveri SV. Idiotypic regulation of anti-factor VIII antibodies. *Haematologica* 2000; **85** (Suppl. to No. 10): 97–9.
5 Scharrer I, Bray GL, Neutzling O. Incidence of inhibitors in haemophilia A patients: a review of recent studies of recombinant and plasma-derived FVIII concentrates. *Haemophilia* 1999; **5**: 145–54.
6 Gilles JG, Jaquemin MG, Saint-Remy J-MR. FVIII inhibitors. *Thromb Haemost* 1997; **78**: 641–6.
7 Oldenburg J, Brackmann HH, Schwaab R. Risk factors for inhibitor development in haemophilia A. *Haematologica* 2000; **85** (Suppl. to No. 10): 7–14.
8 Lorenzo JI, López A, Altisent C, Aznar JA. Incidence of factor VIII inhibitors in severe haemophilia: the importance of patient age. *Br J Haematol* 2001; **113**: 600–3.
9 Nilsson IM, Jonsson S, Sundqvist SB, Ahlberg Å, Bergentz SE. A procedure for removing high titer antibodies by extracorporeal protein A sepharose adsorption in haemophilia: substitution therapy and surgery in patients with haemophilia B and antibodies. *Blood* 1981; **58**: 38–44.
10 Knöbl P, Derfler K, Lorninger L *et al*. Elimination of acquired FVIII antibodies by extracorporeal antibody-based immunoadsorption (Ig-TheraSorb). *Thromb Haemost* 1995; **74**: 1035–9.
11 Brackmann HH, Gormsen J. Massive FVIII infusion in a haemophilic with FVIII inhibitor: high response. *Lancet* 1977; **2**: 933.
12 Mauser-Bunschoten EP, Nieuwnhius HK, Roosendaal G, van den Berg HM. Low dose immune tolerance induction in haemophilia A patients with inhibitors. *Blood* 1995; **86**: 983–8.
13 Unuvar A, Warrier I, Lusher JM. Immune tolerance induction in the treatment of paediatric haemophilia A patients with FVIII inhibitors. *Haemophilia* 2000; **6**: 150–7.
14 Freiburghaus C, Berntorp E, Ekman M *et al*. Immunoadsorption for removal of inhibitors: update on treatments in Malmö-Lund between 1980 and 1995. *Haemophilia* 1998; **4**: 16–29.
15 Knöbl P, Derfler K. Extracorporeal immunoadsorption for the treatment of hemophilic patients with inhibitors to FVIII or XI. *Vox Sang* 1999; **77** (Suppl. 1): 57–64.
16 Jansen M, Schmaldienst S, Banyai S *et al*. Treatment of coagulation inhibitors with extracorporeal immunoadsorption (Ig-TheraSorb). *Br J Haematol* 2001; **112**: 91–7.
17 Borberg H, Stoffel W, Oette K. The development of specific plasma immunoadsorption. *Plasma Ther Transfus Technol* 1983; **4**: 459–66.
18 Stoffel W, Borberg H, Geve V. Application of specific extracorporeal removal of low density lipoprotein in familial hypercholesterolemia. *Lancet* 1981; **2**: 1005–7.
19 Koll RA. Ig-Therasorb immunoadsorption for selective removal of human immunoglobulins in diseases associated with pathogenic antibodies of all classes and IgG subclasses, immune complexes, and fragments of immunoglobulins. *Ther Apher* 1998; **2**: 147–52.
20 Richter WO, Jacob BG, Ritter MM *et al*. Three-year treatment of familial hypercholesterolemia by extracorporeal low-density lipoprotein immunoadsorption with polyclonal apolipoprotein B antibodies. *Metabolism* 1993; **42**: 888–94.
21 Toepfer M, Spannagl M, Sitter T *et al*. Successful reduction of acquired high titer FVIII antibodies by extracorporeal antibody-based immunoadsorption. *Nephrol Dial Transplant* 1997; **12**: A170.
22 Tribl B, Knöbl P, Derfler K *et al*. Rapid elimination of a high-titer spontaneous factor V antibody by extracorporeal antibody-based immunoadsorption and immunosuppression. *Ann Hematol* 1995; **71**: 199–203.
23 Zeitler H, Unkrieg C, Brackmann H *et al*. An immunmodulatory treatment of acquired haemophilia A with long-term IgG-immunoadsorption, immunosuppression and antigen substitution: a modified Bonn protocol inducing immuntolerance. *Blood* 1997; **90** (Suppl. 1): 36A.
24 Julius U, Patzak A, Schaich M, Ehninger G, Kamin G. Immunogene Thrombozytopenie, Anämie und Leukopenie während Schwangerschaft: Erfolgreiche extrakorporaltherapie mit Immunadsorption. *Deutsch Med Wochenschr* 1997; **122**: 220–4.
25 Swoboda K, Derfler K, Koppensteiner R *et al*. Extracorporeal lipid elimination for treatment of gestational hyperlipidemic pancreatitis. *Gastroenterology* 1993; **104**: 1527–31.
26 Aledort LM. Immune tolerance induction: is it cost effective? We know too little. *Semin Thromb Hemost* 2000; **26**: 189–93.
27 Colowick AB, Bohn RL, Avorn J, Ewenstein BM. Immune tolerance induction in haemophilia patients with inhibitors: costly can be cheaper. *Blood* 2000; **96**: 1698–702.
28 Feldmann M, Marini JC. Cell cooperation in the antibody response. In: Roitt IM, Brostoff S, Male D, eds. *Immunology*, 6th edn. London: Mosby, 2001: 131–46.
29 Cooke A. Regulation of immune response. In: Roitt IM, Brostoff S, Male D, eds. *Immunology*, 6th edn. London: Mosby, 2001: 173–89.
30 Sultan Y, Maisonneuve P, Kazatchkine MD, Maionneuve P, Nydegger UE. Anti-idiotypic suppression of autoantibodies to FVIII (anti-haemophilic factor) by high dose intravenous gamma globulin. *Lancet* 1984; **2**: 765–81.

CHAPTER 33

Standard, high and mega bolus doses of rVIIa or recombinant VIIa: comparison of three different treatment regimens

G. Kenet and U. Martinowitz

The first report on the use of recombinant activated factor VII (rFVIIa) together with tranexamic acid during surgical synovectomy in a patient with severe haemophilia A and inhibitor was published in 1988 [1]. It has taken 8 and 11 years since that report for the product to be approved for the prevention and treatment of bleeding episodes in haemophilia patients with inhibitors in Europe and the USA, respectively. Nevertheless, in spite of extensive usage of the drug for this indication [2–6] including some dose finding studies [7,8], the optimal dose, therapeutic levels, mode of administration and monitoring are still controversial. Today the most commonly used initial dose for both treatment of bleeding episodes and surgery is 90 µg/kg. Bolus injections repeated at 2–3 h intervals followed by longer intervals until the bleeding subsides are the most common mode of administration in most centres, although a few reports have been published on continuous infusion of the product [9,10].

This chapter briefly reviews the mechanism of action of rFVIIa as well as issues regarding modes of administration, dosage and safety.

Mechanism of action

Activated FVII circulates in small quantities in normal conditions in the plasma but it is biologically inactive. Only upon complex formation with tissue factor, is it able to generate small amounts of factor Xa, the prothrombinase complex and, most importantly, thrombin which activates platelets [11–14] at the site of vascular injury. These activated platelets then serve as a template for the binding of factors IXa, VIIIa, Xa and Va. Following this, larger amounts of thrombin are generated [13]. As more thrombin is generated, feedback loops with factors V, VIII, XI and XIII occur and cross-linked fibrin is formed producing a stable covalently linked clot. Higher doses of rFVIIa (50 nM or higher) can directly activate FX in the absence of either FVIII or FIX [12,13], creating a huge increase of factor VIIa level, compared with the physiological state, leading to faster and higher thrombin generation [12,13]. *In vitro* analysis of the fibrin clots formed in the presence of a high thrombin concentration have shown that such clots have a different type of architecture; they are far more resistant to degradation by fibrinolytic enzymes compared to normal clots [15,16]. One of the mechanisms for this resistance to fibrinolysis is via activation of thrombin-activated fibrinolysis inhibitor (TAFI), which occurs only in the presence of high thrombin concentrations [17]. *In vitro* experiments, together with animal and clinical studies, support the compartmentalized effect of rFVIIa at the site of injury without systemic activation of coagulation. Administration of a high dose of rFVIIa caused no significant changes in activation markers of coagulations in haemophilia patients [18].

Drug pharmacokinetics and other dosage considerations

Phase I investigations with rFVIIa indicated that the drug has a mean half-life of 2.7 h in adult patients with haemophilia [19]. Similar data were recently reported in a study comparing the effect of bolus injections in adults with haemophilia, liver cirrhosis, or healthy volunteers [20]. The mean clearance values for adult patients were 33–35 mL/h/kg. Shorter half-lives and more rapid clearance values were obtained for paediatric patients [21,22].

Other considerations that should be taken into account with regard to rFVIIa regimens are the interpatient variability in thrombin generation at the same levels of rFVIIa, making it difficult to recommend broad dosage guidelines [23]. Additionally, there may be a slightly increased rate of clearance during bleeding episodes but more data are needed [24]. Higher doses of rFVIIa can illicit a more rapid thrombin burst that has been shown to produce a more stable fibrin clot. Such clots are less prone to fibrinolysis [15,16,25].

Modes of administration

Bolus injections

Because of the very short half-life, rFVIIa has mostly been administered within treatment regimens consisting of frequent recurrent bolus injections. A regimen of boluses (90–120 µg/kg), given initially every second hour, and followed by same doses at increasing intervals, was effective for prophylaxis of bleeding in elective major surgical procedures, whereas for minor surgery the same doses were used for shorter periods and time intervals extended earlier [26,27]. Lower bolus doses were proven to be less effective in surgical patients [27].

Lusher et al. [26] reviewed the experience with rFVIIa in 103 surgical patients, utilizing conventional interval dosage (90 µg/kg every 2–3 h), having an excellent haemostatic response of 81%. Shapiro et al. [8] made a randomized comparison of two regimens in surgical patients: 35 vs. 90 µg/kg given every 2 h for the first 48 h and then every 2–6 h for the next 3 days. Although haemostasis was determined to be effective overall, the lower bolus group had less than optimal control (12 of 15 vs. 15 of 15) of bleeding compared to the higher bolus group. Adjuvant haemostatic agents such as fibrinolytic inhibitors and/or fibrin glue were not used in these patients. Both agents, known to enhance local haemostasis, are used more frequently within the European community [27] and may increase safety and perhaps allow some dosage reduction.

For non-surgical bleeding episodes in haemophilia patients with inhibitors, recurrent intravenous bolus doses of rFVIIa represent a common mode of treatment. Recently, an overall success rate of 92% was reported following home treatment of haemarthroses with 90 µg/kg bolus doses at 3 h intervals [28]. Effective home treatment, achieving haemostasis in most patients after 2–3 bolus injections, was also reported by other groups [29–31]. All studies utilized 90 µg/kg every 2–3 h for primarily treating joint bleeding. Hemostasis was good–excellent (79–92%) after a mean of two doses. Overall, treatment was initiated quite early, usually 1–3 h after first noticing the bleed. Better results and faster haemostasis were achieved in these studies, in concordance with previous reports, when early intervention was feasible [32]. The median total dose to achieve haemostasis in patients treated by bolus injections at home was 180–270 µg/kg each bleeding episode.

Continuous infusion

Administration of coagulation factors VIII and IX by continuous infusion has proven to be safe, convenient and cost effective [33]. The extension of this therapeutic approach to rFVIIa appears logical in view of its pharmacokinetic profile. The inconvenience of frequent bolus injections and the risk of bleeding complications (especially when prolonged treatment is necessary for surgical patients) because of a missed or delayed dose have stimulated us to set up the use of continuous infusion of rFVIIa [34]. Recombinant FVIIa is stable in a minipump for 72 h [35], after reconstitution at room temperature. To avoid frequent local thrombophlebitis, low molecular weight heparin (LMWH) at a concentration 5 U anti-Xa/mL solution is added. The addition of regular heparin used with continuous infusion of all other coagulation products is unsuitable with rFVIIa because of its effect on the stability of this product [34]. Constant plasma FVII:C levels were maintained above a predetermined haemostatic trough. This was achieved by administering an initial bolus dose of 90 µg/kg, followed by continuous infusion according to the patient's individual pharmacokinetics. Daily calculations of individual clearance rates are determined using the steady-state equation:

Infusion rate (U/kg/h) = clearance (mL/h/kg) × desired concentration (U/mL)

Following the first report from our centre [36], the use of continuous infusion of rFVIIa has expanded to the treatment of a variety of surgical procedures [37–40]. Interestingly, the same phenomenon of decrease in clearance values during continuous infusion reported with continuous infusion of FVIII, FIX and vWF has been observed with rFVIIa. This may further reduce the total dose required to achieve haemostasis. The successful use of continuous infusion has also been reported for the treatment of haemarthroses and muscle bleeds [36,37,41,42]. Experience with continuous infusion in Australia and Thailand have demonstrated cost savings of up to 25% during the first 12 h of treatment compared to bolus dosage [41,42]. However, a dose reduction of 35% in one of these cases was complicated by bleeding, suggesting that there might be some lower limit of haemostatic control with this mode of administration [42].

Most continuous infusion regimens have used an initial bolus dose of 90 µg/kg followed by continuous rFVIIa administration at initial rate of 15–16.5 µg/kg/h and thereafter adjusting rates to the individual patient's clearance rate to maintain FVII:C levels above the haemostatic trough. Plasma FVII activity level of at least 6 U/mL have been reported to adequately maintain haemostasis for bleeding events at closed sites [43]. In a recent UK study, rFVIIa was given by continuous infusion to surgical patients to maintain FVII activity levels of 10 U/mL. This was associated with an increased incidence of bleeding [44]. Targeting FVII:C levels to 30–40 U/mL in the immediate postoperative period is likely to provide a margin of safety sufficient for most patients [9]. In our centre the target trough FVII level was 10 U/mL and we had good clinical response; however, we used it together with other agents that enhance haemostasis, such as fibrin glue and tranexamic acid. The use of this standard continuous infusion protocol in our centre yielded an average daily rFVIIa consumption of 450 µg/kg for the first day, followed by a total daily amount of 360 µg/kg or less, depending upon the individual clearance adjustment. In contrast, patients treated with bolus injections might require up to 1080 µg/kg of rFVIIa for the first day of treatment, followed by 540–360 µg/kg/day. Continuous infusion may elicit laboratory markers of thrombin activation but does not appear to promote the risk for systemic thrombosis [45]. Because the kinetics of thrombin generation are likely to influence the structure of the fibrin clot as well as its susceptibility to lysis [46], recurrent boluses of rFVIIa may carry a better haemostatic effect than continuous infusion.

The Israeli experience

Continuous infusion: two regimens

In 1996 we reported on the use of continuous infusion of rFVIIa (with tranexamic acid and use of fibrin glue) [34,36] in patients with haemophilia and inhibitors undergoing surgery. The use of rFVIIa continuous infusion has since expanded to treat bleeding episodes too. In treating these bleeding episodes and surgical patients our aim was to maintain FVII:C trough level above 10 U/mL. While this level was satisfactory (together with fibrin glue and tranexamic acid) to prevent peri- and postoperative bleeding, new haemarthroses occurred in these patients while the level of FVII:C was above our target trough level of 10 U/mL.

Prior studies [2–5] have reported high peak levels (≥30 U/mL) of FVII:C to be associated with successful haemostasis. This prompted us to evaluate the efficacy and safety of an augmented dose-shortened rFVIIa continuous infusion protocol in our haemophilia patients with inhibitors treated for bleeding, and to compare treatment results using the two different protocols. Since 1998, an augmented regimen consisting of higher bolus doses and continuous infusion rates (160–180 μg/kg bolus followed by 30 μg/kg/h, respectively) was utilized in an attempt to achieve higher FVII:C steady-state levels in our patients [47].

Home treatment of haemarthroses and muscle bleeds was effective and safe. Haemostasis was achieved in about 70% of bleeding episodes in both groups. Patients treated with the augmented protocol achieved a significantly faster response (within 60–90 min as compared to 4–6 h with the standard protocol) and the proportion of patients responding within 6 h of treatment onset was significantly higher (72 vs. 48%). As the augmented continuous infusion yielded faster pain relief and slightly faster bleeding response rates for patients with haemarthroses and muscle bleeds, it even paradoxically allowed for some dose reduction although the difference was not significant.

Bolus 'mega dose'

Based upon the good results obtained by the use of the augmented continuous infusion protocol [47] and the theoretical benefit anticipated when a high initial thrombin burst is generated, a new protocol composed of a very high single dose ('mega dose') of rFVIIa (300 μg/kg) was introduced. The efficacy and safety of this regimen was compared prospectively with the two continuous infusion protocols previously used in the same patients. The total number of bleeding episodes treated in these patients was 214 episodes (58 and 72 bleeding episodes were treated according to standard or augmented continuous infusion protocols, respectively, whereas 84 bleeding episodes were treated by the 'mega-dose'). Higher efficacy (75% response) and faster response (30–40 min) were obtained by use of the high bolus regimen, together with lower rFVIIa consumption (300 μg/kg) as compared to the augmented continuous infusion protocol [48]. No thrombotic complications were documented.

At the present time, until more safety data can be gathered in adults, we recommend that the use of rFVIIa 'mega dose' should be considered for bleeding episodes of young patients with inhibitors.

Safety

The prevalence of thrombotic events with the use of rFVIIa is extremely low, especially in light of the many high-risk clinical situations in which the drug is administered.

There have been at least three cases describing laboratory evidence of disseminated intravascular coagulation (DIC) with the use of rFVIIa. However, it is unclear whether laboratory markers of activated coagulation are predictive of subsequent thrombotic events. Baudo et al. [45] have recently cited the lack of clinical complications despite laboratory evidence of DIC. Some investigators cite evidence that activated coagulation is not common with the use of rFVIIa. However, further follow-up is needed and caution advised, especially when dealing with disease states where tissue factor is overproduced. All three cases of presumed DIC reported in the literature encompassed concomitant conditions, which might support activated coagulation [49–51].

Roberts's [52] review of the safety of rFVIIa in 1998 cites two cases of angina in elderly people. Recombinant FVIIa should probably be used with caution in cases with known heart disease and other chronic diseases [49–53].

Other thrombotic complications that were reported included a few cases of stroke, deep vein thrombosis and pulmonary embolism (unpublished data). As rFVIIa complexes with tissue factor locally at the site of bleeding, systemic activation of the clotting system is less likely. None the less, caution should be exercised in the use of this agent in states of excessive tissue factor production or in liver disease [54].

Conclusions

At the present time, the relationship between rFVIIa dosage given to any patient, the thrombin burst induced by it—crucial for stable clot formation—and the actual measurable blood FVII:C levels remain to be clarified. Thus, the potential benefit of haemostasis maintained above trough (by continuous infusion) compared to high recurrent thrombin generations (by bolus doses) has not been established. Nevertheless, it has become clear that for short courses of therapy, such as most haemarthroses or muscle bleeds, continuous infusion is hardly required. On the other hand, its use is certainly recommended for prolonged treatments and surgical procedures. The use of rFVIIa 'mega dose' should be considered for bleeding episodes of young patients with inhibitors.

References

1. Hender U, Glazer S, Pingel K et al. Successful use of rFVIIa in a patient with severe hemophilia A subjected to synovectomy. Lancet 1988; 2: 1993.
2. Hender U, Ingerslev J. Clinical use of recombinant FVIIa. Transfus Sci 1998; 19: 163–76.
3. Glazer S, Hender U, Falch JF. Clinical update on the use of recombinant factor VIIa. In: Aledort LM, Hoyer LW, Lusher JM, Reisner HM, White GC II, eds. Inhibitors to Coagulation Factors. New York: Plenum Press, 1995: 163–74.
4. Ingerslev J. Efficacy and safety of recombinant factor VIIa in the prophylaxis of bleeding in various surgical procedures in hemophilia patients with factor VIII or IX inhibitors. Semin Thromb Hemost 2000; 26: 425–32.
5. Negrier C, Hay CR. The treatment of bleeding in hemophilic patients with inhibitors with recombinant factor VIIa. Semin Thromb Hemost 2000; 26: 407–12.
6. Lusher J, Ingerslev J, Roberts H, Hender U. Clinical experience with recombinant activated factor VIIa. Blood Coag Fibrinolysis 1998; 9: 119–28.
7. Lusher JM, Roberts HR, Davignon G et al. A randomized, double-blind comparison of two dosage levels of recombinant activated factor VIIa in the treatment of joint, muscle and mucocutaneous hemorrhages in persons with hemophilia A and B, with and without inhibitors. Haemophilia 1998; 4: 790–8.
8. Shapiro AD, Gilchrist GS, Hoots WK, Cooper H, Gastineau DA. Prospective, randomized trial of two doses of rFVIIa in hemophilia patients with inhibitors undergoing surgery. Thromb Haemost 1998; 80: 773–8.
9. Ewenstein BM. Continous infusion of recombinant factor VIIa: continue or not? Thromb Haemost 2001; 86: 942–5.
10. Schulman S. Continuous infusion of recombinant factor VIIa in hemophilia patients with inhibitors: safety, monitoring and cost effectiveness. Semin Thromb Hemost 2000; 26: 421–4.
11. Rapaport SI, Rao LVM. Initiation and regulation of tissue factor-dependent blood coagulation. Arterioscler Thromb 1992; 12: 1111–21.
12. Hoffman M, Monroe DM, Roberts HR. Human monocytes support factor X activation by factor VIIa, independent of tissue factor: implications for the therapeutic mechanism of high-dose factor VIIa in hemophillia. Blood 1994; 83: 38–42.
13. Monroe DM, Hoffman M, Oliver JA, Roberts HR. Platelet activity of high-dose factor VIIa is independent of tissue factor. Br J Haematol 1997; 99: 542–7.
14. Rauch U, Bonderman D, Badimon J et al. Platelets become tissue factor positive during thrombus formation. Blood 1998; 92: 347A.
15. Blombäck B, Carlsson K, Fatah K, Hessel B, Procyk R. Fibrin in human plasma: gel architectures governed by rate and nature of fibrinogen activation. Thromb Res 1994; 75: 521–38.
16. Gabriel DA, Muga K, Boothroyd EM. The effect of fibrin structure on fibrinolysis. J Biol Chem 1992; 267: 24259–63.
17. Bajzar L. Thrombin activatable fibrinolysis inhibitor and an antifibrinolytic pathway. Arterioscler Thromb Vasc Biol 2000; 20: 2511–18.
18. Macik BG, Lindley CM, Lusher J et al. Safety and initial clinical efficacy of three dose levels of recombinant activated factor VII: results of a phase I study. Blood Coag Fibrinolysis 1993; 4: 521–7.
19. Lindley CM, Sawyer WT, Macik BG et al. Pharmacokinetics and pharmacodynamics of recombinant factor VIIa. Clin Pharmacol Ther 1994; 55: 638–48.
20. Erhardsen E. Pharmacokinetics of recombinant activated factor VII. Semin Thromb Hemost 2000; 26: 385–92.
21. Shapiro AD. Recombinant factor VIIa in the treatment of bleeding in hemophilic children with inhibitors. Semin Thromb Hemost 2000; 26: 413–19.
22. Hedner U, Kristensen HI, Berntorp E et al. Pharmakokinetics of FVIIa in children. Haemophilia 1998; 4: 244A.
23. Sumner WT, Monroe DM, Hoffman M. Variability in platelet procoagulant activity in healthy volunteers. Thromb Res 1996; 81: 533–43.
24. Hedner U. Dosing and monitoring Novoseven treatment. Haemostasis 1996; 26 (Suppl. 1): 102–8.
25. Weisel JW, Veklich Y, Collet J-P, Francis CW. Structural studies of fibrinolysis by electron and light microscopy. Thromb Haemost 1999; 82: 277–82.
26. Lusher JM, Ingerslev J, Roberts H, Hender U. Clinical experience with recombinant factor VIIa. Blood Coag Fibrinolysis 1998; 9: 119–28.
27. Ingerslev J. Efficacy and safety of recombinant factor VIIa in the prophylaxis of bleeding in various surgical procedures in hemophilia patients with factor VIII or IX inhibitors. Semin Thromb Hemost 2000; 26: 425–32.
28. Key NS, Aledort LM, Beardsley D et al. Home treatment of mild to moderate bleeding episodes using recombinant factor VIIa in hemophiliacs with inhibitors. Thromb Haemost 1998; 80: 912–18.
29. Ingerslev J, Thykjaer H, Kudsk Jensen O, Fredberg U. Home treatment with recombinant activated factor VII: results from one center. Blood Coag Fibrinolysis 1998; 9: S107–10.
30. Laurian Y, Goudemand J, Negrier C et al. Use of recombinant activated factor VII as first line therapy for bleeding episodes in hemophiliacs with factor VII or IX inhibitors (NOSEPAC study). Blood Coag Fibrinolysis 1998; 9: S155–6.
31. Santagostino E, Gringeri A, Mannucci PM. Home treatment with recombinant activated factor VII in patients with factor VIII inhibitors: the advantages of early intervention. Br J Haematol 1999; 104: 22–6.
32. Lusher JM. Acute hemarthroses: the benefits of early versus late treatment with recombinant activated factor VII. Blood Coag Fibrinolysis 2000; 11: S45–9.
33. Martinowitz U, Schulman S. Coagulation factor concentrates by continuous infusion. Transfus Med 1997; 11: 56–63.
34. Schulman S, Bech Jensen M et al. Feasibility of using recombinant factor VIIa in continuous infusion. Thromb Haemost 1996; 75: 432–6.
35. Bonde C, Bech Jensen M. Continuous infusion of recombinant activated factor VII: stability in infusion pump systems. Blood Coagfibrinol 1989; 9: S103–5.
36. Schulman S, d'Orion R, Martinowitz U et al. Feasibility if using recombinant factor VIIa in continuous infusion. Thromb Haemost 1998; 9: S97–101.
37. Mauser-Bunschoten EP, de Goede-Bolder A, Wielenga JJ, Levi M, Peerlinck K. Continuous infusion of recombinant factor VIIa in patients with hemophilia and inhibitors: experience in the Netherlands and Belgium. Neth J Med 1998; 53: 249–55.
38. Montoro JB, Altisent C, Pico M et al. Recombinant factor VIIa in continuous infusion during central line insertion in a child with factor VIII high titer inhibitor. Haemophilia 1998; 4: 762–5.
39. Vermylen J, Peerlinck K. Optimal care of inhibitor patients during surgery. Eur J Haematol 1998; 61 (Suppl. 63): 15–17.
40. Tagariello G, De Biasi E, Gajo GB et al. Recombinant factor VIIa continuous infusion and total hip replacement in patients with hemophilia and high titre of inhibitors to FVIII: experience of two cases. Haemophilia 2000; 6: 581–3.

41 McPherson J, Sutcharitchan P, Lloyd J et al. Experience with continuous infusion of recombinant activated factor VII in the Asia-Pacific region. *Blood Coag Fibrinolysis* 2000; **11**: S31–4.

42 Chuamsumrit A, Isarangkura P, Angchaisuksiri P et al. Controlling acute bleeding episodes with recombinant factor VIIa in haemophiliacs with inhibitor; continuous infusion and bolus injection. *Haemophilia* 2000; **6** (2): 61.

43 Hender U. Dosing and monitoring NovoSeven treatment. *Hemostasis* 1996; **26** (Suppl. 7): 102–8.

44 Smith MP, Ludlam CA, Collins PW et al. Elective surgery on factor VIII inhibitor patients using continuous infusion of recombinant activated factor VII: plasma factor VII activity of 10 IU/mL is associated with an increased incidence of bleeding. *Thromb Haemost* 2001; **86**: 949–54.

45 Baudo F, Redaelli R, Caimi TM et al. The continuous infusion of recombinant activated factor VIIa in patients with factor VIII inhibitors activates the coagulation and fibrinolytic systems without clinical complications. *Thromb Res* 2000; **99**: 21–4.

46 Torbert J. The thrombin activation pathway modulates the assembly, structure and lysis of human plasma clots *in vitro*. *Thromb Haemost* 1995; **73**: 785–92.

47 Kenet G, Lubetsky A, Gitel S et al. Treatment of bleeding episodes in patients with haemophilia and inhibitor: comparison of two treatment protocols with recombinant activated factor VII. *Blood Coag Fibrinolysis* 2000; **11**: S35–8.

48 Kenet G, Lubetsky A, Luboshitz J et al. Treatment of inhibitor patients with rFVIIa: continuous infusion protocols as compared to a single, large dose. *Haemophilia* 2000; **6**: 279.

49 Gallistl S, Cvirn G, Muntean W. Recombinant factor VIIa does not induce hypercoagulability *in vitro*. *Thromb Haemost* 1999; **81**: 245–9.

50 Hay CRM, Negrier C, Ludlam CA. The treatment of bleeding in acquired haemophilia with recombinant factor VIIa: a multicentre study. *Thromb Haemost* 1997; **78**: 1463–7.

51 Stein SF, Duncan A, Cutler D, et al. DIC in a haemophiliac treated with recombinant factor VIIa. *Blood* 1990; **76**: 438A.

52 Roberts HR. Clinical experience with activated factor VII: focus on safety aspects. *Blood Coag Fibrinolysis* 1998; **9** (Suppl. 1): S115–18.

53 Peerlinck KK, Vermylen J. Acute myocardial infarction following administration of recombinant activated factor VII (NovoSeven) in a patient with haemophilia A and inhibitor. *Thromb Haemost* 1999; **82**: 1775–6.

54 Aledort LM. Recombinant factor VIIa is a pan-haemostatic agent? *Thromb Haemost* 2000; **83**: 637–8.

Index

Page numbers in *italics* indicate figures, and those in **bold** indicate tables

acquired haemophilia A 98–108, 199
  associated diseases 98–100
  bleeding episodes 78–9, 92, 101, 102–6
  clinical features 100
  immunoadsorption 34, 105, 201–2
  laboratory diagnosis 101–2
  physiotherapy 165–7
  treatment 102–8
  see also autoimmune anti-factor VIII antibody inhibitors
activated factor VII, recombinant (rFVIIa) *see* recombinant activated factor VII
activated partial prothrombin time (APPT) 74
activated partial thromboplastin time (aPTT) 101
activated protein C (APC) 101
activated prothrombin complex concentrates (aPCC, Feiba™, Autoplex™) 146, 149, 159, 169
  in acquired haemophilia 100, 102, **103**, 105–6, 166
  adverse effects 46, 79, 106, 180
  for bleeding episodes **47**, 75–6, **149**
    in children 70
    factor V inhibitors 202
    in FIX allergy 89, 90
    with immunoadsorption 202
    life- or limb-threatening 78, 79, 80, 81, 197
  in mild/moderate haemophilia 96
  for central venous line insertion 40
  for compartment syndrome 139
  economic aspects 204
  for general and emergency surgery 180–1, 182
  in haemarthrosis 120
  in immune tolerance induction **47**
    Bonn protocol 45–6, 49, 56–7, 200
    low-dose protocol 50, 51–2
  MRI evaluation of treatment 135
  protein A immunoadsorption with 29
  for synoviorthesis 129
  in synovitis 153
  vs. extracorporeal inhibitor removal 34
  vs. rFVIIa 72
acute bleeding episodes *see* bleeding episodes
acute renal failure 107
African descent, patients of
  immune tolerance induction 62
  natural history of inhibitors 4–5
  risk of inhibitors 3, 24, 196

age
  central line infection and 39
  at inhibitor development 51, 92, 199
  success of immune tolerance induction and 62
AIDS *see* HIV infection
Air Cam Walker 158
Airsplint devices 158
albumin, in FVIII concentrates 11
alcohol abuse 171–2
allergic/anaphylactic reactions
  aPCCs 180
  COX-2 selective NSAIDs 147
  factor VIII immune tolerance induction 63–4
  factor IX products 21, 71–2, 87–90, 197–8
    immune tolerance therapy 64, 89
    management 88–90
    natural history 7
    pathogenesis 24, 88
    see also nephrotic syndrome
  porcine factor VIII 79
alloantibodies 98
ε-aminocaproic acid (Amicar) **103**, 184
amputations, for pseudotumours 143, 144
anaesthesia *see* general anaesthesia; local anaesthesia
analgesia
  in acute haemarthrosis 123
  in chronic arthropathy 147
  postoperative 179, 184
anaphylactic reactions *see* allergic/anaphylactic reactions
angina 208
angular deformities 116
ankle
  arthrocentesis 121–2, 123, Plate 18.2
  arthrodesis 172, *175*
  chemical synoviorthesis 127
  fixed flexion deformity 116
  haemarthrosis 158
  ligament injuries 158
  MRI *133*, 134
  orthotics 158
  radiosynoviorthesis **130**, 131
  rehabilitation 158
  valgus deformity 116
antibiotics 38, 39, 99
anti-CD20 monoclonal antibody 107
anti-CD40 ligand 107–8
anticonvulsants 99
anti-factor VIII antibodies
  autoimmune *see* autoimmune anti-factor VIII antibody inhibitors

  ELISA 10
  human monoclonal 11–12, Plate 2.1
  immunopathology 199–200
  inhibitory *see* factor VIII inhibitors
  non-inhibitory 10
  two populations 200
anti-factor IX antibodies, inhibitory *see* factor IX inhibitors
antifibrinolytic agents
  in acquired haemophilia 105
  in mild/moderate haemophilia A 198
  with rFVIIa 82
  for surgery 96, 169, 180–1, 184
  see also tranexamic acid
antihaemophilic globulin 9, 14
anti-idiotypic antibodies 107, 199, 200
  immunoadsorption and 11, 205
anti-immunoglobulin antibodies
  immunoadsorption with 31–2, 196, 199–205
  sheep antihuman 200–1
antiretroviral therapy 179
antithrombin concentrate 129–30
aPCC *see* activated prothrombin complex concentrates
apolipoprotein B-100 200
arterial embolization, pseudotumours 143, 144
arterial puncture 31, 33
arterial thrombosis 33
arthrocentesis 116, 120, 121–3
  ankle 121–2, 123, Plate 18.2
  elbow 122, 123, Plate 18.4
  hip 122, 123, Plate 18.3
  knee 121, 123, Plate 18.1
  recommendations 124
  shoulder 122, 123, Plate 18.5
arthrodesis, ankle 172, *175*
arthropathy, haemophilic 117, 121, 152–3
  in children with inhibitors 69
  immune tolerance induction and 46
  medical management 147
  MRI 132–4, 135–6
  MRI evaluation scale 135
  neurological complications 118
  orthopaedic surgery 169
  pathogenesis 115–17, Plates 17.1–4
  physiotherapy 164–5
  pseudotumours and 142
  quality of life aspects 187
  radiographic classification **126**
  rehabilitation 153–4
  see also haemarthrosis; synovitis, haemophilic

assays, FVIII inhibitor 9–10
  in acquired haemophilia 101
  Bethesda method *see* Bethesda assay
  cut-off values 5, 50
  Oxford and new Oxford method 9
  plasma sampling 50
autoimmune anti-factor VIII antibody inhibitors 98–108
  disease associations 98–100
  eradication therapy 106–8
  immunological characteristics 100–1
  laboratory diagnosis 101–2
  see also acquired haemophilia A
autoimmune diseases **98, 99**
Autoplex™ *see* activated prothrombin complex concentrates
azathioprine 107

bed rest 123
Bethesda assay 9–10, 74, 195
  in acquired haemophilia 101
  cut-off values 5
  Nijmegen modification 5, 50
Bethesda unit (BU) 10, 74, 101, 195
Bisanect™ 16, 17
bleeding episodes
  in acquired haemophilia 78–9, 92, 101, 102–6
  in children with inhibitors 69–72
    aims of treatment 69
    characteristics 69
    home treatment 72
    individualized treatment scheme 70–2
    preventive treatment 72
    rFVIIa *vs.* aPCC 72
    treatment options 69–70
  during immune tolerance therapy 50, 51–3, 64, 71
  in factor IX allergy 89, 90
  frequency 149, 188, 189
  German treatment guidelines 47
  immunoadsorption 47, 71, 76, 200, 201–2, 203
  immunoglobulin depletion therapy 200
  in immunotolerized patients 53
  inhibitor development and 92, **93**
  life- and limb-threatening 78–82, 92, 102, 197
    side-effects of treatment 79–80
    treatment options 78–9
  medical management 74–7
  in mild/moderate haemophilia A 96
  musculoskeletal *see* musculoskeletal bleeding
  psychological management 190
  rFVIIa therapy *see* recombinant activated factor VII, for bleeding episodes
  surgical *see* intra-operative bleeding; postoperative bleeding
blood cyst, haemophilic *see* pseudotumour, haemophilic
blood transfusions, pseudotumour excision 143
B lymphocytes 199
BO2C11 monoclonal antibody 11, Plate 2.1
bone
  cysts 134, 135
  grafts, pseudotumours 143
  hypertrophy 106
  lesions, MRI 133, 134
  surgery, control of bleeding 170

Bonn protocol (Brackmann) 45–7, 55, 196, 200
  choice of concentrates 56, 57
  clinical results 46, 56–7
  development 45
  indications 32–3
  regimen 45–6, 49, 76
  side-effects 46
BPL8Y 18
braces 164, 165
Broviac catheters 36
Bundesärztekammer guidelines 46–7
bypass agents 34
  in acquired haemophilia 102, **103**
  for bleeding episodes 70, 71, 72, 75–6
  for central venous line insertion 40
  in mild/moderate haemophilia A 96
  see also activated prothrombin complex concentrates; prothrombin complex concentrates; recombinant activated factor VII

calcification, pseudotumours 118
calcium gluconate, intravenous 105
cancer, acquired haemophilia **98**, 99–100, 166
Candida albicans 38
cartilage damage
  MRI 132, 133, 134
  pathogenesis 115–16, Plates 17.1–4
celecoxib (coxib) 147, 153
central nervous system haemorrhages 78, 80, 81, 197
central venous access devices (CVADs) 36–40, 196
  fibrin clot in tip 37, 39, 40
  haemostasis during insertion 40, 181
  infections complicating 37–9, 196
    causative pathogens 36
    during immune tolerance induction 52–3, 64
    management 39
    risk factors 36, 37
  other complications 37, 39–40, 64
  position of tip 39–40
  saline flush/heparin lock 40
  studies on record 36–7
  types 36
  see also Port-A-Cath™ devices
children
  bleeding episodes *see* bleeding episodes, in children with inhibitors
  haemophilic pseudotumours 143
  inhibitor development 3, 92–3
  physiotherapy 161–2
  preventative treatment 72
  psychosocial impact of inhibitors 187–91
chronic obstructive pulmonary disease (COPD) 166
Citem 10/Immunosorba R™ system 30
collagen vascular disorders 98–9
communication 190
compartmental pressure, measurement 139, 140–1
compartment syndrome 78, 117, 139–41
  after protein A immunoadsorption 33
  decompression 140
  management 139–40, 197
  postoperative 171–2
  rFVIIa therapy 81, **82**, 139
contractures
  flexion *see* flexion contractures
  Volkmann 140

contrast agents 122, 134
corticosteroids
  in acquired haemophilia 99, 100, 107, 108, 166, 167
  in immune tolerance induction 33, 49, 54, 203
  intra-articular injection 147
  in mild/moderate haemophilia 96
  in synovitis 147
costs, economic 69, 204
counselling 161
COX-2-selective NSAIDs 147, 153
creatine kinase, serum 139
crural nerve palsy 118
Cryocuff™, in acute haemarthrosis 123
cryoprecipitate, lyophilized 18
cryotherapy *see* ice therapy
Cyclokapron *see* tranexamic acid
cyclophosphamide 169
  in acquired haemophilia 107, 108, 166, 167
  for bleeding episodes 71
  in Hannover–Heidelberg protocol 203
  in immune tolerance induction 49, 202
  in Malmö protocol 29, 32, 33, 51, 77, 196
    in FIX allergy 89
    in mild/moderate haemophilia 96
    in synovitis 129–30
cytokines, in FVIII concentrates 15, 16–17
cytotoxic chemotherapy, in acquired haemophilia 99, 100, 107

DDAVP *see* desmopressin
dental extraction 183–4
dermatoses, autoimmune **98, 99**
desensitization *see* immune tolerance induction
desmopressin (DDAVP)
  in acquired haemophilia A 103, 104–5
  adverse effects 104
  FVIII genotype and responsiveness 25
  in mild/moderate haemophilia A 92, 93, 96, 198
  in musculoskeletal bleeding 147
  tachyphylaxis 104–5
dextropropoxyphene 123
disability 165
disseminated intravascular coagulation (DIC)
  immune tolerance induction and 46
  PCC and aPCC-associated 79, 106, 120, 180
  rFVIIa-associated 106, 121, 169, 208
drug-induced acquired haemophilia 99

economic costs 69, 204
education, patient 164, 165
elbow
  angular deformity 116
  arthrocentesis 122, 123, Plate 18.4
  chemical synoviorthesis 127
  haemarthrosis 155
  orthotics 156
  radiosynoviorthesis 130, 131
  rehabilitation 155–6
  synovectomy **172**
  ulnar nerve compression 118
elderly, acquired haemophilia **98, 99**, 166–7
electric stimulation, functional **152**
electromyography 139
electrotherapy **152**, 153, 158, 162, 164
ELISA (enzyme-linked immunosorbent assay) 5, 10
embolization, pseudotumours 143, 144
emergency surgery 179–82
EMLA™-anaesthetic cream 38

INDEX

empowerment, personal resources 190–1
endocarditis, bacterial 39
enzyme-linked immunosorbent assay (ELISA) 5, 10
epiphyseal overgrowth 116–17
exercises
   graded 162
   in subacute haemarthrosis 123
   see also physiotherapy
extension–desubluxation cast, hinged 123
extracorporeal techniques *see* immunoadsorption; plasmapheresis

factor V inhibitors, acquired 32, 202
factor VII
   congenital deficiency 81
   plasma 82
   recombinant activated (rFVIIa) *see* recombinant activated factor VII
factor VIII concentrates 14–15
   in acquired haemophilia 102–4, 105, 166
   antigenicity 15–18, 61, 195–6
      epidemiological studies 17–18
      factors affecting 16–17, 19
      in mild/moderate haemophilia 95–6
   for bleeding episodes 47, 74–5, **149**
      in children 70–1
      with immunoadsorption 202
      in mild/moderate haemophilia 96
   continuous infusion 95–6
   for dental extraction 183, 184
   economic aspects 204
   for general and emergency surgery 179
   in immune tolerance induction 55–8, 163, 196–7
      in acquired haemophilia 108
      Bonn protocol 45–6, 49, 56–7, 76, 200
      German recommendations **47**
      in Hannover–Heidelberg protocol 203
      low-dose (Oxford) protocol 76
      low-dose (Van Creveld) regimen 49, 50, 51, 54
      Malmö protocol 33
      North American Registry 61, 62
   inhibitor development after exposure 3
   for orthopaedic procedures 120
   phospholipid-containing ('protected') 105
   plasma-derived *see* plasma-derived factor VIII concentrates
   porcine *see* porcine factor VIII
   purity 15, 16–17, 18, 19, 195–6
   recombinant *see* recombinant factor VIII (rFVIII) concentrates
   switching between 18, 19, 95
   in synovitis 153
factor VIII gene
   defects, inhibitor development and 5, 15, 22–4, 93–5, 196
   intron 22 inversions 22, 23, 24–5
   large deletions 22, 23, 196
   missense mutations 22, 23, 93–5
   nonsense mutations 22, 23
   small deletions/insertions 22, 23
   splice site mutations 22, 23
factor VIII inhibitors 9–12
   acquired *see* autoimmune anti-factor VIII antibody inhibitors
   assays *see* assays, FVIII inhibitor
   epitopes 3, 10, 199
      in acquired haemophilia 100
      genetic factors 25
      immune tolerance induction and 57, 58
   extracorporeal removal methods 29–34

FVIII product-related risk 15–18, 19, 55, 195–6
   genetic factors 21–5
   history 9, 14
   immunoadsorption 29–34
   immunopathology 199–200
   importance of early detection 69
   incidence 3–7, 17, 59, 120
   isotypes 9
   management 120–1
   mechanisms of FVIII inactivation 10–11
   in mild and moderate haemophilia A 92–6
   monoclonal 11–12, Plate 2.1
   natural history 3–7
   pathogenesis 15–17
   plasmapheresis 29
   prevalence 3–7, 17, 23–4
   risk factors **15**, 19
   study design issues 17
   type I (A, classic) *see* type I (A) inhibitors
   type II *see* type II (B) inhibitors
factor VIII protein
   in acquired haemophilia A 101–2
   assays 50
   conformational change in mutated 93, 94
   human monoclonal antibodies 11–12, Plate 2.1
   maternal 23
   mechanisms of inhibition 10–11, 195
   structure–function relationship 10
factor IX concentrates
   anaphylactic/allergic reactions *see* allergic/anaphylactic reactions, factor IX products
   for bleeding episodes 70–1, 74–5
   for general and emergency surgery 179
   in immune tolerance induction 33, 61, 76
   inhibitor development after exposure 3, 7
   for orthopaedic procedures 120
   in synovitis 153
factor IX gene
   defects, inhibitor development and 7, 21–4, 87, 196
   large deletions 22, 71, 87, 88
   missense mutations 22, 23
   nonsense mutations 22
   small deletions/insertions 22, 23
factor IX inhibitors 87–90
   characteristics 88
   genetic factors 21–5, 87
   immunoadsorption 29–31, 32
   importance of early detection 69
   international registry 88
   management 120–1
   natural history 3, 7
   prevalence 23–4, 59, 195
   reasons for low incidence 87–8
factor IX protein
   in acquired haemophilia A 101
   activated (FIXa), FVIIIa interaction 10, 100
   homology to vitamin K-dependent factors 24, 87
   non-plasma levels 87–8
   Sepharose™ coupled 32, 105
   size 87
factor X, activated (Xa) 105, 206
factor XI, in acquired haemophilia A 101
factor XII, in acquired haemophilia A 101
family attitudes 188–9
family history of inhibitors 3, 24, 196
   in mild/moderate haemophilia A 93, **94**
Fanhdi™, for immune tolerance induction 57
fasciotomy 140, **172**

Feiba™ *see* activated prothrombin complex concentrate
femoral neck fracture 171, 172
femoral nerve compression 117, 139–40, 146
fibrin clot
   rFVIIa-generated 206
   tip of central line **37**, 39, 40
fibrin glue 105, 207, 208
   in decompression surgery 140, 141
   in dental surgery 184
   in orthopaedic surgery 171, 172
   in pseudotumour excision 143
fibrinogen, plasma 80
fibrinolytic inhibitors *see* antifibrinolytic agents
FK506 (tacrolimus) 107
flexion contractures 116, 123
flexion deformity, fixed 116
fludarabine 99
footwear 158, 164, 165
forearm, muscle haematomas 155–6
friends, relationships with 188, 189
fungal infections 38
FVIII:CPS-P 16
FVIII:P 16

gadolinium contrast agents 134
gait re-education 157–8
Gammagard™ 200
gammaglobulin, intravenous *see* intravenous immunoglobulin
gastrocnemius haematomas 158
gastrointestinal bleeding 147
general anaesthesia 122, 140
genetics 3, 21–5
   FVIII antibody epitopes and 25
   FVIII/FIX genotypes 22–3, 196
   haemophilia A *vs*. B 23, 87
   immune system 24–5
   mild/moderate haemophilia A 93–5
German Medical Association guidelines 46–7
glomerulonephritis, membranous 89
glucosamine sulphate 153
gluteus haematoma 156
glycosylation, FVIII concentrate antigenicity and 16
gold, radioactive (^{198}Au) 127, 128, 129–31
gradient echo pulse MRI sequences 132, *133*
Gram-negative organisms 38

haemarthrosis 152–3
   acute 121–3, **124**
   ankle 158
   elbow 155
   hip 117, 123, 156
   knee 157
   medical management 74–7, 207, 208
   MRI 116, 134, 135
   orthopaedic management 120–4
   pathogenic effects 115–16
   physiotherapy **152**, 162–3
   postoperative 171
   prevention 129–31, 153
   shoulder 154–5
   sites **150**
   subacute 121, 123
   types 121
   ultrasound imaging 116, 152–3
   see also arthropathy, haemophilic
haematomas
   muscle *see* muscle haematomas
   postoperative 170–1
   superficial 153

haematuria, medical management 74–7, 81
Haemoctin SDH™, for immune tolerance induction 57
haemolysis
  complicating immune tolerance induction 46
  FVIII concentrate-induced 103
Haemophil M™ 17, **18**
haemophilia A 14–19
  acquired *see* acquired haemophilia A
  current treatment regimens 14–15
  FVIII inhibitors *see* factor VIII inhibitors
  history of treatment **14**
  immune tolerance induction 30–4, 45–7, 49–54, 59–64
  mild and moderate *see* mild/moderate haemophilia A
  natural history of inhibitors 3–7, 195
haemophilia B
  anaphylactic reactions *see* allergic/anaphylactic reactions, factor IX products
  FIX inhibitors *see* factor IX inhibitors
  immune tolerance induction (ITI) 24, 72
    with FIX allergy 89
    Malmö protocol 32, 33, 34
    North American Registry 59–64
  prevalence of inhibitors 23–4, 59, 195
  severe 87, 90
Haemo-QoL questionnaire 188
haemosiderin 115, 134, Plates 17.2–4
head trauma 80
health
  perceived 188, 189
  physical 188, 189
heat treatment, FVIII antigenicity and 16, 17
heparin 40, 82
  low molecular weight (LMW) 207
hepatitis A vaccination 179
hepatitis B 46, 51, 147, 179
  vaccination 179
hepatitis C 46, 51, 147, 179
Hickman catheters 36
high responders 3, 4, 160–1
  bleeding episodes **47**, 71, 74, 75
  general and emergency surgery 179–82
  immune tolerance induction 33, 46, **47**, 56–7, 60, **61**
  intracranial bleeds 80
  orthopaedic surgery 169–76
  physiotherapy 162
high titre inhibitors 160–1
  bleeding episodes 71
  immune tolerance induction 32–3, 34, 45, 51, **52**, 55
  natural history 5
  orthopaedic procedures 120
  plasmapheresis 29
  porcine factor VIII therapy 70
hinged extension desubluxation cast 123
hip
  arthrocentesis 122, 123, Plate 18.3
  closed reduction and internal fixation (CRIF) 171, **172**
  flexion deformity 117
  haemarthrosis 117, 123, 156
  MRI 135
  orthotics 156–7
  osteonecrosis 135
  rehabilitation 156–7
  replacement, total (THR) 96, **172**
Hispanic patients 5

HIV infection 46, 147
  central venous line infections and 36, 37
  immune tolerance induction 51, 53, **61**, 62, 64
  musculoskeletal complications 117
  surgery and 179
home adaptations 165
home treatment
  bleeds in children with inhibitors 72
  rFVIIa 72, 76, 79, 207, 208
Humate p™, for immune tolerance induction 56–7
hyaluronic acid, intra-articular 135, 153
Hyate C™ *see* porcine factor VIII
hydrotherapy 147, 157, 158, 164
ice therapy 152, 162
  in acute haemarthrosis 123, 124, 155, 158
  muscle haematomas 153, 155, 156
IgG
  autoimmune factor VIII inhibitors 100
  extracorporeal removal *see* Ig-TheraSorb™ immunoadsorption
  factor VIII inhibitors 9, 195
  factor IX inhibitors 88
  intravenous therapy *see* intravenous immunoglobulin
  protein A binding 29, 30
IgG4 subclass antibodies 9, 88, 195
Ig-TheraSorb™ immunoadsorption 31–2, 34, 196, 199–205
  in acquired haemophilia 105, 201–2
  in acute bleeding 201–2
  economic aspects 204
  in elective surgery 202
  Hannover and Heidelberg experience 202–3
  in pregnancy 202
  principle 200–1
  side-effects 201
iliopsoas haematoma 117, 118, 149
  diagnosis 139
  management 139–40, 197
  rehabilitation 153, 156–7
immobilization 152, 153, 162
  see also rest
Immunate™, for immune tolerance induction 57
immune system, genetics 24–5
immunoadsorption 29–34
  in acquired haemophilia 34, 105, 201–2
  with anti-immunoglobulin antibodies 31–2, 196, 199–205
  in bleeding episodes **47**, 71, 76, 200, 201–2, 203
  Hannover–Heidelberg protocol 202–3
  in immune tolerance induction 32–3, 34, 202–3
  other methods 31–2
  protein A *see* protein A immunoadsorption
immunoglobulins 199
  rationale for depletion 200
  see also IgG
Immunosorba™ columns 30, 31–2, 105
immunosuppressive therapy
  in acquired haemophilia 99, 100, 106, 107–8
  for bleeding episodes 71, 202
  in immune tolerance induction 203
  see also corticosteroids; cyclophosphamide
immune tolerance induction (ITI, ITT) 3, 76–7, 120, 146, 149
  in acquired haemophilia 34, 106, 108
  acute bleeding during 50, 51–3, 64, 71

adverse effects 63–4
Budapest protocol 108
choice of concentrates 55–8, 196–7
  at Frankfurt haemophilia centre 56–7
  hypothetical aspects 57
definitions of success 50, 56, 61
duration 61, **62**
for elective surgery 179
German recommendations 46–7
in haemophilia A 30–4, 45–7, 49–54, 59–64
in haemophilia B *see* haemophilia B, immune tolerance induction
Hannover–Heidelberg protocol 202–3
high-dose protocol *see* Bonn protocol
immunoadsorption with 32–3, 34, 202–3
low-dose (Oxford) protocol 76
low-dose (Van Creveld) regimen 49–54, **55**, 56, 196
  clinical results 51–3
  complications 51–3
  indications 54
  methodology 49–51
  in mild/moderate haemophilia A 96
  pros and cons 53–4
  treatment of bleeds during 50
maintenance 63, **64**
Malmö protocol *see* Malmö protocol
in mild/moderate haemophilia A 96
natural history of inhibitors and 5, 6
nephrotic syndrome complicating 24, 64, 89, 90
North American Registry (NAITR) 59–64, 197
outcome predictors 55, 59, 62–3, 64
plasmapheresis with 33, 34
psychological management 190
rationale 69, 200
success rates 59, 61–2
venous access for 46, 49, 52–3, 163
incidence of inhibitors 74, 195
  FVIII products and 17–18
  in haemophilia B 120
  in mild/moderate haemophilia A 92, 95, 96, 120
  in severe haemophilia A 3–7, 17, 59, 120
infections
  central venous lines *see* central venous access devices (CVADs), infections complicating
  complicating immune tolerance induction 46
  peripheral intravenous access devices 37
  postoperative 169
  soft tissue haematomas 117
inferior alveolar nerve block 183
information, giving and eliciting 190
infrared therapy **152**
inhibitors (inhibitor antibodies)
  characterization 9–12
  natural history 3–7
  see also factor VIII inhibitors; factor IX inhibitors
inhibitor titres 160–1
  in acquired haemophilia A 102
  bleeding episode management and 70–1, 74
  general and emergency surgery 179
  immunoadsorption and 30, **31**, 202
  immune tolerance induction and
    Bonn protocol 45–6, **47**
    low-dose protocol 49, 51, **52**, 53
    North American Registry 62, 64

INDEX

measurement *see* assays, FVIII inhibitor
see also high titre inhibitors; low titre inhibitors
insoles 158, 164, 165
interferon-alpha 99, 107
internal bleeds 78, 81–2, 197
International Immune Tolerance Registry (IITR) 59, 64
International Society on Thrombosis and Haemostasis (ISTH) 59, 88
intra-abdominal haemorrhages 78, 81–2
intracranial haemorrhages 78, 80, 81, 197
intra-operative bleeding 169, 170–2
intravenous immunoglobulin (gammaglobulin, IVIG)
 in acquired haemophilia 106, 107, 108, 167
 in Hannover–Heidelberg protocol 203
 in immune tolerance induction 49, 54, 205
 in Malmö protocol (Nilsson) 32, 51, 129, 196
  dosage 33
  in FIX allergy 89
  in mild/moderate haemophilia 96
 sheep antibody production 200
iontophoresis 152
iron deposits 115, 116, 147, 153
ischial weight-bearing brace 156–7
ITI, ITT *see* immune tolerance induction

joint
 aspiration *see* arthrocentesis
 bleeding into *see* haemarthrosis
 examination 150, 161
 mobilization 164–5
 pain 121, 153, 165
 protection 147
 range of motion (ROM) 150, *151*
 replacement surgery 148, 169, **172**, *173–4*
 sepsis 121, 147
 target 115, 134–5, 146

kinesiohydrotherapy 154, 155, 156
kinesiotherapy 151, 152, 153, 154
 for elbow problems 155, 156
 for hip problems 156
 in shoulder haemarthrosis 154–5
knee
 arthrocentesis 121, 123, Plate 18.1
 arthroplasty, total (TKA) 169, 171, 172, *173–4*, 181
 flexion deformity 116, 170
 haemarthrosis 157
 MRI 133, 134
 orthotics 157–8
 Oxford splinting technique 123
 radiosynoviorthesis 130, 131
 rehabilitation 157–8
 synoviorthesis 127–8
 valgus deformity 116
Kogenate™ 15
 in acquired haemophilia **103**
 for immune tolerance induction 56
 natural history of inhibitors 4–5, 6
Kogenate™ SF **4**, 5
Kryobulin–VH™ 18

lactate dehydrogenase, serum 139
La Paz University Hospital Haemophilia Centre children's inhibitor study 187–9
laser therapy **152**, 153
Latin Americans, immune tolerance induction 62

LDL-Therasorb™ system 200, 201, **202**
LE2E9 monoclonal antibody 11–12
leg length discrepancies 116–17
life-threatening bleeding episodes 78–82, 197
limb
 amputations, for pseudotumours 143, 144
 -threatening bleeding episodes 78–82, 197
local anaesthesia 122, 140, 183
low responders 3, **4**, 74, 160
 in acquired haemophilia A 102
 bleeding episodes 47, 70–1, 74, **75**
 immune tolerance induction 46, **47**, 56–7
low titre inhibitors 5–6
 bleeding episodes 70–1
 immune tolerance induction 32–3, 51, **52**
lupus anticoagulant 101
lupus erythematosus, systemic (SLE) 98–9, 166
lymphoproliferative malignancies **98**, 99–100

magnetic resonance imaging (MRI)
 arthropathy scale 135
 in haemarthrosis 116, 134, 135
 musculoskeletal 132–7
magnetic waves, low frequency **152**, 153
major histocompatibility complex *see* MHC
malignant tumours
 acquired haemophilia **98**, 99–100
 vs. pseudotumours 118
Malmö protocol (Nilsson *et al*) 29, 32–3, 34, 55, 77
 choice of concentrate 56
 in FIX allergy 89
 in mild/moderate haemophilia A 96
 modifications 33, 34, 108, 202–3
 in North American Registry 61
 protein A immunoadsorption 32–3, 34, 129, 130, 196
 switching to 49, 51
 in synovitis 129–30
manual therapy 164–5
membranous nephropathy 89, 90
Memory™ 162
6-mercaptopurine, in acquired haemophilia 107
methylprednisolone 108
MHC
 in antigen presentation 199
 genotype, inhibitor formation and 24–5, 196, 199–200
mild/moderate haemophilia A 92–6, 198
 clinical presentation 92
 genetic predisposition 93–5
 incidence of inhibitors 92, 95, 96, 120
 inhibitor characteristics 92–5
 natural history of inhibitors 92–3
 prevention of inhibitors 96
 treatment of inhibitors 96
 type and delivery of treatment 95–6
Monoclate™ 17, **18**
MRI *see* magnetic resonance imaging
mucocutaneous bleeding
 in acquired haemophilia 100, 101
 medical management 74–7
muscle
 balance, evaluation 150, 151
 intra-operative bleeding 170–1
 problems 117–18
 re-education techniques 165
 strengthening exercises 154
 strength testing 161
 testing 161

muscle haematomas 117
 ankle 158
 complications 117–18
 encapsulated 142
 forearm 155–6
 limb-threatening 78, 79, 81
 medical management 74–7, 79, 146, 207, 208
 physiotherapy **152**, 153, 162–3, 166
 postoperative 170–1
 sites **150**
 thigh 157
musculoskeletal bleeding 146–7
 incidence 149
 sites **150**
 see also haemarthrosis; muscle haematomas
musculoskeletal system
 complications of haemophilia 115–18
 physical examination 150–1
musculotendinous lengthening procedures 170
mycophenolate mofetil 202
myocardial infarction 33, 79, 106

natural history of inhibitors
 in mild/moderate haemophilia A 92–3, **94**
 in severe haemophilia A and B 3–7, 195
nephrotic syndrome 89–90, 197–8
 complicating immune tolerance induction 24, 64, 89, 90
 indicating Malmö protocol 33, 34
 pathogenesis 71, 89–90
nerve conduction studies 139
neurological problems 118
NiaStase *see* recombinant activated factor VII
non-steroidal anti-inflammatory drugs (NSAIDs) 147, 153, 179
North American Immune Tolerance Registry (NAITR) 59–64, 197
 data collection 59–60
 definition of variables 60
 results 60–4
NovoSeven *see* recombinant activated factor VII

occupational therapy 165
Octavi SDPlus™ 16, 17
Octostim *see* desmopressin
oral surgery 183–4
orthopaedic surgery/procedures 148, 169–76
 in compartment syndrome 139, 140, **172**
 in haemarthroses 120–4
 indications 123, 169
 operations, results and complications 170–2, *173–5*
 planning 135–6, 169–70
 for pseudotumours 143–4, 171, **172**
orthotics 153
 ankle 158
 in arthropathy 165
 in chronic synovitis 164
 elbow 156
 hip 156–7
 knee 157–8
 shoulder 155
 see also splinting
ossification, pseudotumours 118
osteoarthritis 135
osteonecrosis, hip 135
osteoporosis 148, 153
osteotomies, in dental surgery 184
Oxford/new Oxford assay 9
Oxford splinting technique 123

215

pain
  joint 121, 153, 165
  relief see analgesia
paracetamol 123, 184
P.A.S. Ports™ 36
PCC see prothrombin complex concentrates
Percuseal™ device 37
peripheral intravenous access 36
  complications 64
  devices 36–7
phospholipids
  in factor VIII concentrates 105
  FVIII binding 10, 11–12, 100, Plate 2.1
physiotherapy 147, 160–7
  in acquired haemophilia 165–7
  acute joint and muscle bleeds 162–3
  annual reviews 162
  in arthropathy 164–5
  in chronic synovitis 163–4
  comprehensive care approach 161–2
  initial examination 160–1
  outcome measures 167
  in subacute haemarthrosis 123, 124
  techniques 151–2
  see also rehabilitation
plasma-derived factor VIII concentrates 14–15
  in acquired haemophilia 102–4
  antigenicity 4, 16–18, 55
  contaminants 15, 16–17, 18
  high purity 15, 17, 18
  for immune tolerance induction 46, 56, 57
  intermediate purity 15, 18
  low purity 57
plasmapheresis 29, 33–4
  in acquired haemophilia A 105, 108
  for bleeding episodes 71, 76
  immune tolerance induction with 33, 34, 49, 54
PlasmaSelect AG 200
plasminogen, plasma 80
platelet(s)
  activation 206
  porcine factor VIII and 79
  transfusions, in acquired haemophilia 105
porcine factor VIII (Hyate C™) 146, 149
  in acquired haemophilia 100, 101, **103**, 104, 106, 166
  adverse effects 79, 104, 180
  antigenicity 70
  availability 79
  in Bethesda assay 101
  for bleeding episodes **47**, 75
    in children 69–70
    life- and limb-threatening 78–9, 80, 81, 197
    in mild/moderate haemophilia 96
  for central venous line insertion 40, 51
  for compartment syndrome 139
  cross-reactivity 70, 75
  for general and emergency surgery 179–80
  for orthopaedic procedures 120
  preventative treatment 72
Port-A-Cath™ devices 36, 162
  for immune tolerance induction 49, 51, 52–3, 163
  infections 37–9, 52–3
  other complications 37, 39–40
postoperative analgesia 179, 184
postoperative bleeding
  dental surgery 184
  general and emergency surgery 181
  orthopaedic surgery 169, 170–2

postpartum period, acquired haemophilia 99
precipitin reaction 9
prednisone/prednisolone
  in acquired haemophilia 107, 108, 166, 167
  in immune tolerance induction 202, 203
  in synovitis 147
pregnancy
  acquired haemophilia 99, 166
  Ig-Therasorb immunoadsorption 202
preoperative assessment 179
prevalence of inhibitors
  FVIII products and 17–18
  haemophilia A/B difference 23
  in haemophilia B 7, 59, 195
  in severe haemophilia A 3–7, 17, 195
Profilate™, for immune tolerance induction 57
prophylactic treatment, children with inhibitors 72
Proplex™ 78
Prosorba columns 105
protein A immunoadsorption 29–31, 32–4, 196
  in acquired haemophilia 105
  for bleeding episodes 71, 76
  clinical experience 30–1
  limitations 31
  in Malmö protocol 32–3, 34, 129, 130, 196
  side-effects 33
  two-column systems 29–30
protein C, activated (APC) 101
proteins, residual, in FVIII products 15, 16–17, 18
proteoglycans 115, 116
proteolysis
  FVIII, inhibitor-mediated 11
  FVIII concentrate antigenicity and 16, 17
prothrombin 105
prothrombinase complex 206
prothrombin complex concentrates (PCC) 146, 165
  in acquired haemophilia 100, 102, **103**, 105–6
  activated see activated prothrombin complex concentrates
  adverse effects 46, 79
  for bleeding episodes 70, 75–6
    factor V inhibitors 202
    in FIX allergy 89
    life- and limb threatening 78, **79**, 197
    in mild/moderate haemophilia 96
  for compartment syndrome 139
  for dental surgery 184
  for general and emergency surgery 180
  for orthopaedic procedures 120
  for radiosynoviorthesis 129–30
prothrombin time 50, 101
Prothromblex™
  for bleeding episodes 78
  for central venous line insertion 40
pseudotumour, haemophilic 78, 118, 142–4, Plates 17.5–6
  complications 118, 143
  lower jaw 183
  management 142–4, 171, **172**
  MRI 134, 136
  pathogenesis 142
  postoperative 170–1, **172**
psoas haematoma see iliopsoas haematoma
psychological management 149, 189–91
psychosocial impact 187–91
pulmonary embolism 39, 208
pulsed short-wave diathermy (PSWD) 163

quadriceps
  atrophy/weakening 157, 170
  haematoma 157
quality of life 187–91

racial differences
  immune tolerance therapy outcome 62
  inhibitor development 3, 4–5, 24, 196
radial artery, puncture 31, 33
radiation therapy, pseudotumours 143, 144
radiographs, plain film
  classification of haemophilic arthropathy **126**
  vs. MRI 132, 134–5
radiosynoviorthesis (radionuclide synovectomy) 127–8, 129–31, 135, 136, 147
range of motion (ROM) 150, *151*
rapamycin 107
recombinant activated factor VII (rFVIIa, NovoSeven) 146, 149, 159
  in acquired haemophilia 100, 102, **103**, 106, 108
  adverse effects 79, 106, 182, 208
  for bleeding episodes **47**, 76, 200, 207
    in children 70, 71, 72
    continuous infusion 207, 208
    in FIX allergy 89, 198
    with immunoadsorption 202
    in mild/moderate haemophilia 96
    threatening life or limb 78, 79, 81–2, 197
  bolus injections 206–7, 208
  bolus 'mega dose' 208
  for central venous line insertion 40, 181
  for compartment syndrome 81, **82**, 139
  continuous infusion (CI) 81–2, 121, 181–2, 197, 207–8
  for dental extraction 183, 184
  dosage regimens 197, 206–8
  economic aspects 204
  for general and emergency surgery 181–2
  in haemophilia B 7, 72
  home treatment 72, 76, 79, 207, 208
  in immune tolerance induction 50, 52
  for internal bleeds 81–2, 197
  for intracranial bleeding 80, 197
  Israeli experience 208
  mechanism of action 206
  in mild/moderate haemophilia A 96, 198
  modes of administration 206–7
  MRI evaluation of treatment 135–6
  for orthopaedic surgery 120–1, 124, 144, 169–70
  pharmacokinetics 206
  for physiotherapy 163
  preventative treatment 72
  to prevent inhibitor development 96
  protein A immunoadsorption and 29, 33
  quality of life aspects 187
  for synoviorthesis 127, 128, 129, 130
  in synovitis 153
  vs. aPCC 72
  vs. extracorporeal inhibitor removal 34
recombinant factor VIII (rFVIII) concentrates 15
  in acquired haemophilia 102–4
  antigenicity 17, 55, 96, 195
  for immune tolerance induction 56
  natural history of inhibitors 4–5, 6
  second-generation 15
  truncated B-domainless 15
  see also ReFacto™
  see also Kogenate™; Recombinate™

Recombinate™ 15
   in acquired haemophilia **103**
   for immune tolerance induction 56
   natural history of inhibitors 4–5
ReFacto™ (truncated B-domainless FVIII)
   in acquired haemophilia **103**
   for immunotolerance induction 56
   natural history of inhibitors **4**, 5
rehabilitation 149–59
   aims 150
   common problems 152–3
   haemophilic arthropathy and synovitis 153–4
   in haemophilic synovitis 126–7
   physical examination for 150–1
   specific joints 154–9
   techniques 151–2
   see also physiotherapy
renal failure, acute 107
rest 123, 124, 162
   see also immobilization
retroperitoneal haematoma 92
retropharyngeal haemorrhage 81, **82**
rFVIIa *see* recombinant activated factor VII
rFVIIa-CI Group 81, 82, 197
rhenium, radioactive (^{189}Re) 127–8
rheumatoid arthritis 98–9, 146, 166
rheumatological problems 146–8
rifampicin, intra-articular 127, 147
rituximab (Rituxan) 107
rofecoxib 126–7, 147, 153

Scandicain™ 122
school activities 188, 189
seizures 80
self-esteem 188, 189
Sepharose™ 30, 31–2, 105, 200
   see also Ig-Therasorb™ immunoadsorption
septic arthritis 121, 147
sheep antihuman immunoglobulin antibodies 200–1
shoulder
   arthrocentesis 122, 123, Plate 18.5
   examination 154
   haemarthrosis 154–5
   orthotics 155
   radiosynoviorthesis **130**, 131
   rehabilitation 154–5
siblings 24, 196
SIMS 36
sirolimus 107
Sjögren's syndrome 99
skin erosion, over Port-a-Cath™ ports **39**, 40
Slim-Ports™ 36
snooker 162
snorkelling 159
soft tissue bleeds 162
   in acquired haemophilia 100
   infections 117
   pseudotumour development 142
   see also muscle haematomas
spinal extradural haemorrhage 80
splinting 147, 162, 163
   in acute haemarthrosis 123, 155
   in arthropathy 165
   in chemical synoviorthesis 126–7
   in chronic synovitis 164
   in dental surgery 184

elbow 156
lower limb 157, 158
   see also orthotics
sports 158–9, 162, 188, 189
Staphylococcus aureus 38, 196
Staphylococcus epidermidis 38, 196
sternoclavicular joint 154
steroids *see* corticosteroids
Stimate *see* desmopressin
stroke 208
subarachnoid haemorrhage 80
subchondral cysts 134
subclavian thrombosis 39
subdural haematoma 80
supportive relationship 190
surgery
   general and emergency 179–82
   Ig-Therasorb immunoadsorption 202
   in immunotolerized patients 53
   inducing acquired haemophilia 99
   inhibitor development after 92, 95
   MRI-based planning 135–6
   orthopaedic *see* orthopaedic surgery/procedures
   preoperative assessment 179
   prevention of inhibitor development 96
   rFVIIa therapy 207, 208
   therapeutic strategies 179–82
swimming 147, 157, 159, 162
synovial hypertrophy/hyperplasia 115–16
   management 147
   MRI 132–3, 135, 136
synoviorthesis/synovectomy 123, 153–4
   chemical 126–8
   MRI for 135, 136, *137*
   radionuclide 127–8, 129–31, 135, 147
   surgical 136, 172
synovitis, haemophilic 152–3
   in acute haemarthrosis 122–3
   chronic 120, 163–4
   clinical classification **126**
   management 126–8, 129–31, 147
   pathogenesis 115–16, Plates 17.1–4
   physiotherapy 163–4
   rehabilitation 153–4
systemic lupus erythematosus (SLE) 98–9, 166

T2-weighted MRI images 132
tacrolimus (FK506) 107
t'ai chi ch'uan 159, 162
target joints 115, 134–5, 146
team, multidisciplinary 161–2
tetracycline, intra-articular 147
thigh, muscle haematomas 157
thrombin 206, 208
thrombin-activated fibrinolysis inhibitor (TAFI) 206
thrombocytopenia, porcine factor VIII-induced 79, 180
thromboembolism
   aPCC and PCC-associated 106, 180
   immune tolerance induction and 46
   rFVIIa-associated 79, 106, 182, 208
thrombosis
   central venous catheter 39–40, 64
   peripheral intravenous access devices 36–7
tissue factor 76, 206
tissue plasminogen activator (t-PA) 40

titres, inhibitor *see* inhibitor titres
T lymphocytes 199
tooth extraction 183–4
tranexamic acid
   in acquired haemophilia **103**
   in dental surgery 184
   for intracranial bleeds 80, 197
   for major surgery 169, 181
   mouthwash 184
   with rFVIIa 76, 82, 182, 206, 207, 208
transcutaneous electrical nerve stimulation (TENS) **152**, 153, 164
transforming growth factor β (TGF-β) 15, 17
transient inhibitors 3, 5–6, 50, 200
   see also type II (B) inhibitors
trauma 82, 92, 136
trichloroacetic acid 184
type I (A) inhibitors **16**, 50
   in mild/moderate haemophilia A 92
   vs. type II autoantibodies 101–2
type II (B) inhibitors 10, 11, 50
   in acquired haemophilia 101–2
   in mild/moderate haemophilia A 92
   product-related **16**
   see also transient inhibitors

UK Haemophilia Centre Directors Organization (UKHCDO) database 17, 92, 96
ulnar nerve compression 118
ultrasound
   in haemarthrosis 116, 152–3
   muscle haematomas 117, 139, 142
   in synovitis 126
   therapy **152**, 154, 156
urokinase 40

vaccinations, hepatitis A and B 179
Vacutainer 50
vascular access
   protein A immunoadsorption 31
   see also arterial puncture; venous access
venous access 36–40, 162, 196
   for immune tolerance induction 46, 49, 52–3, 163
   see also central venous access devices (CVADs)
venous thrombosis, deep 39, 208
vincristine, in acquired haemophilia 107, 108
viral infections
   inducing acquired haemophilia 99
   see also hepatitis B; hepatitis C; HIV infection
viruses
   inactivation procedures 17, 18
   transmission risks 64, 103
Volkmann contracture 140
von Willebrand factor (vWF)
   FVIII interaction 10, 11–12, 100–1
   in FVIII products 15, 18, 103, 195
   success of immune tolerance induction and 55–6, 57, 58

walking aids 157–8
wheelchairs 165

X-rays, plain film *see* radiographs, plain film

yttrium, radioactive (^{90}Y) 128, 129–31, 147